Theoretical Studies on
Sex Ratio Evolution

MONOGRAPHS IN POPULATION BIOLOGY

EDITED BY ROBERT M. MAY

THEORETICAL STUDIES

ON

Sex Ratio Evolution

SAMUEL KARLIN AND

SABIN LESSARD

PRINCETON UNIVERSITY PRESS

PRINCETON, NEW JERSEY

1986

Contents

CONTENTS

vi

CONTENTS

Preface

The variety of sexual systems and mechanisms in the living world is prodigious. Bacterial strains usually reproduce asexually. Parthenogenesis is found in apomictic plant species and in certain insect, fish, and reptile populations. Some fish in their lifetimes undergo sex conversions (i.e., they are sequential hermaphrodites); but most never change gender. Most plant species are simultaneous hermaphrodites; many partially self-fertilize, but in others incompatibility mechanisms compel outcrossing. Most species of snails are hermaphrodites with self-fertilization prevented, but two such snails in reproduction act as both male and female with respect to each other. There are the haplodiploid eusocial *Hymenoptera* composed of fertile males (haploid), fertile females, and sterile females. Haplodiploidy is not limited to social *Hymenoptera* although sterility is. Sterile classes of males and females also occur in termite species, which are diploid.

Many organisms encompass in their life cycles both a sexual stage involving recombination events and an asexual stage. Tied to the recombination process is the extent of ploidy. Fungi live most of their lives as haploids and take excursions as diploids. Mammals are invariably diploids. Higher ploidy abounds in plant populations; it is rare in the animal kingdom. Mammals and birds are primarily dimorphic: the two sexes display outwardly different physical forms, some distinct hormonal facilities, differences in growth characteristics, and so on.

Mating behavior patterns are also wide-ranging. The degree of outcrossing as against self-fertilization in plant populations varies across a broad spectrum. There are high selfers in some species, but other closely related species show substantial outcrossing. The dispersal of seed and/or pollen can be propagated by wind currents, insect vectors, contact with roaming animals, and cataclysmic events (e.g., floods).

xi

Some organisms pair off (most bird species, some mammals) and care for their offspring. Sea bass, which are hermaphrodites, not only form pairs but take turns laying eggs and providing sperm. At the other extreme, some fish eject eggs and sperm randomly into the water. Some species mate preferentially in response to chemical and sensory stimuli (e.g., pheromones, vocalization), behavioral rituals, or physical attributes. Among some *Drosophila* species rare types have mating advantages. Aspects of hierarchical and social structures—pecking order, territoriality, aggregation, varying group associations, nest-building, feeding, protection, etc.—may influence mating choice. In all these systems there are interactions and balances between individuals and groups.

To explain the kinds of balances we find in the evolutionary process, it is tempting to seek optimization principles and underlying strategies. Do such principles and strategies exist, or is there a wide tolerance and a confounding redundancy even in adaptation to a particular environment? The large number of different solutions to seemingly similar problems is striking. The compromise between short- and long-term benefits, the balance between much and little recombination, the roles of mutation and randomness, the choice between inbreeding and outbreeding, the adaptation of various selection mechanisms to changing environments, critical adjustments in behavior patterns, all these are arrived at in many ways. The evolution and role of sex ratio is one of the issues and not the least important.

The sex ratio of offspring is a complex matter, especially in reptiles and invertebrates. The progeny of invertebrates range from all male to all female, depending on ambience. Among plants the forms and levels of sex allocation are even more varied. The high incidence of a 1:1 sex ratio at birth in mammal populations is striking compared to the frequent deviations from this ratio in invertebrate species. An approximate 1:1 sex ratio appears to predominate in most vertebrates in the early stages of development, even though a smaller number of males would seem a priori more advantageous in polygynous species. Sex chromosomes generally induce a 1:1 sex ratio at conception according to Mendelian schemes, but there are counterexamples. What, then, are the conditions that favor the evolution of strong sex determiners and 1:1 sex ratios?

There are two main approaches to understanding the causes and effects of sex ratio. One emphasizes the optimization and adaptive functions of sex allocation, the other the consequences of genetic sex-determination mechanisms. Many authors take the first approach to parental and population strategies, often relying on the concept of evolutionary stable strategy (ESS) discussed in Chapter 1. The basic models of our book derive from the second approach. The sex determination mode is specified explicitly in terms of underlying multilocus, multiallelic components; the offspring express male or female sex with probabilities that depend on the zygotic or parental genotype but may take into account the effects of modifier genes coupled to prenatal and/or neonatal interactions and other environmental factors.

Chapter 1 reviews some background facts and issues of sex ratio evolution and various models from the recent literature; Section 1.7 presents a précis of the main results in this book. Chapter 2 formalizes several models of the dynamics of sex ratio evolution for one-locus multiallele genotype (zygotic or parental) sex determination systems in diploid and haplodiploid species. Dichotomous systems and strong sex determiners are given special attention. The models encompass the various sex allocation processes by which plant populations dispense seed and pollen, and they take account of environmental influences that cause temporal and spatial variation in sex expression and differences in gender fitness. Differential fertilities lead to more general models.

The main results obtained from working with the models of Chapter 2 are presented and interpreted in Chapter 3 and various optimality principles are discussed. In Chapter 4 we give a complete analytic delineation of two-allele sex determination models, and characterize all qualitative possibilities in the three-allele case. Special forms of sex expression based on dichotomous partitioning of genotypes are also described in detail. Several multilocus sex determination systems are examined in Chapter 5.

Chapter 6 considers a number of nonrandom mating models bearing on sex ratio evolution, including partial sib mating or selfing practices, breeding structure, and influences of population regulation. We further examine the fate of meiotic drive modifiers causing sex ratio distortions. Chapter 7 provides a complete analysis of a general sex-differentiated two-allele viability selection model and some related models pertinent to sex determination. Chapter 8 presents several theoretical studies of multisex classes arising from incompatibility and self-sterility mechanisms. Chapter 9 treats general models of multifactorial sex determination that are also relevant to environmental sex determination and sex conversion systems. Chapter 10 presents briefly some open problems.

Some notable recent books on the evolution of sex systems are Williams (1975), Maynard Smith (1978), Bell (1982), Charnov (1982), and Bull (1983). The first three are concerned with general issues of sexuality such as comparisons and contrasts in reproduction processes, advantages of recombination, monomorphism versus dimorphism. The last two concentrate mainly on sex allocation phenomena, mechanisms, and controls. Our book represents an effort to synthesize and extend previous studies of sex ratio theory. We set forth a general approach that can be applied to a wide range of problems having to do with genetic and environmental mechanisms of sex determination and sex ratio evolution within the constrains of those mechanisms. The results of our research generalize and unify existing work. In some cases previous conceptions of the theory are strengthened, but in other cases the results may be surprising. On the basis of our findings we venture several interpretations, hypotheses, and speculations.

PREFACE

This book developed from our attempts to understand in
what way a 1:1 sex ratio could be considered optimal and then
why biased sex ratios occur. Although we have learned much
from previous writings on sex ratio theory, more than 60 percent of our material embodies new research.

The authors are happy to acknowledge the continuing support provided by NIH Grant 2R01 GM10452-21, NSF Grant
MCS 82-15131, the Direction Générale de L'Enseignement
Supérieur du Québec, and the Natural Sciences and Engineer-
ing Research Council of Canada, without which this book
would have been much longer in the making. We are indebted
to many colleagues for valuable discussions, notably Suresh
Jayakar, Carlo Matessi, Ilan Eshel, and Marc Feldman; to
Eric Charnov and Marcy Uyenoyama for providing comments
on several chapters that helped us to clarify a number of formulations and arguments, and especially to Jon Seger, who
read the whole manuscript and offered many useful suggestions.
Finally, we are grateful to two groups of students whose questions about our lectures on this material not only helped improve
our presentation but frequently stimulated further research.

January 15, 1985   *Samuel Karlin*
                                    Sabin Lessard

Theoretical Studies on
Sex Ratio Evolution

CHAPTER ONE

Problems, Background, and Summary

1.1. INTRODUCTION

The high variability in sex ratios and sex determination mechanisms is documented and discussed in numerous articles and books; see especially Charnov (1982) and Bull (1983) for much recent information and references. Several kinds of questions arise in considerations of sex ratio evolution. First, what are the stable equilibrium sex ratios for a dioecious or haplodiploid species and a specified sex-determining mechanism? How are they maintained? When does a progeny sex ratio or a population sex ratio achieve stability at 1:1, and in what sense is this "optimal"? Second, when can we expect an individual to be able to alter its allocation to male versus female functions in response to environmental or life history conditions? What is the equilibrium sex order and time of sex change for sequential hermaphrodites? What is the equilibrium sex allocation of resources for simultaneous hermaphrodites? Why is simultaneous hermaphroditism so common in higher plant populations? What is the connection between sex allocation and the degree of selfing compared to outcrossing? Third, what are the advantages of dioecy versus hermaphroditism, and diploidy versus haplodiploidy? When is a mixture of multiple incompatibility classes favored? Fourth, when would environmental sex determination be predominant over genetic sex determination? How is sex ratio affected by fluctuating environments? How is sex ratio affected by mating patterns and population structure?

Many authors discuss sex ratio determination in terms of parental strategies of resource allocation. In many situations the "best" strategy (the one favored by natural selection) is a

1 : 1 sex ratio. But there are exceptions. For example, biased sex ratios can evolve where there is parent-offspring conflict, local mate competition (among males), or local resource competition (among females).

An alternative view is that particular sex ratios occur as an evolutionary concomitant of a genetic sex determination system—simply the consequence of the cytological machinery. For example, XY/XX systems and the formal Mendelian rules help to maintain a 1 : 1 sex ratio. Generally, sex chromosome systems of many kinds induce a 1 : 1 sex ratio at conception, though there are counterexamples. Sex determination under two blocks of genes should ordinarily yield a 1 : 1 sex ratio even with haplodiploid organisms, and X-linked genes that modify the sex of offspring in XY/XX systems are expected to produce a 1 : 1 sex ratio, as elaborated in Chapters 2 and 3.

Thus there are two main approaches (not mutually exclusive) to understanding sex ratios under various systems of sex determination. One emphasizes the optimization and adaptive functions of sex allocation at a phenotypic level, and the other emphasizes the consequences of genetic sex determination mechanisms. In discussing the utility of these two approaches, we shall focus on three topics: the principal sex-determining mechanisms, the various genetic and environmental factors that contribute to adaptive sex expression, and the evolution of different sex-determining systems. For the reader's convenience, representative references for various theoretical formulations and concepts from the recent literature are given in Tables 1.1 and 1.2 of Section 1.6. Then in Section 1.7 we summarize and interpret the most important of the theoretical results that emerge from our own investigations, as detailed in Chapters 2 through 9.

1.2. SEX DETERMINATION SYSTEMS AND SEX RATIOS

Most animal and plant species produce two gamete types that fuse for reproduction. Hermaphroditic organisms can produce both kinds of gametes, and in dioecious organisms the male and

4

female sexes are separated throughout their lives, each providing a single gamete type. Significant numbers of species, especially in the plant kingdom, are subject to incompatibility constraints (e.g., self-sterility classes) based on genotypes. Incompatibility systems, in essence, define multisex types. This theme is elaborated in Chapter 8.

The variety and complexity of sex-determining systems are manifold. Such systems can be classified into seven modes: (1) one-locus multiallele systems; (2) multigene (and polygenic) determination systems with modifier gene effects; (3) chromosomal heteromorphism systems, including distinguished XY/XX, ZZ/ZW, or balanced XO/XX forms; (4) hermaphroditic systems (simultaneous or sequential); (5) mixed parthenogenesis (e.g., haplodiploid) systems; (6) environmental sex-determination systems (e.g., where sex is influenced by cytoplasmic milieux or by endogenous conditions at birth); (7) systems involving extrachromosomal factors (e.g., contagious viral particles, conditions fostering meiotic drive).

The controls on sex expression may depend on the genotype of the offspring or of the parents (e.g., they may reflect intrauterine and cytoplasmic environment and related elements of the mother's genetic and physiological state), or on brood size and sibling interactions. In broad terms, there are distinguished genotypic and environmental determinants subject to zygotic, parental, or population controls. Even completely genetic controls of sex expression can be manipulated by hormonal and physiological covariates.

Most mammals have a 1:1 (progeny) sex ratio. Fisher (1930) proposed that such a ratio should obtain in the long run at least in random mating populations, since an individual increases its contribution to succeeding generations by "investing" more of its reproductive potential in the sex that is less numerous at the time of investment. However, anomalous sex ratios are widespread, especially among invertebrates, and this has raised challenging evolutionary questions since the early works of Gershenson (1928) and Sturtevant and Dobzhansky (1936). Populations with sex-linked sex ratio distorters would not evolve

5

to a 1:1 sex ratio according to Fisher's rule because the genetic materials transmitted through sons and daughters differ. Such distorters would rather maximize their own representation in the next generations, leading to monomorphic (all female or all male) populations unless viability or fertility differences (Edwards, 1961; Thomson and Feldman, 1975) or population-structure factors (Hamilton, 1967; Colwell, 1981) compensate for the driven effects. Hamilton's theory of local mate competition predicts a female-biased sex ratio in species in which sons of a parent must compete with each other for mates (for elaborations, see Sections 1.4, 6.6, and 6.7). Sex allocation in plants is even more complicated, involving geographical covariates (e.g., Lloyd, 1974; Charlesworth and Charlesworth, 1981; Charnov, 1982; Lloyd and Bawa, 1984). Genetic, behavioral, physiological, and environmental factors may all affect an individual's relative allocation of resources to male and female functions.

Sex Partitions Based on Multiple Alleles and Loci

In this category, we may include heterogamety against homogamety or, more generally, sex classes corresponding to a partition of genotypes into two classes. In most animals sex determination is governed by the XX/XY (or ZZ/ZW) chromosomal mechanism, allowing in some cases for multiple X or Y chromosomes or their absence. It should be emphasized that even with the XX/XY system the final representation of the sexes in the population may be affected by further autosomal or sex-linked mutations and modifier genes (see Ohno, 1967, 1979; Bull, 1983). Other agents influencing sex ratio include segregation distortion (meiotic drive) and extrachromosomal contingencies, which are discussed later in this section.

Sex forms in plants are varied. They can incorporate a hierarchy of modifier genes that regulate the allocation of resources to male versus female functions (e.g., Robinson et al., 1976; Frankel and Galun, 1977; and Lewis, 1979) or produce significant qualitative transformations (e.g., incompatibility, male

6

sterility). Examples of self-incompatibility classes in plants are given in Chapter 8.

A documented example of a one-locus multiallele sex determination system occurs in the platyfish (Kallman, 1973; Orzack et al., 1980). The alternative alleles are commonly labeled X, Y, W; females correspond to genotypes WX, WY, or XX and males to genotypes XY or YY. This locus appears to be linked to pigment color genes (and these can relate to differential viability effects). In lemmings, males are heterogametic XY, but some females are also heterogametic with a variant X-chromosome denoted X* (Fredga et al., 1976). Therefore, females may be XX, X*X, or X*Y. Moreover, it has been observed that X*Y females produce mainly daughters, causing a female-biased population sex ratio. A population genetic model for the wood lemming, incorporating fertility differentials and segregation distortion effects, was proposed by Bengtsson (1977). Maynard Smith and Stenseth (1978) conceived an alternative scenario based on inbreeding.

Multifactor (multilocus) systems of sex determination have been observed in the housefly and other Diptera (midges, mosquitos, phorids). In many species, females are homozygous at every relevant locus, whereas males are heterozygous at one or more of the loci (see Milani et al., 1967; Bull, 1983, and references therein). A four-locus version was reported in a population of houseflies (Tsukamoto et al., 1980). Other systems involve combined chromosomal and autosomal sex factors.

In the housefly (*Musca domestica*) Franco et al. (1982) have documented the existence of at least three systems of sex determination: (1) chromosomal (XX/XY); (2) a single autosomal locus of male heterogamety (*Mm* ♂; *mm* ♀); (3) a second independent autosomal locus of female heterogamety (Ff ♀; *ff* ♂). These loci seem to show a dominance hierarchy associated with a temperature-altitude gradient. In northern Europe the standard XX/XY mechanism of sex determination is in force, whereas in southern Italy all flies are XX and the autosomal

7

locus Ff/ff of female heterogamety determines sex expression. Moreover, there are intermediate zones where all extant flies are XX and ff, and then the genotype of the Mm/mm locus governs the sex phenotype. It is hypothesized that the selective agents maintaining this cline of sex-determining loci are associated with the climatic state and the presence of insecticides. Jayakar (1982) has formulated some theoretical selection models to account for the cline.

Haplodiploidy

Haplodiploidy commonly involves a mixed parthenogenic and random mating reproductive process. In most cases unfertilized eggs develop into haploid males (arrhenotoky), the other case of parthenogenesis (theletoky, usually producing female diploids from unfertilized eggs) being rare in haplodiploid species. Control of sex expression can be based on the zygote's or the mother's genotype. Of special interest is the case where the haploid fraction is determined either maternally or to some extent through interaction with sisters. The proportion of haploids to diploids may vary in response to endogenous and exogenous environmental conditions. In some cases the proportion of haploids to diploids may be under facultative parental control. The sex ratio among the diploid representatives may also be genotype-dependent (subject to zygotic and/or mother control), as occurs with the wasp *Habrobracon* (Whiting, 1943) and in other hymenotera including some species of bees and ants (in which homozygote diploids become males, heterozygote diploids develop into females, and unfertilized eggs develop into males). We refer to Bull (1983) for more details on this system, which is known as complementary sex determination.

Another important consideration for many haplodiploid organisms is the degree of inbreeding in mating and the timing in eclosion. Haploid males usually emerge before sisters (protandry), wait for them to eclose, and mate with them. Some solitary wasps (*Eumenes esuriens* and *Sceliphron madraspatanum*— both haplodiploid species) have a pattern of egg-laying that suggests that during a breeding season either they lay eggs of one sex only, or if they lay eggs of both sexes, they lay all male eggs before all female eggs. The population sex ratios are close to 1:1 (Jayakar and Spurway, 1968). One species of solitary wasps (*Stenodynerus miniatus*) has evolved behavior such that every cell of its nest has two compartments, in one of which a male is laid and in the other a female. The two offspring of a cell emerge simultaneously and copulate before leaving the nest (Jayakar and Spurway, 1966). This is complete sib mating with a blatant numerical 1:1 sex ratio.

Models for the evolution of protandry in insects are presented in Bulmer (1983) and Iwasa et al. (1983). Various environmental factors have been identified that affect sex ratio, including crowding of larvae, population size, nutritional state of the mother, and presence of other females (for a review see Charnov, 1982).

Hermaphroditism

In sequential hermaphroditism an individual functions early in life as one sex and then changes to the other sex for the remainder of its life. This pattern is found among invertebrates and fish (reported in Bacci, 1965), and in some plant species (Heslop-Harrison, 1972). Most invertebrates convert from male to female: a number of coral reef fish (e.g., wrasses) reverse from female to male. The timing of the switch may depend on growth rate, size, population density, social milieu, and/or environmental cues. Further, there is a hypothesis proposed by Ghiselin (1974) (see also Policansky, 1982; and Charnov, 1982, chap. 9) that postulates male as the sex for smaller-size individuals, which change to female as they grow larger. Sex change in mollusks, among other invertebrates, also appears to be mediated by size. There are cases of sex reversal (even multiple changes) that are socially controlled, e.g., in *Labroides dimidiatus*, a coral reef fish (Robertson, 1972).

Simultaneous hermaphroditism (an individual produces both kinds of gametes in each breeding season) is widespread in

9

plants and in certain invertebrates (e.g., snails and slugs). The analog of sex ratio is sex allocation to male and female functions. Sex allocation in plants can be modified by many influences including climate conditions, the kinds and numbers of coexisting flora and fauna, and the density of pollinating vectors (e.g. bees).

An extensive review of various cases of sex reversal is presented by Policansky (1982). Policansky also attempts to unify the terminology and classifications. Thus in plant species protandry designates sequential maturation producing first pollen and then ova, and protogyny designates the reverse order. Protandrous mechanisms in animals are those in which an individual begins life as a male, later becoming a female, while protogynous signifies the opposite sequence.

The distribution of gender in higher plants is complex, as attested by the wide spectrum of flower morphology. Plants differ in synchrony versus asynchrony of pollen and seed production and in incompatibility constraints. The flowers of plants can be hermaphrodite (producing both pollen and ova), monoecious (male and female flowers distinct on the same plant), or dioecious (male and female flowers on separate plants). In a survey of over 120,000 angiosperm species, Yampolsky and Yampolsky (1922) (see also Charnov, 1982, table 16.1) estimated the proportion of species of different sex types in the following order: hermaphrodite, 70 percent; monoecious, 5 percent; dioecious, 5 percent; andromonoecious (plant with hermaphrodite and male flowers), 2 percent; gynomonoecious (plant with hermaphrodite and female flowers), 3 percent; mixtures of the above, about 15 percent.

Multifactorial (Polygenic) Sex Determination

Sex determination is said to be multifactorial when a large number of loci (factors) are assumed to contribute to sex expression. Bull (1983) discussed three criteria for assessing multifactorial sex determination: a large variance of between-family sex ratio; asymmetric paternal versus maternal effects on family

10

sex ratio; and a sex ratio response to selection. The prime examples are poeciliid fish, swordtails, and medakas (Kosswig, 1964; Bacci, 1965; Kallman, 1983). The additive-value model of Winge (1934) is usually used in theoretical studies: in this model a numerical value is assigned to each sex factor, and the sum of the values (plus an environmental addend) is compared to some threshold to determine sex. Bulmer and Bull (1982) formalized this approach. Various extensions are developed in Chapter 9.

Environmental Sex Determination (ESD)

In certain species the controlling agent is some environmental factor external to the individual's gamete or zygote. Sex expression is considered to have a strong environmental component when sex is not necessarily determined at conception (Bull, 1983, and references therein). There may exist some interactive genetic influences, but these alone are not decisive. ESD can occur at different life stages. The earliest is an embryonic response to the environment (e.g., to incubation temperature, maternal physiology, or cytoplasmic milieu). Or the response can be to neonatal conditions. Nutrients available after birth are decisive for some parasites. Orchids germinating in sunlight tend to become female, whereas those in shade tend to produce male flowers. Alternatively, sex conversions can occur at later ages, depending on growth rates and local population densities. For example, Werren (1983) noted a positive correlation between the parasitoid wasp (*Nasonia vitripennis*) population size and proportion of males produced.

In parasitic nematodes (roundworms, *Mermethids*), in which XX individuals are hermaphrodite and XO individuals are always male (e.g., Bull, 1983, chap. 13), the proportion of females increases when there is more nourishment available in the host (e.g., fewer worms in the host). Traut (1970), through laboratory experiments on the archiannelid worm (*Dinophilus*) observed a female-biased sex ratio between 2:1 and 3:1 depending on the nutritional state of the mother, reaching at most

1:1 in a situation of severe hunger stress. Since under natural conditions good and hard times generally alternate, the average population sex ratio in this case is expected to be female-biased. Offspring of the marine worm *Bonellia viridis* (see, e.g., Bacci, 1965; Crew, 1965) become females if raised in isolation and males if raised with adult females. In crocodiles and some lizards the embryos turn out to be females if the incubation temperature is low and males if it is high, while the reverse association holds for some turtle species (e.g., Bull, 1980; Bull et al., 1982).

One general argument for environmental control of sex expression is that selection would favor the ability of a female parent to manipulate the sex ratio to improve the sex-specific fitness of her offspring in varying environments. This point is discussed in Section 1.5. The annelid worm *Dinophilus apatrix* determines its sex prior to fertilization, since large eggs develop into females and small ones into males. Both genetic and non-genetic factors can alter the sex ratio (Bacci, 1965).

Sex Ratio Distorters and Extrachromosomal Mechanisms

There are established cases of sex distorter genes inducing progeny of only one sex or a strongly biased sex ratio. In many species of Drosophila, the heterogametic males carrying certain X chromosomes denoted X_r produce mainly daughters (Gershenson, 1928; Sturtevant and Dobzhansky, 1936). In the mosquito *Aedes aegypti*, the Y-linked gene M^D causes its male carrier to produce an excess of sons (Hickey and Craig, 1966; Wood and Newton, 1976). But there are also resistant chromosomes and/or autosomal suppressors able to cancel these effects. Autosomal modifiers having a direct or indirect influence on progeny sex ratio are known in the house mouse (Lewontin, 1968), and in *Drosophila* (see, e.g., Uyenoyama and Feldman, 1978, and references therein). Uyenoyama and Feldman focus on cases in which the progeny sex ratio is affected partly by factors (or viruses) inherited through the cytoplasm and transmitted from mother to daughter. These extrachromosomal (and/or contagious) agents usually confer on their carriers a

12

fertility advantage or disadvantage (e.g., killing zygotes of one sex). Broods of one sex are known in flies and barnacles (see, e.g., Bull, 1983, chap. 14).

1.3. ADAPTATION AND STRATEGY IN SEX RATIO DETERMINATION

The idea of optimal strategies for genetic systems has been common in discussions of evolutionary theory since 1920. The compromise between short- and long-term fitness has been discussed by Fisher, Darlington, Mather, Haldane, and many others. The balance between much and little recombination, the virtues of inbreeding versus outbreeding, the response to changing environments, and the adjustments occasioned by diverse behavioral patterns are other subjects of debate.

The role of randomness in all its guises and the inherent complexity of genetic-ecological systems suggest that optimality criteria are not easily determinable and indeed may be irrelevant to a great deal of evolution. With these caveats in mind, we describe briefly several population and individual strategies, seeking to elucidate the broad spectrum of sex ratio phenomena occuring in natural populations.

Some typical considerations in this literature are: (i) *Parental resources:* How does a mother decide, or when does the situation dictate, the extent of her investment among her offspring, taking into account her reproductive potential, her physiological capability, and the available resources? (ii) *Litter size:* What is the "optimal" litter size in terms of sex ratio control? Matching maternal capacities with progeny production cost may require regulation of both litter size and litter sex ratio. (iii) *Conflict situations for sex ratio:* Various conflict situations arise, including the familiar worker-queen conflicts in haplo-diploid organisms (Trivers, 1974; Trivers and Hare, 1976); conflicts induced by intergenerational differences when a mother delivers several rounds of progeny (in this case conflicts can arise among siblings of the same or different generations, between parents and offspring, or between offspring of different

13

genotypic constitution); and conflicts between competing elements of cytoplasmic DNA mediated in part by extrachromosomal agents, which can result in progeny consisting exclusively of sons or daughters or exhibiting drastic shifts from the normal sex ratio. Such alterations can be mediated by viral particles, infections, and environmental factors.

Some typical conclusions from models of adaptive sex expression follow. The fitness rationale in each case is presumably the intuitive one.

(1) Simultaneous hermaphroditism is generally associated with organisms having a sessile or relatively sedentary lifestyle.

(2) Parasitoid organisms nurtured in small hosts tend to produce more sons, which presumably need less sustenance than daughters to attain an appropriate fitness. Predictions of sex expression based on the host size are reviewed in Charnov (1982) for some wasp (insect parasitic) and nematode (plant parasitic) populations. For hosts receiving egg deposits, sex realization may relate to the choice of insect host made by the mother.

(3) A number of studies (e.g., Trivers and Willard, 1973) present evidence for mammals correlating the sex of offspring with mother's physiological capabilities. Males tend to be produced by females in good physical condition, it being argued that in polygynous species strong males are more effective sexual competitors. Since females virtually always get mated, mothers in poor condition profit more by producing daughters.

(4) Delayed mating and/or fertilization depending on the condition of the mother may suggest some effective sex ratio control. In haplodiploid Hymenoptera the mother may restrict sperm to produce male haploids. This presumably facultative action provides a physiological mechanism for a mother's regulation of the sex ratio in her progeny.

(5) Plants may vary sex expression by way of adapting to the physical and/or nutritional characteristics of their microhabitat. Many kinds of environmental conditions affect sex allocation in plant populations, including light exposure, soil composition, moisture, and stress conditions. Stress apparently

14

causes a shift of more sex allocation to the male function. Further questions on labile sex distribution in plants relate to the degree and nature of synchrony in the male and female functions.

(6) Sex ratio as well as sex-dependent viability pressures may vary temporally and spatially (Bull, 1981a; see also Chapter 2, Model IV). Differential migration based on sex is widely recognized. In higher mammal populations (especially polygamous ones) the males apparently move more, and in bird populations females often nest relatively far from mating areas. Migration rates, apart from being sex-dependent, can also be genotype-dependent.

(7) Among the behavioral and population-structure influences on sex ratio realizations, the principle of local mate competition (LMC) is presumed to be an important factor operating in sparsely distributed founding colonies. LMC applies most directly to organisms such as female parasitoids that search out hosts and produce a brood relatively isolated from broods produced by conspecifics. A simple model of this situation due to Hamilton (1967) (see Section 1.4 for technical details) envisions colonies of n inseminated females each of which produces a brood of N offspring. The offspring mate locally and subsequently the inseminated females of the next generation disperse to found new colonies. The "optimal" proportion of males (in the ESS-evolutionary stable sense) is $(n - 1)/2n < 1/2$. (An ESS property refers to a monomorphic population state invulnerable to mutational events; see Section 3.7 for more elaborations.) With sib mating prohibited, the ESS sex ratio is less female-biased, namely $(n - 2)/(2n - 3)$ (Charnov, 1982) exceeding $(n - 1)/2n$ for $n > 3$. With increasing numbers of foundress females per colony, the male sex ratio approaches $1/2$ from below in both cases.

For a variable (random) group size the corresponding ESS sex ratio depends on the distribution of n. In the case of a Poisson distribution with parameter v, the corresponding ESS sex ratio is $(e^{-v} + v - 1)/2v$. In general, the ESS sex ratio is

15

$(v^* - 1)/(2v^*)$ where v^* is the expected number of founders in occupied colonies. The mathematical arguments are presented in Section 6.5.

(8) The sex ratio bias associated with LMC can be construed as a concomitant of inbreeding (e.g., because of small local population size and sib mating). Inbreeding versus LMC models have been presented by Taylor and Bulmer (1980), Alexander and Sherman (1977), Wilson and Colwell (1981), and Uyenoyama and Bengtsson (1982). Werren (1980) stressed that parental inclusive fitness increases as fewer males are produced and competition among brothers for mates declines. Colwell (1981) argued that LMC is better interpreted as a form of group selection in which demes producing more females have a higher rate of increase.

(9) The relationship among the evolution of sex ratio, kin selection, and measures of genetic relatedness was considered in Uyenoyama and Bengtsson (1981), Hamilton (1972), Trivers and Hare (1976), Oster et al. (1977), and others. It is suggested that the sex ratio should be adjusted according to the degree of relatedness.

(10) Charlesworth (1977) and Charnov (1979a) studied demographic effects on sex ratio. Sex ratio is quite variable in this respect, generally showing a negative (female-biased) correlation with age cohorts caused by higher death rates among males, which, however, are partly compensated for by parental replacement of lost offspring (cf. Edwards, 1962). In human populations, a male-biased sex ratio at conception is generally the rule, suggesting some genetic control. There also certainly exist cultural and/or physiological determinants (e.g., infanticide, family planning, mother's condition). Data collected by Edwards (1966, 1970) suggest that both sexes tend always to be represented in a family. Moreover, the proportion of males generally declines as the mother's age increases (Schuster and Schuster, 1972; Teitelbaum, 1972). Other factors (e.g., age difference at marriage, population growth, mortality) are considered in Beiles (1974).

16

(11) The high frequency of 1:1 sex ratios at birth in mammal and bird populations contrasts strikingly with the significant deviations from that ratio in invertebrate species. Williams (1979) and Charnov (1982) surveyed the literature on litter sex ratio in mammals including deer, mice, rabbits, mink, and diverse livestock and in a number of bird populations (e.g., geese, blackbirds). They found that the evidence strongly supports a 1:1 litter sex ratio according to a Mendelian scheme; it is not clear, however, how and why this ratio evolved. Moreover, at least in one case (domesticated cattle), Bar-Anon and Robertson (1975) observed a standard deviation of 1.5 percent among the male progeny sex ratios and a significant correlation of 0.5 from father to son. A 1:1 population sex ratio per se does not preclude a large amount of variability in individual brood sex ratios (more than what would be expected with a binomial distribution), as noted by Kolman (1960).

(12) In finite or structured populations, the number of possible encounters between males and females may have to be increased in order to maximize the number of offspring produced (Crew, 1937; Kalmus and Smith, 1960). Verner (1965) (see also Bodmer and Edwards, 1960; Taylor and Sauer, 1980) proposed that any deviation from a 1:1 brood sex ratio would tend to be reduced and even eliminated. Cannings and Cruz Orive (1975) evaluated sex ratio patterns in terms of individual encounter schemes based on movement, signaling, and spatial structures. They concluded that the sex ratio that maximizes the expected number of fertilized females is generally biased toward the more mobile sex.

1.4. WHEN AND WHY IS SEX RATIO ONE-TO-ONE?

Optimality Criteria

Assuming random encounters, MacArthur (1965) proposed that the optimum sex ratio was the one that would maximize $m \times f$ where m is the number of males and f the number of females. With a constant brood size $m + f = N$, clearly $m(N - m)$ is maximized for $m = f = N/2$. For elaborations on this theme, we refer to Charnov's (1982) graphical approach in terms of fitness sets.

A mixed qualitative-quantitative argument predicting a $1:1$ sex ratio attributed to Fisher (cf. Eshel, 1975) goes as follows. Consider a three-generation population—I, grandparents; II, parents; III, children—in which generation II consists of n_1 males and n_2 females who produce an aggregate of N offspring in generation III. Then the average number of children for a male of generation II is N/n_1, and the average number of children for a female of generation II is N/n_2. Say the grandparents of generation I have a total of m offspring consisting of x males and $m - x$ females. In this situation the expected number of descendants in generation III for a grandparent of generation I is

$$T = x(N/n_1) + (m - x)(N/n_2) = mN/n_2 + (1/n_1 - 1/n_2)Nx.$$

T is an increasing function of x when $n_2 > n_1$, but a decreasing function of x when $n_2 < n_1$. Therefore the expected number of children of a grandparent in generation I increases if he (or she) produces more offspring of the underrepresented sex. If each grandparent makes an equal genetic contribution to each of its progeny, the expected sex ratio is $1:1$. When males "cost" ϕ compared to 1 for females (ϕ is analogous to a viability differential factor for males as against females) and the total "parental expenditure" is constant, the expected male to female sex ratio at conception is $1:\phi$ such that parental expenditure is equalized between the sexes for disparate costs in rearing males and females. The sex ratio at the end of parental care remains $1:1$ (cf. Chapter 2, p. 56).

Shaw and Mohler (1953) gave an elegant early demonstration of the ESS principle in the context of sex ratio evolution (cf. Charnov, 1982). They consider a dioecious random mating population with normal brood sex ratio M. If a deviant female adopts a sex ratio m, the relative number of her genes transferred to the third generation through her male offspring is

18

m/M compared to 1 for a typical female and the number transferred through her female offspring is $(1 - m)/(1 - M)$ compared to 1; the total number is then $m/M + (1 - m)/(1 - M)$ compared to 2 for a typical female. Elementary analysis of this quantity shows that for $M \neq 1/2$ there exists a value m closer to $1/2$ that gives a greater contribution of genes to the subsequent generations, and that for $M = 1/2$ there is no such m. This attribute of $M = 1/2$ is the ESS property. The expression

$$m/M + f/F \qquad (1.1)$$

where $m(f)$ is interpreted as a male (female) brood size with average value $M(F)$ in the population, is known as the Shaw-Mohler formula.

Uyenoyama and Bengtsson (1979), following Speith (1974), investigated models in which the success of a litter may depend on sex ratio. They interpret the quantity (1.1) (or equivalently $M \cdot F$ in MacArthur's formulation) as an adaptive function that should increase or be maximized at a stable equilibrium. Actually, they propose a necessary and sufficient condition reflecting some kind of symmetry for this to be true in a general two-allele one-locus autosomal setting. This topic is elaborated in Chapter 7. The quantity (1.1) certainly plays a role in the *initial increase* of a mutant allele in pure sex determination models, as will be illustrated in Chapter 3.

Uyenoyama and Feldman (1981) constructed an extended version of (1.1) called "effective fitness" that involves relatedness between individuals defined in terms of regression coefficients. This analysis depends also on how the parents, the individual himself, and his siblings affect sex expression and viability. Theoretical results for sex ratio evolution in social Hymenoptera were discussed by Oster et al. (1977), who derived mean sex ratios in the range $1/4 \leq m \leq 1/2$. The extreme case $m = 1/2$ results if the sex ratio is fully governed by the queen; $m = 1/4$ is the ESS sex-ratio under complete daughter (i.e., worker) control; $1/4 < m < 1/2$ results from mixed controls. These results are obtained by maximizing inclusive fitness functions, taking into account the proportion of genes shared by relatives. (See, e.g., Trivers and Hare, 1976; Charnov, 1978; Stubblefield, 1980; and Iwasa, 1981 for further developments.)

Local Mate Competition (LMC) Models

As a further elementary illustration of an ESS analysis in sex ratio theory, we present a model of local mate competition following Hamilton (1967). There is no genetic structure (i.e., detailing the genotypic frequency changes) in the formulation. Consider fertilized females, distributed across numerous islands of local size n. After reproduction and local mating the male offspring die, but the fertilized females recolonize in groups of size n. Let the common brood sex ratio be r. Consider a mutant female with altered sex ratio $\hat{r}$. Her fitness (reproductive success) is measured by the number of migrants to which she contributes genes through sons and daughters. Suppose each mother produces b offspring. The fitness of the mutant mother through daughters is $b(1 - \hat{r})$ and through sons (those mated to other females) is

$$\frac{\hat{r}b}{b\hat{r} + (n - 1)rb} \left[b(1 - \hat{r}) + (n - 1)(1 - r)b \right].$$

The combined fitness of the mutant mother (W) is

$$b \left[(1 - \hat{r}) + \frac{\hat{r}}{\hat{r} + (n - 1)r} \left\{ (1 - \hat{r}) + (n - 1)(1 - r) \right\} \right]$$

(1.2)

We now seek a value r^* of r such that $W \le 2b(1 - r^*)$ for all $\hat{r}$, that is,

$$\left. \frac{\partial W}{\partial \hat{r}} \right|_{\hat{r} = r} \ne 0 \quad \text{for } r \ne r^*$$

and

$$\left. \frac{\partial W}{\partial \hat{r}} \right|_{\hat{r} = r^*} = 0 \quad \text{for } r = r^*$$

20

This value $r*$ is an ESS in the same sense as $M = 1/2$ in (1.1). Simple calculus shows that $r* = (n - 1)/2n$. More rigorous arguments are given in Chapter 6 in the general situation of groups of random size. By contrast, the case of the wasp, *Stenodynerus miniatus* (Spurway et al., 1964), although practicing full sib mating, adheres strictly to an even sex ratio.

If sib mating is disallowed, then the contribution through male offspring must be replaced by

$$b\{\hat{r}(n - 1)(1 - r)\}/\{\hat{r} + (n - 2)r\}$$

which yields an ESS sex ratio $r* = (n - 2)/(2n - 3)$ (Charnov, 1982).

Taylor and Bulmer (1980) extended these models to allow a finite number of colonies and incorporated genetic structures for diploid and haplodiploid populations (see Chapter 6). A more realistic model would also incorporate effects due to migration depending on genotype, variation in clutch size, differential viability, etc.

The haystack model (Bulmer and Taylor, 1980a; Wilson and Colwell, 1981) envisions, in addition to population subdivision, several generations of breeding and population growth between migration events. Variants of the model consider contingencies in which females disperse with males after mating or females mate after dispersal.

The Local Resource Competition (LRC) model places daughters in territories with sons dispersing. Competition occurs among daughters for local resources. The expectation is that a male-biased sex ratio will result since competition occurs between female sibs but not between males (Clark, 1978).

Other examples of ESS analysis pertaining to sex ratio strategies are reviewed by Charnov (1982), who discusses population subdivision structures (e.g., involving an array of foundress colonies), varying cost-benefit sex-dependent parameters over space and time, aspects of sex reversal, demographic effects, and mixed hermaphroditic systems, among other matters.

Population Genetic Models

Modifiers of genetic sex ratio determination are mainly of three kinds: autosomal, sex-linked, and extrachromosomal. The first generally produces a 1:1 sex ratio, the other two a skewed sex ratio. In early theoretical papers, Scudo (1964, 1967a) set forth a wide spectrum of simple sex-ratio models (with one- and two-locus diploid sex determination) incorporating viability differentials, factors of meiotic drive, gametic selection, potential for sex reversal, encounter processes, and mixed systems of positive and negative assortative mating. According to Scudo (1964), pairs of genes at two independent loci, determining sex, can lead to a stable non-even sex ratio equilibrium in the case in which double heterozygotes characterize one sex and all other genotypes the other sex (compare with Model I of Chapter 5). All such models occur as special cases of multiallele systems of exact (dichotomous) sex determination as studied in Chapters 2–5. The role of recombination in multilocus sex-determination systems is studied in Chapter 5.

Theoretical genetic models accommodating incompatibility and self-sterility mechanisms were developed early by Wright (1921) and Fisher (1941). Finney (1952) distinguished zygote and gamete elimination effects and introduced the notion of isoplethy (multiple phenotypic classes equally represented). General models are considered in Speith (1971) and especially in Heuch (1979). Mating systems entailing different classes of prohibited matings are elaborated in Workman (1964) and Karlin and Feldman (1968a, b). Some theoretical results in these areas are developed in Chapter 8.

Sex-linked modifiers of progeny sex ratio are usually studied in concert with viability and/or fertility differences (e.g., Edwards, 1961; Thomson and Feldman, 1975; Bengtsson, 1977), population subdivision (Hamilton, 1967), degree of inbreeding (Maynard Smith and Stenseth, 1978), and recombination rate (Maffi and Jayakar, 1981) in order to explain

22

observed polymorphisms in natural populations. The effects of these influences are further analyzed in Chapter 6.

Autosomal modifiers of sex expression are considered in Shaw (1958), and a two-allele model is studied in Eshel (1975). Nur (1974), Speith (1974), and Uyenoyama and Bengtsson (1979), among others, consider the case of progeny sex ratios and fertility differences determined at an autosomal locus in the female parent in diploid and haplodiploid populations. Eshel and Feldman (1982a, b) find conditions for the initial increase in frequency of a new modifier in multiallele contexts.

1.5. THE EVOLUTION OF SEX-DETERMINING MECHANISMS

Most animals are dioecious diploids and transmit their maternal and paternal genes in equal frequencies. Plants are overwhelmingly hermaphrodites, and a majority allow some degree of self-fertilization. Sex conversions related to size and environmental ambience occur in many fish populations. More complex organisms rarely undergo a sex change. Some plants change their relative expenditure between seed (or fruit) and pollen production depending on environmental shifts. When and to what extent is sex environmentally determined? Which came first, environmental sex determination (ESD) or genetic sex determination (GSD)?

Regular alteration between haploid and diploid phases is common in eucaryotes; cellular fusion is followed by nuclear fusion and recombination, the latter events usually delayed until meiosis. Procaryotes ordinarily do not undergo cellular fusion. The evolution of sexual dimorphism in eucaryotes may have entailed a selective process through intermediate steps involving a multiplicity of haploid sex genotypes representing several partially self-sterile mating types. One possible order of events starts from hermaphroditism and evolves to sex determination based on two alternative blocks of genes, followed by a loss of segregational machinery leading ultimately to

23

a chromosomal sex determination system such as XX/XY (ZW/ZZ) or XX/XO (balanced). The two kinds of gametes— male and female—can be more efficiently produced, i.e., pre- serve sex differentiation, by being distinct in form and size.

Chromosomal Sex Determination

Bull (1983) provides a recent summary of sex-determining mechanisms emphasizing the vast diversity across animal and plant species. Sex chromosomes in mammals are generally thought to have evolved in two stages out of an initial X-Y pair of homologous autosomes: the first stage involves mech- anisms for suppressing crossing over between X and Y (or Z-W in avian species), and in the second stage there is degeneration of Y (or W) gene functions. Moreover, in most invertebrates with chromosomal sex determinants Y lacks many of the genes on X and YY becomes inviable (demonstrated for *Drosophila* [Bridges, 1916] and guppies [Winge, 1934]). Sex factors presumably concentrate tightly on Y (or W) operating early in development, while the X chromosome retains the initial gene functions of the progenitor autosomes. In many cases, weak sex determinants are retained by X with covariate autosomal genes. White (1973) surveys many cases of a variable number of X chromosomes.

The X and Y chromosomes of vertebrates tend to be hetero- morphic, of different size and shape. Sex chromosome hetero- morphism in hermaphroditic species is documented for only a few flowering plants (Westergaard, 1958). Several mosses and liverworts also have heteromorphic sex chromosomes, but these species in general show multiple (often haploid) sex factors. Amphibians and reptiles usually show only slightly differen- tiated sex chromosomes.

In mammals the Y chromosome is largely heterochromatic and genetically inert. Dosage compensation (inactivation of one of the X chromosomes in females) guarantees equal gene ac- tivity in the two sexes. Ordinarily the specification of which X (maternal or paternal) becomes heterochromatic is random

24

over all cell types for placental mammals. In marsupials it is usually the paternal contribution that becomes heterochromatic. Autosomal gene modifiers can affect which X is inactivated, and this can depend also on the allele content.

Absence of dosage compensation is virtually universal in ZZ/ZW systems. Ohno (1967, pt. 2) reviews possible theories of the evolution of dosage compensation mechanisms and highlights contrasts between mammalian genes and X-linked genes in *Drosophila*. The absence of dosage compensation for Z-linked genes of avian species is also discussed. Ohno envisions polyphyletic origins emanating from the reptile line such that birds, snakes, and lizards branched off first and much earlier, establishing female heterogamety (ZZ/ZW)without dosage compensation, whereas the placental mammal lineage developed or branched off much later, at the beginning of the Cenozoic era ($80–140 \times 10^6$ years ago), developing male heterogamety (XX/XY) and dosage compensation.

Most fish species are without chromosomal sex differentiation and it is hypothesized (e.g., Ohno, 1979) that prior to 300 million years ago chromosomal sex determination did not exist. In many lower vertebrates sex chromosomes exist in a primitive state of differentiation accommodating a multiplicity of sex tendencies.

It is often hypothesized that the primitive state of sex expression was rather randomly set, often acting in response to environmental cues. The following evolutionary scenario has been suggested. The earliest fish and invertebrates were hermaphrodites but later specialized to genotypic control of sex expression. From single-gene sex determination, or determination by sets of genes, the cumulation of mutations (multiple alleles) and the reduction of linkage associated with homozygote lethals could have given rise to chromosomal sex determination.

A survey of the literature confirms that male heterogamety or female heterogamety are highly common forms of chromosomal sex determination. In some taxonomic groups such as birds and mammals, a single form of heterogamety uniformly occurs. In various invertebrate groups, e.g., salamanders and midges, the nature of heterogamety varies. Bull (1983, chaps. 16–17; see also Bull and Charnov, 1977) proposes some multi-locus models that may mediate transition from male to female heterogamety or the other way around; see Chapter 5.

The progenitors of XX/XY (and ZZ/ZW) sex determination are generally postulated to be autosomes: that is, it is argued that the Y (and W) eliminated all their Mendelian autosomal genes with progressive genetic deterioration. There is some evidence that these constrictions were accelerated by substantial chromosomal rearrangements. In fact, a pericentric inversion on the Y (and W) has been postulated as a first step toward heteromorphism. It has been suggested (see Charlesworth, 1978) that the development of chromosomal sex determination may lead to gene expression with different effects in males and females, or conversely. Concomitantly, sex differences in fitness could lead to X-Y crossover suppression. Moreover, Y degeneration and X inactivation may mutually accelerate the evolution of both.

No precise knowledge of the male- or female-determining factors associated with the X and Y chromosomes is available. But, it is known that in mammals Y contributes decisively to sex determination, apparently in early ontogeny, and that in most vertebrates Y plays a dominant role, as demonstrated by deletion experiments. It appears that the X of mammals does carry some female-determining factors (Ohno, 1967, pt. 3). In invertebrates a balance effect on the number of female factors often operates in sex determination, so that XO in Drosophila generally functions as male. Autosomal modifiers can confound the outcome. Each X-linked gene in mammals has adjusted to a hemizygous state (as a consequence of dosage compensation), but this is often compensated for by an increased rate of product. Lyon (1974) points out that several related genetic functions are duplicated on the mammalian X chromosome, strengthening the stability of sex-linked traits.

26

As pointed out earlier, the traditional model of the evolution of the sex-determining Y chromosome, or W in birds and snakes, following Muller (1918) (see also Charlesworth, 1978), suggests a gradual evolution to specialization accompanied by the accumulation of lethals and followed by condensation. The compensatory inactivation of one of the X chromosomes in mammalian species is considered to have evolved by an independent mechanism.

Jones (1983) proposes that evolution of sex chromosomes was an abrupt singular event emerging as a consequence of chromosomal inactivation. He suggests a possible mechanism of insertional mutagenesis involving specific genes and transposons, a mechanism that failed to occur in snakes and certain bird species in which chromosomal sex determination did not evolve. Some yeast Ty-insertion elements that act under influences of mating types support the feasibility of this scenario. In this view, conserved repeated sequences are associated with sex chromosomes, probably resulting from transposition, interspersion, and amplification. Evolutionary changes within the sex-determining chromosomes possibly involving conserved transposable sequences could supply an important mechanism for speciation. Genetic isolation of the sex-determining chromosomes may cause rapid evolution of developmental incompatibilities.

Haplodiploidy

The formulation of genetically based models designed to explain the evolution of haplodiploidy is an intriguing problem with a voluminous literature. Analytic models are found in Hartl and Brown (1970), Bull (1979, 1981c), Borgia (1980), and others.

Hartl and Brown (1970) present two models for the evolution of haplodiploidy from a precursor diploid situation. Their environmental model envisions various external stimuli that permit unfertilized eggs to develop. A fraction $1 - u$ of eggs are fertilized and the remainder unfertilized. Stipulating a sufficient fitness s for haploid males relative to a value 1 for diploid males and females, Hartl and Brown show the attainment of a haplodiploid mechanism. The exact conditions leading to such an attainment in a multiallele context are presented in Chapter 3 (see also Bull, 1981c; Eshel and Feldman, 1982b). Their second model is a genetic single-locus sex determination system in which a recessive allele allows haploid progeny. In this model haplodiploidy evolves through allelic substitution.

Both models are based on random mating and can realize 1:1 equilibrium sex ratios. However, Alstad and Edmund (1983) sampled a host of trees for black pineleaf scale insects (involving haploid males and diploid females) observing a range of sex ratios from 0.005 to 0.320.

Hartl and Brown emphasize the relative paucity of haplodiploid systems attributing the primary obstacle to the nature of the sex-determining mechanism. The Lepidoptera (e.g., butterflies) and Diptera (flies) are pure diploid organisms with sex controlled through ZZ/ZW or XX/XY and XX/XO mechanisms. There are rare cases in these groups in which unfertilized eggs develop into females. White (1973) reviews the vast array of sex-determining mechanisms for insects, precluding any unified theory but stressing the importance of a careful classification.

White (1973) compares male haplodiploidy to an extreme manifestation of sex linkage. It is not well understood why female haploidy is so rare. Bull (1979) emphasizes in support of haplodiploidy the advantage of haploid sons, as compared with diploid biparental sons, of having a twofold representation of maternal alleles in their gametes. He argues that when the fitness of haploid males exceeds half that of diploid males, selection could favor genes that act in the mother to produce haploid sons.

The ability of females in haplodiploid species to control sex ratio suggests a basis for selecting alleles directed to the establishment of male haploidy correlated with inbreeding. Such controls would also facilitate the immediate adjustment of sex ratios in response to environmental conditions. Diploids endowed with precise sex ratio controls (e.g., aphids) often have complex life cycles. There are no known examples of haplodiploid evolution with a heterogametic female progenitor. What are the relative benefits to be derived from interspersing a parthenogenic phase in reproduction (latent or controlled)? Borgia (1980) emphasizes two systems of sex determination that may allow the evolution of male haploidy: an XO female sex-determining system, and a system with overriding cytoplasmic factors.

As with most biological phenomena, there are difficulties and counterexamples for all models proposed to date, and there may be multiple routes. White (1973) in his discussion of the sex determination of haplodiploids (and in general of many insect organisms) suggests that a genetic balance depending on the female-determining X versus male-determining autosomes (akin to XO), rather than a dominant Y, appears to be paramount.

Environmental Sex Determination (ESD)

There is no clear empirical evidence suggesting the evolutionary order of ESD compared to GSD (genotypic sex determination), even though the ESD mechanism is considered very primitive. ESD is often regarded as an adaptation to special life histories, e.g., parasitic conditions.

Charnov and Bull (1977) emphasize that in patchy environments, where male and female fitnesses are variable and/or uncertain in space and time, there may be disadvantages in determining sex at conception. In this perspective ESD allows the embryo to adjust its sex according to environmental effects on fitness. This "intuitive" argument asserts that wherever selection favors sex ratio control, it favors a mechanism to promote that control. Hermaphroditism occurring sporadically in reptiles is generally considered a recent evolutionary phenomenon. Bull (1981a, b) examines cases in which the sex ratio and gender fitness vary spatially. If the probabilities of male expression and male fitness vary together and exhibit values opposite those for females, Bull predicts that an ESD system will evolve. He also examines some simple theoretical models starting with XX/XY sex determination that evolve to ESD.

Bulmer and Bull (1982) analyze several theoretical models in which sex determination is dichotomized by a threshold level based on a continuously varying phenotype. They suggest that the mean sex ratio of the population in such cases comes closer to 1:1 over successive generations. However, our analysis of multifactorial sex determination (MSD) models (Chapter 9) shows that a 1:1 population sex ratio outcome is unlikely, especially when environmental factors act asymmetrically on sex expression or maternal versus paternal transmission effects are unequal. Moreover, even when parental transmission is symmetric, the evolution of the index (phenotype) variable(s) governing MSD is sensitive to initial conditions (historical events) and population structure parameters.

When multifactorial sex determination depends on an index of physiological and environmental variables V, the system can be described as follows. There may exist a threshold v^* such that $V < v^*$ induces maleness and $V > v^*$ femaleness. In annelids, for example, large eggs tend to be female and small ones to be male, with subsidiary cytoplasmic controls. Some actual systems (in turtles) have two critical nest temperatures v^* and v^{**} that determine sex expression as depicted below:

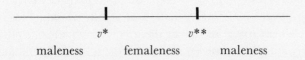

$$v^* \qquad v^{**}$$

maleness   femaleness   maleness

It is reported in Mrosovsky et al. (1983) that hatchling loggerhead turtles produced female brood proportions ranging from 0 female in early May to 0.80 in July which then dropped to 0.10 in August. These data are consistent with the role of temperature in directing sexual differentiation for turtle species as in the diagram.

In our study of theoretical models in which sex expression is governed by a threshold level of some index variable (Chapter 9), it appears that MSD will evolve to a single index value. This cannot be viably maintained unless sex determination ultimately becomes either genic or chromosomal.

Hermaphroditism

Although there are hermaphroditic plants with incompatibility systems, hermaphroditism is generally regarded as an adaptation to permit self-fertilization. Ghiselin (1974) describes three possible models for the origin of hermaphroditism.

(1) The low-density model suggests that where mating encounters are few (e.g., as a result of low mobility or low population density), hermaphroditism can help to ensure reproductive fulfillment. The high frequency of hermaphroditism in plants is often attributed to their sessility. Also hermaphroditism, like parthenogenesis and related phenomena, can have a decisive advantage where there are population bottlenecks, or in highly fluctuating environments.

(2) The size-advantage model is often used to account for sequential hermaphroditism. Reproductive potential can change with size, favoring one sex in early years and the other sex in later years.

(3) The gene-dispersal model postulates that dispersal constraints may affect population structure. Sequential hermaphroditism may increase reproduction by allowing more time for dispersal and by partially circumventing inbreeding effects. In small populations (especially when they are isolated) there often exists an excess of one sex, and in such cases hermaphroditism can increase the average fitness of an offspring.

31

Issues of kin selection and altruism are also sometimes introduced to explain the evolution of hermaphroditism and parthenogenesis. In plants, increased self-fertilization often correlates with diminished pollen production.

1.6. REVIEW OF THE LITERATURE ON SEX RATIO THEORY

Tables 1.1 and 1.2 list some models concerned with sex and sex ratio evolution, and cite sources that can be consulted for further reference. The last column of Table 1.1 relates the various sex-determining systems to the models investigated in this book.

TABLE 1.1. Genetic Models Related to Sex Ratio Evolution

Sex-determining systems	Some references	Relevant chapters in this book
One-locus/dioecious autosomal, X-linked, dichotomous (determined by the offspring genotype)	Shaw (1958), Scudo (1964) Lloyd (1974), Eshel (1975), Eshel & Feldman (1982a), Jayakar (1982), Karlin and Lessard (1983)	Chaps. 2–4
Multilocus/dioecious	Scudo (1967a), Bull & Charnov (1977), Bull (1983)	Chap. 5
Multifactorial ("polygenic")	Kosswig (1964), Bacci (1965), Bulmer & Bull (1982), Kallman (1983)	Chap. 9
Haplodiploid	Hartl & Brown (1970), Charnov (1978), Bull (1979, 1981c), Borgia (1980), Eshel & Feldman (1982b), Charnov (1982)	Chaps. 2–3
Haploid-multiallele, Y-linked, X-linked in the hetero-gametic sex	Edwards (1961), Hamilton (1967), Maynard Smith (1978), Colwell (1981), Gregorius (1982) Lessard (1984)	Chap. 6

(Continued)

TABLE 1.1. (Continued)

Sex-determining systems	Some references	Relevant chapters in this book
Maternally controlled, extra-chromosomal, cytoplasmic, contagion	Nur (1974), Speith (1974), Uyenoyama & Bengtsson (1979, (1981), Uyenoyama & Feldman (1978), Bull (1983)	Chaps. 2, 3, 7
Hermaphroditism: sex allocation to sperm and ova function, sex conversion	Ghiselin (1974), Leigh et al. (1976), Maynard Smith (1978), Charlesworth & Charlesworth (1981), Charnov (1982)	Chaps. 2, 3, 7
Incompatibility systems: multisex classes	Finney (1952), Workman (1964), Karlin & Feldman (1968a, b), Speith (1971), Heuch (1979)	Chap. 8
Environmental sex determination	Charnov & Bull (1977), Charnov (1982), Bull (1983)	Chap. 9
Sex determination with non-random mating, population subdivision, spatial and temporal variation, demographic effects, random drift	Hamilton (1967), Charlesworth (1977), Charnov (1979a), Taylor & Bulmer (1980), Wilson & Colwell (1981), Bull (1981a), Uyenoyama & Bengtsson (1982), Lloyd and Bawa (1984)	Chaps. 2, 6, 7
General two-sex two-allele selection models and related fertility models	Bodmer (1965), Karlin (1972, 1978), Speith (1974), Uyenoyama & Bengtsson (1979)	Chap. 7
Meiotic drive	Thomson & Feldman (1975), Bengtsson (1977), Maffi & Jayakar (1981)	Chap. 6

NOTE: (i) Controls of sex expression: zygotic, maternal or paternal genotypic control, probabilistic versus exact (dichotomous) sex classes. Models of multifactorial sex determination are based on zygote or maternal phenotype or suitable index variables. For sex influenced by cytoplasmic or extrachromosomal elements there can be viral mediators. (ii) Mating systems include: random mating, partial sib mating, partial selfing with possible multiple mating periods, population dispersal between matings, population subdivision. (iii) Selection forces: viability selection; litter size differentials depending on mating type, which may vary spatially and temporally.

TABLE 1.2. Adaptive Concepts Related to Sex and Sex Ratio Evolution

Concepts	Some references
Parental investment (sex allocation)	Fisher (1930), MacArthur (1965), Trivers (1972), Charnov (1982)
Family planning and sex ratio	Edwards (1962, 1966, 1970), Beiles (1974), Teitelbaum (1972)
Local mate competition (LMC) between males	Hamilton (1967, 1979), Alexander & Sherman (1977), Werren (1980), Taylor & Bulmer (1980)
Local resource competition (LRC) between females	Clark (1978), Charnov (1982), Bulmer & Taylor (1980b)
Population structure and sex ratio (group selection)	Bulmer & Taylor (1980a), Colwell (1981), Wilson & Colwell (1981)
Age-dependent sex ratio	Charlesworth (1977), Charnov (1975, 1979a)
Size hypothesis in sex reversal	Ghiselin (1974), Policansky (1981), Charnov (1982)
Encounter and behavioral covariate of sex ratio	Crew (1937), Kalmus & Smith (1960) Cannings & Cruz Orive (1975)
Contagion and sex ratio	Uyenoyama & Feldman (1978), Hamilton (1980)
High variability in progeny sex ratios	Kolman (1960), Bar-Anon & Robertson (1975)
Intergenerational conflicts on sex choice (parent-offspring conflict)	Trivers (1974)
Sib-mating effects on sex ratio	Maynard Smith (1978), Stenseth (1978), Uyenoyama & Bengtsson (1982)
Sex ratio when gender fitness varies spatially and temporally (patchy versus fine environment)	Charnov & Bull (1977), Charnov (1979a), Bull (1981a)
Haplodiploidy	Hartl & Brown (1970), Borgia (1980), Bull (1979, 1981c)
Mother condition and sex of offspring	Schuster & Schuster (1972), Trivers and Willard (1973)
Sex ratio versus relatedness (kin selection)	Hamilton (1972), Trivers & Hare (1976), Uyenoyama & Bengtsson (1981)
Mendelian segregation (non-adaptive sex ratio)	Williams (1979)

(Continued)

TABLE 1.2. (Continued)

Concepts	Some references
Sex-ratio homeostasis in finite populations	Bodmer & Edwards (1960), Verner (1965), Taylor & Sauer (1980)
Inclusive fitness functions in sexual populations	Oster et al. (1977), Stubblefield (1980), Uyenoyama & Feldman (1981)
Selfing versus incompatibility (hermaphroditism versus dioecy)	Charnov et al. (1976), Charlesworth & Charlesworth (1978a, b, 1979), Maynard Smith (1978), Bawa & Beach (1981), Lloyd (1982)
Short-term versus long-term advantages for sex and recombination (population vs. individual fitness)	Crow & Kimura (1965), Maynard Smith (1968, 1978), Eshel & Feldman (1970), Karlin (1973), Williams (1975)
Dimorphism related to sex: anisogamy, sexual selection	Scudo (1967b), Parker et al. (1972), O'Donald (1977, 1980), Maynard Smith (1978), Bell (1982)
Sex chromosomes	Ohno (1967, 1979), White (1973), Bull (1983)
Mating system and hermaphroditism	Robertson (1972), Fischer (1980), Charnov (1979b), Queller (1983)
Cost of meiosis and sex ratio	Williams & Mitton (1973), Maynard Smith (1978), Lloyd (1980), Uyenoyama (1984)

1.7. SUMMARY OF THE MAIN RESULTS OF THIS BOOK

The detailed analyses of Chapters 2–9 have two main objectives. First, a general formalism is established that can be applied to the study of a wide range of theoretical models involving mechanisms of sex determination and the evolution of sex ratio within the constraints imposed by these mechanisms. Second, the analytic models attempt to elucidate functional relationships between parameter specifications of the pertinent biological and environmental factors and the resulting evolutionary patterns. The solution to the models will generally not be quantitatively applicable but may help provide a qualitative conceptual basis for the interpretation of field and laboratory data and suggest further experiments and observations.

Multiallelic Sex Determination and
Associated Two-Sex Viability Selection Systems

The evolutionary dynamics of the basic zygotic multiallelic sex determination model can be identified as an important case of the dynamic system describing the allele frequency changes for a sex-differentiated viability model. Specifically, consider a bisexual population with r possible alleles $A_1, A_2, \ldots, A_r$ at an autosomal locus primarily responsible for sex determination. Let m_{ij} be the probability that an $A_i A_j$ individual is male, and $1 - m_{ij}$ its probability of being female. The matrix $M = \|m_{ij}\|_{i,j=1}^r$ is referred to as the *sex determination coefficient matrix.*

The case $0 < m_{ij} < 1$ may reflect the effects of modifier genes or other exogenous factors. When $m_{ij} = 1$ or 0, the collection of all genotypes $A_i A_j$ divides into two groups $\mathscr{G}_M$ and $\mathscr{G}_F$, where all individuals of type $\mathscr{G}_M$ and $\mathscr{G}_F$ are unambiguously male and female, respectively. This situation is referred to as *dichotomous (exact) genotypic sex determination.*

Consider the model of allele frequency changes for a dioecious panmictic population subject to sex-differentiated viability selection with viability coefficient matrix $M = \|m_{ij}\|$ (interpreting m_{ij} as the fraction of males among $A_i A_j$ zygotes that become reproductive) and female viability matrix $F = \|(1 - m_{ij})\| = U - M$ (U is the matrix of all unit entries). The dynamics in this model summarize precisely the evolution of sex ratio when sex is determined by the zygote genotype at an autosomal locus. In terms of a two-sex viability system with male and female viability matrices M and $U - M$, respectively, a rather complete analysis is possible. A new characterization of the equilibria of the classical multiallele monoecious viability model, based on the critical points of a spectral function (Theorem 3.1), has been found to be decisive for a theoretical analysis of the evolution toward stable sex ratios in panmictic populations.

The models formulated in Chapter 2 concern sex expression determined by the zygote genotype (Model I) or by one or both parental genotypes (Models II, II′, II″). A sex allocation perspective appropriate for simultaneous hermaphroditism, as often occurs with plant or fish species, requires only a reinterpretation of the sex determination parameters m_{ij} as the relative proportions of fixed individual resources allocated to male versus female functions (Model VI). Similar formulations can be applied to sex modification at an X-linked locus with multiple alleles (Model VII, VII′), to mixed sex determination in haplodiploid populations (Model V), and to the analysis of allele frequency changes with general fertility schemes (Models III, VIII, IX). All these models reduce to special classes of two-sex viability models, or variant models in which the viability of offspring depends on one or both parental genotypes.

A mixed regime of probabilistic sex determination and exact sex determination entails a number of strong sex-determining alleles analogous to the mammalian Y chromosome and a hierarchy of alleles X_i, $i = 1, \ldots, r$, bearing female tendencies. In this setting any genotype X_iY carrying Y is invariably male while the genotype X_iX_j will become male with probability m_{ij} and female with probability $1 - m_{ij}$. By relabeling Y as allele X_{r+1} and defining $m_{r+1,i} = 1$ for all i, we reduce theoretical considerations to the study of Model I of genotypic sex determination with $r + 1$ alleles.

When the sex determination coefficients and/or gender fitness parameters vary spatially and/or temporally, the multiallele sex determination model again reduces to a two-sex viability model, but this time with M and F matrices such that the special relationship $F = U - M$ may not hold (e.g., in Model IV of Chapter 2). In Chapter 7 general two-sex viability models are fully analyzed in the case of two alleles.

Genotypic and Phenotypic Equilibria

The models of sex ratio evolution analyzed in Chapter 3 (except Model IV and general fertility schemes) are concerned

37

with two kinds of equilibria, genotypic and phenotypic. A genotypic equilibrium is characterized by identical gamete frequencies for the male and female populations; it tends to show a biased population sex ratio and is vulnerable to allelic mutational substitutions that are generally in the direction of reducing the bias. By contrast, a phenotypic equilibrium persistently shows an adjusted $1:1$ population sex ratio, but this ratio can be realized by many genotypic frequency configurations. Phenotypic equilibria are structurally stable, that is, they cannot be altered by mutational events, although the underlying genotypic frequencies may change. The existence of a continuum of genotypic equilibria yielding a $1:1$ population sex ratio derives from the linear relationship $f_{ij} = 1 - m_{ij}$ for all i,j (it suffices that $f_{ij} = a + bm_{ij}$ to have a similar property, see Chapter 7). For general $M = \|m_{ij}\|$ and $F = \|f_{ij}\|$ where $F \neq aU + bM$, we hypothesize that only finitely many stable genotypic equilibria can exist. In this case convergence of the dynamic system (the genotypic frequencies over successive generations) to one of the stable equilibrium points should occur, the specific equilibrium attained depending on the initial population composition. Spatial and temporal variation in the sex determination coefficients coupled to sex-specific viability selection effects generally correspond to M and F not linearly related (see Section 2.5).

A comprehensive classification of all two-allele sex-ratio equilibria and various equilibrium contingencies for three-allele sex-determining systems are presented in Chapter 4.

"Optimal" Sex Ratios

Optimality properties of phenotypic equilibria are established in Chapter 3 for most of the models of Chapter 2. In the basic sex determination model of r alleles (Model I), the optimality property for an even sex-ratio evolution is understood to mean that, with the introduction of a new allele (from r to $r + 1$ alleles), all possible equilibrium states that are stable for the extended model cannot attain a sex ratio farther from 1:1 than the existing ratio in the r-allele subsystem. This expresses an evolutionary tendency toward an even sex ratio. A sex ratio that is evolutionary "optimal" has the property that under modifications by successive mutant genes any deviant sex ratio would normally evolve to a sex ratio closer to this optimum. The 1:1 population sex ratio is optimal in this sense, following Theorem 3.7. This finding applies to sex determination under zygotic or parental controls in diploid and haplodiploid populations.

Although the Y and W chromosomes of XX/XY and ZZ/ZW systems, respectively, are primarily heterochromatic, it is conceivable that with respect to sex determination they function as a large multiallele system, thus approaching a 1:1 sex ratio in line with Theorem 3.7.

Some authors distinguish between the numerical sex ratio (the head count for a particular cohort) and the sex allocation of parental investment (the relative cost of producing male versus female offspring). Differences in cost can, for example, reflect differences in gender fitness or relate to sex-specific environmental factors. The tendency toward a 1:1 sex ratio with parental controls applies for variable costs of male and female offspring provided the total parental investment is constant and the population is censused at the appropriate life stage, namely, at the end of parental care. Sex ratio counts at different life stages with associated sex-specific fitness differences will vary as well. However, when genotypic and gender viability or fertility differential selection effects are absent, the sex ratio at conception will essentially coincide with the adult sex ratio.

In haplodiploid populations, the sex ratio is naturally partitioned among male haploids, male diploids, and female diploids. In this context, the sex ratio can be appraised either by contrasting male diploid with female diploid numbers or by cumulating haploid and diploid males into a total haploid number and contrasting this number with the haploid number accruing from female diploids. Sex ratio counts would take into account selection differences that can exist between haploid and diploid males.

Model V of Chapter 2 on sex ratio determination in a haplodiploid species involves parameters α and s such that α indicates the fraction of unfertilized eggs (haploid males) per generation and s is the selection coefficient that measures the relative viability of haploid males compared to one for diploid males, and a general probabilistic genotypic sex determination coefficient matrix $\|m_{ij}\|$ for diploid offspring. Assuming equal virility for haploid and diploid males, the condition for only haploid males to be "optimal" is $\alpha \geq 1/(1 + 2s)$. Thus we might expect the population to retain only haploid males if either the proportion of unfertilized eggs is relatively large or selection strongly favors haploid over diploid male types. In other words, even when haploid males are much disadvantaged (s is small) the diploid male population is ultimately eliminated provided the proportion of unfertilized eggs per generation is sufficiently large. This result suggests that the wasp species *Habrobracon*, in which both haploid and diploid males are represented, has α and s parameters such that $\alpha < 1/(1 + 2s)$. If the virility of haploid males is one-half that of diploid males then this condition becomes $\alpha < 1/(1 + s)$ leading to a $1:1$ population sex ratio at maturity.

Are There Simple Optimality Principles Underlying Sex Ratio Realizations?

A number of authors have proposed simple optimality principles governing sex ratio evolution. On the basis of the Shaw-Mohler formula (1.1), MacArthur (1965) proposed that populations should evolve to maximize the product $m \cdot f$, where $m(f)$ is a suitable quantitative measure of maleness (femaleness) (see Section 7.4 for details). Charnov (1982) presents a variety of graphical displays based on male and female "fitness sets" which are presumed to summarize spatial, temporal, demo-

40

graphic, mating pattern, and other characteristics. Postulating that a stable population sex ratio state corresponds to a fitness configuration that maximizes $m \cdot f$, he examines cases of fitness sets bounded by concave and convex curves and other regular shapes, describing in these cases the qualitative directions of sex ratio evolution.

The Shaw-Mohler equation, which comes into play when conditions favor an initial increase in frequency of new alleles, can serve to determine *fixation states* (i.e., sex allocation strategies practiced by all members of the population) that can be stable against any new mutant stategy satisfying specific constraints on the individual resources allocated to sex. This leaves open the question of evolutionary patterns in *polymorphic* diploid and haplodiploid populations.

For general two-sex viability models there are no demonstrated optimality properties over successive equilibria. Even in the case of two alleles, the quantity $m \cdot f$ is not necessarily maximized at a stable equilibrium (see, e.g., Speith, 1974; Uyenoyama and Bengtsson, 1979).

When one-locus sex allocation matrices F and M, designating female and male functions, respectively, obey a linear relationship in the form $F = aU + bM$ leading to $f = a + bm$ on the average in the population, the "optimal" allocation m to the male function in the population maximizes the product $f \cdot m$ in the sense that this quantity increases from one equilibrium to the next following the introduction of a new allele (see Theorem 7.2). However, this quantity does not necessarily increase over successive generations, allowing some oscillations even near a stable equilibrium.

Maximization characterizations of polymorphic sex ratio equilibrium states are generally not available in models allowing for spatial variation in sex expression, or in the presence of sex-specific fitness differences. Optimality expressed by $\max(m \cdot f)$ intrinsically reflects some kind of symmetry and is generally restricted to one-locus controls. Where sex-determining genes are linked to viability factors, a spectrum of sex ratios

41

can be expected, and simple optimization criteria that would characterize the corresponding stable polymorphic equilibria are unknown (see Chapter 7).

Exact Genotypic Sex Determination

When the genotype-phenotype classes of sex determination at one locus are not absolute, that is, for $0 < m_{ij} < 1$, a stable non-even sex ratio is possible according to Theorem 3.3. This may be the case if a probabilistic mechanism is imposed by virtue of endogenous or exogenous genetic and environmental co-variates. Theorem 3.5 establishes that if exact genotypic sex determination ($m_{ij} = 0$ or 1 for all i, j)delimits the sex dimorphism of the two phenotype classes in such a way that $\mathcal{G}_M$ consists of all genotypes with $m_{ij} = 1$ and $\mathcal{G}_F$ consists of all genotypes with $m_{ij} = 0$, then *only an even population sex ratio can be stable*. Global convergence to such equilibria is demonstrated in the case of three allele dichotomous sex determination systems and for the general multiallelic model where heterozygotes determine one sex and homozygotes determine the other sex. By contrast, as shown in Chapter 5, this outcome of an even sex ratio generally does not occur in the case of multilocus exact genotypic sex determination with loose linkage.

Chapter 8 examines some models on incompatibility systems with several phenotypic classes. General conditions for the existence of isoplethic equilibrium states (the multisex analog of an even sex ratio) are ascertained. A dichotomy of genotypic versus phenotypic equilibrium realizations emerges in this context, as with two-sex determining systems. It is expected that the "optimality" characteristic of equal phenotypic frequencies will prevail, but a rigorous proof seems difficult.

Multilocus and Multifactorial Sex Determination

The study of multifactorial sex determination is relevant in at least two contexts. First, the sex phenotype may be influenced by many loci (e.g., mediated through some developmental or physiological process), as is probable with various fish, amphibian, reptile, and invertebrate species. Second, environmental sex determination (e.g., where sex depends on nesting temperature) probably reflects a complex situation of gene-environment interactions. In this perspective the environmental variable can be regarded as an autocorrelated multifactorial trait obeying certain familial phenotypic transmission rules.

The concept of a multifactorial (polygenic) model is intrinsically unclear. Many genes are interacting, and their interactions may be coupled to environmental stimuli in complex ways. In the models of Chapter 9 we assume that sex espression depends on a phenotype value V (e.g., some physiological covariate) or is determined in response to a variable V that indexes an environmental state (e.g., temperature, humidity) such that an offspring encountering the state $V = v$ becomes male with probability $p(v)$ and female with probability $1 - p(v)$. We do not attempt to make precise the underlying genetics. The changes of the variable V are described by a transmission process, in which the offspring value receives some "blend" of the parental values modified by independent individual residual terms. Section 9.6 considers a mixed model in which a major gene combines with multifactorial influences to control sex expression.

The main conclusion that emerges from our studies of multifactorial sex determination (Chapter 9) is that a non-1:1 population sex ratio is expected when environmental pressures affect the sexes asymmetrically, or when paternal and maternal genetic or phenotypic contributions are unequal.

Non-even sex ratios are also promoted by multilocus determinants of sex expression under loose linkage (Chapter 5). The prevalence of non-even sex ratios among invertebrates leads to the hypothesis that the sex factors in many fish, amphibian, reptile, and invertebrate species are not confined to a single gene but are distributed over the genome.

The dichotomy of one locus as against polygenic inheritance excludes a multitude of natural alternatives in seeking a genetic mechanism for the determination of continuously distributed traits. It may be reasonable to envision genotypic-phenotypic associations based on 1–10 loci contributions coupled to environmental factors that may underline phenotypic expression. In this perspective, two or more loci models may be construed to represent polygenic inheritance (cf. Lewontin, 1974, chap. 5). Classifications and characterization pertaining to the dynamics and equilibrium behavior in multilocus theory incorporating components of selection-recombination events, nonrandom mating patterns, and facets of population structure are becoming increasingly amenable to theoretical analyses, numerical simulation, and interpretations, e.g., see Karlin (1979a), Karlin and Liberman (1979a–d), and Karlin and Avni (1981).

The models of Chapter 5 having the sex phenotype determined by the genotypic composition of several loci are analyzed using the general theory of sex-differentiated models allowing for general recombination schemes. It is striking that already for two (or more) unlinked genes of dichotomous *exact* (nonprobabilistic) sex determination the stable sex ratio is generally not 1:1. This contrasts sharply with the results of Chapters 2–4, attesting to the tendency for an even sex ratio when sex expression is controlled at a single multiallelic locus.

Biased Sex Ratios

Why is the sex ratio not 1:1 for many invertebrates? This question can be partly answered in terms of the following factors and patterns: the effects of population subdivision accompanied by differential gender selection over space and time (Chapter 7); the consequences of inbreeding, local mate competition, and life history strategies (Chapter 6); sex-differentiated viability and fertility forces (Chapter 7); X- or Y-linked modifiers of sex determination (Chapter 6); meiotic drive and extrachromosomal factors (Chapter 6); multilocus and multifactorial sex determinations (Chapters 5 and 9); and behavioral adaptations and manipulations.

44

Chapter 6 presents several models of sex allocation in structured populations, allowing for nonrandom mating patterns embracing partial selfing and sib mating. We have concentrated on population-genetic models governing sex ratio evolution. Population structures and regulations causing local mate competition or local resource competition tend to bias the sex ratio in order to equalize male and female competition for reproduction independently of the breeding regime.

Meiotic drive, which effectively induces sex-differentiated selection, also promotes non-even sex ratio outcomes.

For general two-sex models, the sex ratio is entirely dependent on the viability and/or the fertility regime and the genetic constraints. Some fertility schemes and consequences of sex conversion possibilities are analyzed in Chapters 7 and 9.

CHAPTER TWO

Concepts and Parametrizations for One-Locus Multiallele Sex-Determining Systems and Related Systems

Sex determination governed at an autosomal locus, with probabilities of being male or female depending on the zygote genotype, was first given a theoretical formulation in Shaw (1958). Eshel (1975) studied a two-allele model. A variant of this model dealt with in Nur (1974), Speith (1974), Uyenoyama and Bengtsson (1979) among others, involving maternal (or paternal) genotype controls on the expected brood sex ratio and brood size, produced similar qualitative conclusions to Eshel's. Eshel and Feldman (1982a, b) considered generalizations to multiallelic cases including haplodiploid models. Bull (1981a) and Charnov (1979a) studied spatial and temporal variations on sex expression and intersex viability.

In this chapter we set forth, for random mating populations, a series of sex-determining mechanisms involving an autosomal or sex-linked locus that entails multiallelic variants allowing sex expression to depend on the zygote genotype, or one or both of the parental genotypes. These mechanisms relate to general two-sex viability models as elaborated later. We subsume in the same framework the context of sex allocation to seed and pollen functions in plant populations. The effects of differential fertilities are examined separately. We also incorporate environmental influences on sex expression and fitness differences between the sexes allowing for spatial and temporal variations. Lastly, the corresponding models for sex ratio evolution in haplodiploid populations are set forth. Results and proofs can be found in Chapter 3 while specific examples are presented in Chapter 4. Multilocus sex determination is investigated in Chapter 5. The effects of inbreeding (e.g., sib-mating, selfing, and population subdivision influences) and meiotic drive are considered in Chapter 6. Chapter 7 presents a complete analysis for general two-sex two-allele models.

MODEL I. AN AUTOSOMAL, MULTIALLELE SEX-DETERMINATION SYSTEM BASED ON THE GENOTYPE OF THE OFFSPRING

Consider a bisexual infinite population with r possible alleles $A_1, A_2, \ldots, A_r$ at an autosomal locus primarily responsible for sex determination. Let us denote the frequency of genotype $A_i A_j$ in the female population by $2p_{ij}$ when $i \neq j$ and by p_{ii} when $i = j$. The frequency of allele A_i is $p_i = \sum_{j=1}^{r} p_{ij}$. The quantities $2q_{ij}$, q_{ii}, and q_i are defined analogously with respect to the male population. We assume discrete generations, random mating, Mendelian segregation, and equal fertility for all mating types. Let m_{ij} be the probability for an $A_i A_j$ individual to be a male, and $1 - m_{ij}$ that to be a female. Clearly, $0 \leq m_{ij} = m_{ji} \leq 1$. We refer to the non-negative symmetric matrix $M = \|m_{ij}\|_{i,j=1}^{r}$ as the *sex-determination coefficient matrix*.

The case $0 < m_{ij} < 1$ may reflect the effects of modifier genes coupled to prenatal and/or neonatal interactions with the milieu. Where $m_{ij} = 1$ or 0, the sex phenotype is determined such that the collection of all genotypes $A_i A_j$ partition into two groups, $\mathscr{G}_M$ or $\mathscr{G}_F$, where every individual of type $\mathscr{G}_M$ and $\mathscr{G}_F$ is unambiguously male and female, respectively, creating a dioecious population. We refer to this situation as *dichotomous (exact) genotypic sex-determination*. Corresponding multitype decompositions are appropriate for studies of incompatibility systems and for elucidating consequences of nonrandom mating patterns.

47

TABLE 2.1 Dichotomous Sex Partitions
with Two-Alleles at One Locus

	$\mathcal{G}_F$	$\mathcal{G}_M$
(i)	AA	AB BB
(ii)	AA BB	AB

With two alleles at a locus, modulo relabeling of alleles, there are two sex partitions possible, as shown in Table 2.1. In case (i), BB is eliminated in one generation, which corresponds to the classic XX/XY system $(X = A, Y = B)$. With three alleles $\{A, B, C\}$ at a locus there are nine distinguishable partitions apart from the relabeling of the alleles. All possible cases are listed in Table 2.2.

The system known in the platyfish (Kallman, 1973) corresponds to case (VI). The number of inequivalent dichotomous subdivisions with four alleles is 255 (Cotterman, 1953; Scudo, 1967a). With increasing numbers of alleles the number of dichotomous partitions grows more than exponentially. These dichotomous models are analyzed in Chapter 4.

TABLE 2.2. Dichotomous Sex Partitions with Three Alleles at One Locus

	$\mathcal{G}_M$			$\mathcal{G}_F$			
(I)	AA			AB	AC	BB	BC CC
(II)	AB			AA	BB	AC	BC CC
(III)	AA	BB		AB	AC	BC	CC
(IV)	AA	BC		AB	AC	BB	CC
(V)	AB	AC		AA	BB	CC	BC
(VI)	AA	AB		BB	AC	BC	CC
(VII)	AA	BB AB		AC	BC	CC	
(VIII)	AA	BB AC		AB	BC	CC	
(IX)	AA	BB CC		AB	AC	BC	

Returning to the general model, the genotype frequencies over two successive generations obey the recursion relations

$$q'_{ij} = \frac{m_{ij}(p_i q_j + p_j q_i)}{2w}, \tag{2.1a}$$

$$p'_{ij} = \frac{(1 - m_{ij})(p_i q_j + p_j q_i)}{2(1 - w)}, \quad i,j = 1, \ldots, r, \tag{2.1b}$$

where $w = \sum\limits_{i,j=1}^{r} m_{ij} p_i q_j$ is the proportion of males, the *sex ratio*, in the total population for the given allelic frequency state.

If gender male and female (but genotype-independent) viabilities act in the ratio of s to 1, the transformation equations (2.1) are unaltered although the juvenile and adult sex ratios differ. The juvenile (primary) sex ratio, as previously noted, is

$$w = \sum_{i,j=1}^{r} m_{ij} p_i q_j \tag{2.2}$$

but the adult (secondary) sex ratio is

$$\tilde{w} = \frac{sw}{sw + 1 - w}. \tag{2.3}$$

A higher mortality rate in males ($s < 1$ leading to $\tilde{w} < w$) is not uncommon in natural populations. It has often been observed that such differences should not alter the primary sex ratio (e.g., Shaw and Mohler, 1953; Bodmer and Edwards, 1960; Edwards, 1962; Leigh, 1970; Crow and Kimura, 1970).

It will be convenient for later purposes to employ the Schur product operation of two vectors $\mathbf{x} = (x_1, x_2, \ldots, x_r)$, and $\mathbf{y} = (y_1, y_2, \ldots, y_r)$, symbolized by $\mathbf{x} \circ \mathbf{y} = (x_1 y_1, x_2 y_2, \ldots, x_r y_r)$, and the inner product denoted by $\langle \mathbf{x}, \mathbf{y} \rangle = \sum\limits_{i=1}^{r} x_i y_i$. Further, if M is any matrix the representation $\mathbf{x} \circ M$ stands for the matrix product $D_{\mathbf{x}} M$ where $D_{\mathbf{x}}$ designates the diagonal matrix with entries $x_1, x_2, \ldots, x_r$ down the main diagonal. In this notation, the system (2.1) can be converted into the following recurrence equations for the frequency vectors $\mathbf{p} = (p_1, p_2, \ldots, p_r)$ and $\mathbf{q} = (q_1, q_2, \ldots, q_r)$.

$$q' = \frac{\mathbf{p} \circ M\mathbf{q} + \mathbf{q} \circ M\mathbf{p}}{2w} \qquad (2.4a)$$

$$\mathbf{p}' = \frac{\mathbf{p} \circ (U - M)\mathbf{q} + \mathbf{q} \circ (U - M)\mathbf{p}}{2(1 - w)} \qquad (2.4b)$$

where $w = \langle \mathbf{p}, M\mathbf{q} \rangle = \sum_{i=1}^{r} p_i (M\mathbf{q})_i = \sum_{i,j=1}^{r} m_{ij} p_i q_j$ and U denotes the matrix with all unit entries.

MODEL I'. MODEL OF SEX DETERMINATION IN RELATION TO SEX-DIFFERENTIATED GENOTYPE VIABILITY EFFECTS

The transformation equations (2.4) represent an important case of the dynamic system describing the allele frequency changes for a two-sex viability model. To this end, consider a diploid sex-differentiated population characterized by r alleles $A_1, \ldots, A_r$ at an autosomal locus with associated genotypes $A_i A_j$ under the effects of differential viability selection, random mating, and Mendelian segregation. The parameters of the model are as follows. The viability fitness matrix for females is $F = \| f_{ij} \|_1^r$ and for males $M = \| m_{ij} \|_1^r$, where the quantity $f_{ij}(m_{ij})$ is interpreted as the relative number of $A_i A_j$ female (male) zygotes that survive to contribute to the next generation. The frequency of the genotype $A_i A_j$ in the female population is denoted by $2p_{ij}$ when $i \neq j$ and p_{ii} for $i = j$, and the frequency of allele A_i in this population is thereby $p_i = \sum_{j=1}^{r} p_{ij}$. The corresponding frequencies for the male population are denoted by $2q_{ij}$, q_{ii}, and q_i.

The frequencies p_i' and q_i' in the next generation are determined by the transformation equations (cf., e.g., Karlin, 1978).

$$p_i' = \frac{\frac{1}{2}\left[p_i \left(\sum_{j=1}^{r} f_{ij} q_j \right) + q_i \left(\sum_{j=1}^{r} f_{ij} p_j \right) \right]}{\sum_{i,j=1}^{r} f_{ij} p_i q_j}, \qquad i = 1, 2, \ldots, r.$$

$$(2.5a)$$

$$q_i' = \frac{\frac{1}{2}\left[p_i\left(\sum_{j=1}^{r} m_{ij}q_j\right) + q_i\left(\sum_{j=1}^{r} m_{ij}p_j\right)\right]}{\sum_{i,j=1}^{r} m_{ij}p_iq_j}, \quad i = 1,2,\ldots,r.$$

(2.5b)

In the notation of (2.4) the recursion relations (2.5) take the form

$$\mathbf{p}' = \frac{\mathbf{p} \circ F\mathbf{q} + \mathbf{q} \circ F\mathbf{p}}{2\langle \mathbf{p}, F\mathbf{q}\rangle}, \mathbf{q}' = \frac{\mathbf{q} \circ M\mathbf{p} + \mathbf{p} \circ M\mathbf{q}}{2\langle \mathbf{p}, M\mathbf{q}\rangle}. \quad (2.6)$$

The above expression displays a nonlinear transformation T of $2r$ variables, ($(2r - 2)$ independent ones, since $\sum_{i=1}^{r} p_i = \sum_{i=1}^{r} q_i = 1$) which we abbreviate as

$$\{\mathbf{p}',\mathbf{q}'\} = T\{\mathbf{p},\mathbf{q}\}.$$

In the case $M = F$ (such that selection operates on the male and female genotypes symmetrically) then $\mathbf{p}' = \mathbf{q}'$ and in all subsequent generations the transformation T can be reduced to the single set

$$\mathbf{p}' = \frac{\mathbf{p} \circ M\mathbf{p}}{\langle \mathbf{p}, M\mathbf{p}\rangle}, \quad (2.7)$$

where the sexes need not be distinguished.

The analysis of the transformation (2.7), its convergence and equilibrium properties, has received much attention (e.g., Kingman, 1961a, b; Crow and Kimura, 1970; Nagylaki, 1977; Ewens, 1979; Karlin, 1978, and references therein). For this model (the classical one-locus multiallele viability selection model), the mean fitness function

$$w(\mathbf{p}) = \sum_{i,j=1}^{r} m_{ij}p_ip_j \quad (2.8)$$

provides a strict Lyapounov function for the mapping (2.7), such that

$$w(\mathbf{p}') \geq w(\mathbf{p}) \text{ with equality if and only if } \mathbf{p}' = \mathbf{p}. \quad (2.9)$$

51

This remarkable property is sometimes referred to as the discrete form of the fundamental theorem of natural selection (Crow and Kimura, 1970, chap. 5). For the general sex-differentiated model (2.6) (allowing $M \neq F$), there can exist multiple interior equilibria and therefore any associated Lyapounov function cannot be a multivariate quadratic polynomial, or a ratio of quadratic forms. Numerical studies virtually always exhibit global convergence of the iterates of (2.6). Nevertheless, we have not yet ascertained a usable Lyapounov function for the general sex-differentiated viability model.

The case

$$F = aU + bM \qquad (2.10)$$

(where U is the $r \times r$ matrix of all unit entries and a,b are two constants) is of special interest since any equilibrium $\mathbf{p}^*$ of (2.7) engenders an equilibrium pair for the equations (2.6) in the form

$$\{\tilde{\mathbf{q}}, \tilde{\mathbf{p}}\} = \{\mathbf{p}^*, \mathbf{p}^*\} \qquad (2.11)$$

referred to henceforth as a *symmetric* (in the sexes) *equilibrium*. There may also exist *nonsymmetric equilibrium* combinations, that is, equilibrium pairs $\{\tilde{\mathbf{q}}, \tilde{\mathbf{p}}\}$ with $\tilde{\mathbf{q}} \neq \tilde{\mathbf{p}}$.

Comparing (2.4) and (2.6) we see that the sex determination model, where an offspring of genotype $A_i A_j$ is male (female) with probability $m_{ij}((1 - m_{ij}))$, is equivalent to a sex-differentiated viability model with viability matrices $M = \|m_{ij}\|$ and $F = \|1 - m_{ij}\| = U - M$ for males and females, respectively. The fact that M and $F = U - M$ (as in (2.10)) are generated from a single matrix M allows a more complete analysis of the system (2.4) as we shall see in Chapter 3.

Fitness differences are sometimes linked to sex-determining genes and differ from one sex to the other (for example, pigment genes in some fish and bird populations are beneficial in males, serving objectives of sexual selection, while disadvantageous to females because of increased predation). Male and female platyfish express many color genes considered to be sex-linked (Kallman, 1970). Female guppies appear to have some Y-linked color genes (Haskins et al., 1970). In such circumstances, we must resort to the general formulation (2.6). A complete analysis in the case of two alleles is elaborated in Chapter 7.

MODEL II. MULTIALLELIC SEX RATIO DETERMINATION MODEL UNDER MATERNAL (OR PATERNAL) GENOTYPIC CONTROL

Consider a dioecious population with a sex-determining mechanism entailing r alleles $A_1, A_2, \ldots, A_r$ where the sex of the offspring is governed by its maternal genotype. Documentation and interpretations of this mechanism are amply illustrated in Chapter 1. In the present context we prescribe m_{ij} *to be the expected proportion of males in the progeny of an A_iA_j female parent.* The symmetric matrix $M = \|m_{ij}\|_1^r$ will be called the *maternal sex ratio determination matrix*. Probabilistic sex ratio determination with $0 < m_{ij} < 1$ subject to maternal control can result from influences of modifier genes with interactions to the cytoplasmic milieux. The maternal physiological capabilities and neonatal responses to environmental conditions can also be contributory factors.

Let $\tilde{p}_{ij}$ and $\tilde{q}_{ij}$ be the frequencies of the ordered A_iA_j genotype (the first allele A_i deriving from the maternal side and A_j from the paternal) in the female and male populations, respectively. The corresponding frequencies of the unordered A_iA_j genotypes are $2p_{ij} = \tilde{p}_{ij} + \tilde{p}_{ji}$ and $2q_{ij} = \tilde{q}_{ij} + \tilde{q}_{ji}$ for $i \neq j$ (p_{ii} and q_{ii} when $i = j$).

An A_i gamete is contributed from a maternal parent of genotype A_iA_l with probability $(\tilde{p}_{il} + \tilde{p}_{li})/2 = p_{li}$. Note that $\frac{1}{2} \sum_j (\tilde{p}_{jk} + \tilde{p}_{kj}) = p_k$ is the kth allele frequency in the female population and the corresponding male allele frequency is $\frac{1}{2} \sum_l (\tilde{q}_{kl} + \tilde{q}_{lk}) = \sum_l q_{kl} = q_k$.

53

Under *random mating*, the expected proportion of male offspring in the next generation when sex determination is subject to maternal control as delineated above is

$$\tilde{q}_{ij} = \sum_k \sum_l m_{il} \frac{(\tilde{p}_{il} + \tilde{p}_{li})}{2} \frac{(\tilde{q}_{jk} + \tilde{q}_{kj})}{2}. \qquad (2.12)$$

It is convenient to introduce the quantity $z_i = \sum_l m_{il} p_{il}$, $i = 1, 2, \ldots, r$, such that z_i/p_i, $i = 1, \ldots, r$, can be interpreted as the probability that an offspring receiving a maternal allele A_i is a male. Obviously, $w = \sum_{i=1}^{r} z_i = \sum_{i,l} m_{il} p_{il}$ is the *sex ratio* in the next generation.

The recursion equations can be compactly written in the form

$$q'_{ij} = \frac{\tilde{q}_{ij} + \tilde{q}_{ji}}{2} = \frac{z_i q_j + z_j q_i}{2w}. \qquad (2.13a)$$

and

$$p'_{ij} = \frac{(p_i - z_i) q_j + (p_j - z_j) q_i}{2(1-w)}. \qquad (2.13b)$$

The above system can be reduced to the component variables of **p**, **q**, and **z** since only these variables appear on the right of (2.13) leading to

$$\mathbf{q}' = \frac{\mathbf{q}}{2} + \frac{\mathbf{z}}{2w}, \qquad (2.14a)$$

$$\mathbf{p}' = \frac{\mathbf{q}}{2} + \frac{\mathbf{p} - \mathbf{z}}{2(1-w)}, \qquad (2.14b)$$

$$\mathbf{z}' = \frac{(\mathbf{p} - \mathbf{z}) \circ M\mathbf{q} + \mathbf{q} \circ M(\mathbf{p} - \mathbf{z})}{2(1-w)} \qquad (2.14c)$$

with $w = \sum_{i=1}^{r} z_i$.

MODEL II'. UNEQUAL COSTS IN PRODUCING OFFSPRING OF DIFFERENT SEXES

We now consider the model of maternal control acting both on progeny sex ratio and on brood size. More specifically, for a maternal genotype $A_i A_j$ let n_{ij} be the number of offspring and m_{ij} the brood sex ratio. Let the cost of producing a male offspring be c compared to 1 for a female offspring. Assuming a constant total (for convenience, normalized to unity) expenditure per family (see, e.g., Speith, 1974) an $A_i A_j$ mother bears progeny consisting of males and females of relative costs $n_{ij} c m_{ij}$ to $n_{ij}(1 - m_{ij})$ subject to the constraint

$$n_{ij} c m_{ij} + n_{ij}(1 - m_{ij}) = 1 \qquad (2.15)$$

In this context, Fisher (1930) predicted that selection will result in a collectively equal expenditure on daughters and sons. The foregoing model reduces precisely to Model II, the maternal sex-control inheritance model, identifying the maternal sex-determination matrix to be $\tilde{M} = \|\tilde{m}_{ij}\|$ where $\tilde{m}_{ij} = n_{ij} c m_{ij}$.

Without the restriction (2.15), admitting general fertility and sex ratio parameters n_{ij} and m_{ij}, respectively, the analysis is complicated and usually not tractable except for special relations on the parameter sets or when restricted to the stability analysis of fixation events. Uyenoyama and Bengtsson (1979) studied the two-allele maternal control model for the parameter scheme

maternal genotype	$A_1 A_1$	$A_1 A_2$	$A_2 A_2$
progeny sex ratio proportions	m_{11}	m_{12}	m_{22}
progeny relative fertility numbers	n_{11}	n_{12}	n_{22}

Partial results pertaining to the stability conditions for the equilibrium states were obtained and interpreted for various cases including (2.15). In Uyenoyama and Bengtsson's formulation, the fitnesses of the male and female offspring of an $A_i A_j$

55

mother are $\tilde{m}_{ij} = n_{ij}m_{ij}$ and $\tilde{f}_{ij} = n_{ij}(1 - m_{ij})$, leading to a general two-sex viability model. Such models are analyzed in Section 7.3 in relation to our study of the general two-sex model (2.6).

A more general model would consider the case where brood size and progeny sex ratio are determined by the genotypes of both parents. More specifically for the mating type

$$\begin{array}{cc} \female & \male \\ A_iA_j & \times & A_kA_\ell \end{array}$$

a fertility rate $n_{ij;\,kl}$ is effective with sex ratio $r_{ij;\,kl}$ under joint parental determination. The case $n_{ij;\,kl} = 1$ and $r_{ij;\,kl} = r_{ij}s_{kl}$ can be reduced to a variant of model (2.6). Compare with the case $n_{ij;\,kl} = \mu_{ij}v_{kl}$ and $r_{ij;\,kl} = \frac{1}{2}$ treated in Bodmer (1965) and Karlin (1978). (See Appendix A later in this chapter.)

MODEL II″. DIFFERENTIAL VIABILITY WITH REPLACEMENT UNDER MATERNAL CONTROL

It is instructive to reverse the order of events underlying (2.15). Suppose that males and females have relative viabilities s to 1, respectively, from conception to maturity. Suppose that lethal (lost) offspring are replaced resulting in *equal brood sizes* at maturity. Let m_{ij} be the proportion of males in a typical brood reared to maturity by an A_iA_j female. Model II is applicable at maturity with sex ratio $w_+ = \sum_{i,j} m_{ij}p_{ij}$ where p_{ij} is the frequency of A_iA_j mothers. The sex ratio at conception can be deduced by extrapolation to be (compare with (2.3))

$$w_- = \frac{\displaystyle\sum_{i,j} \frac{m_{ij}p_{ij}}{s}}{\displaystyle\sum_{i,j} \left[\frac{m_{ij}}{s} + (1 - m_{ij}) \right]p_{ij}} = \frac{w_+}{w_+ + (1 - w_+)s}.$$

In particular, a 1:1 sex ratio at maturity corresponds to a 1:s sex ratio at conception.

56

Under differential viability without replacement the corresponding results hold at conception since the recursion equations are actually independent of viability differences between the sexes, in agreement with Leigh (1970) and Crow and Kimura (1970), among others. The sex ratio at any stage is easily determined from the primary sex ratio.

MODEL III. ONE-LOCUS MULTIALLELE VIABILITY MODEL OF MATERNAL INHERITANCE

The methods of the model of (2.14) can be adapted to a maternal inheritance model where the viability of the offspring depends only on the mother's genotype. This is not sex-ratio determination but the formulation is closely allied. Let m_{ij} be the expected number of offspring (assuming $1:1$ sex ratio) of an $A_i A_j$ female parent. In this context, we refer to $M = \|m_{ij}\|$ as the *maternal inheritance viability matrix*. The frequency of genotype $A_i A_j$ (males or females) is denoted by $2p_{ij}$ when $i \neq j$ and p_{ii} for $i = j$ such that $p_i = \sum_{j=1}^{r} p_{ij}$ is the frequency of allele A_i. We introduce the quantity.

$$z_i = \sum_{j=1}^{r} m_{ij} p_{ij}$$

meaning that z_i / p_i is the expected number of offspring in the next generation emanating from a maternal gene A_i. Assuming random mating we have recursively

$$p'_{ij} = \frac{z_i p_j + p_i z_j}{2w}, \quad i,j = 1,2,\ldots,r$$

where $w = \sum_{i=1}^{r} z_i$. The pertinent recursion relations can be completely summarized in terms of the vectors $\mathbf{z}$ and $\mathbf{p}$ and their successors $\mathbf{z}'$ and $\mathbf{p}'$ leading to

$$\mathbf{p}' = \frac{\mathbf{p}}{2} + \frac{\mathbf{z}}{2w}, \quad \mathbf{z}' = \frac{\mathbf{z} \circ M\mathbf{p} + \mathbf{p} \circ M\mathbf{z}}{2w}. \tag{2.16}$$

57

MODEL IV. SEX RATIO EVOLUTION UNDER SEX-SPECIFIC SPATIAL FITNESS VARIATION

Several discussions of sex ratio evolution have emphasized the idea that offspring fitness and sex expression may vary with environmental circumstances, e.g., thermal or chemical characteristics, or resource availability of the habitat. Size as an adult may be correlated with size at weaning, which differs between sexes. Fitness can depend on size, which can also reflect habitat quality. The expression of maleness versus femaleness, depending on its genotypic composition, may also respond differentially to the environmental conditions.

In order to study the problem of sex ratio evolution when fitness between sexes varies spatially we consider the simplest population subdivision structure, the Levene model (cf. Karlin, 1977, 1982a; see also Bull, 1981a). The Levene subdivision model is characterized by three main features: (1) Numerous microhabitats are available for the population; (2) mating occurs at random across the local habitat structure; and (3) the output from each site is locally set. Some classes of organisms that possibly fit this lifestyle include the polychaetes, which are principally sessile but during mating engage in swarming maneuvers. A number of fish populations breed together in spawning areas and then disperse back to habitats located up various streams, and thus resemble the Levene population structure. Other examples approximating the Levene subdivision model include seabird populations that nest in large rookeries. There are also parasitic species including wasps and nematodes whose fitness and sex expression may depend on the host condition (see, e.g., Charnov and Bull, 1977; Charnov, 1979a), which in turn depends on the environmental state.

Suppose that the population is subdivided into N natural habitats $\mathscr{D}_v$, $v = 1, \ldots, N$, where the viability differentials operating on male versus female zygotes is σ_v to τ_v, $v = 1, 2, \ldots, N$

across the different habitats. Assume that random mating occurs in a common area and subsequently a fraction c_v of the total population settles in habitat $\mathcal{D}_v$, $c_v > 0$, $\sum c_v = 1$. The parameter c_v can alternatively be interpreted as the fraction of individuals contributed by habitat $\mathcal{D}_v$ to the mating pool of the next generation. This population structure may be appropriate for a species whose numbers are regulated within each of the separate habitats but not within the whole population.

We assume sex determination to be genotypically determined depending on the environmental milieux. We can envision the population to mate in a central area and then distribute itself among the N habitats to give birth (e.g., to nest) and for the offspring to mature within a habitat prior to congregating again in the mating area. Let $m_{ij}^{(v)}$ be the probability that a zygote of genotype $A_i A_j$ becomes male in habitat (environment) $\mathcal{D}_v$ and $1 - m_{ij}^{(v)}$ that it becomes female. We denote the sex determination coefficient matrix for habitat $\mathcal{D}_v$ by

$$M^{(v)} = \left\| m_{ij}^{(v)} \right\|_{i,j=1}^{r}.$$

Let $\mathbf{p} = (p_1, \ldots, p_r)$ be the global population frequencies of the alleles $(A_1, \ldots, A_r)$ in the female population when assembled at the mating area and $\mathbf{q} = (q_1, \ldots, q_r)$ the corresponding adult male allelic frequency vector.

Following mating and dispersal, the relative numbers of male zygotes as against female zygotes in the vth habitat at conception is

$$[\mathbf{p} \circ M^{(v)} \mathbf{q} + \mathbf{q} \circ M^{(v)} \mathbf{p}], \text{ and}$$
$$[\mathbf{p} \circ (U - M^{(v)}) \mathbf{q} + \mathbf{q} \circ (U - M^{(v)}) \mathbf{p}],$$

respectively.

These numbers are modified as a result of selection to

$$[\mathbf{p} \circ M^{(v)} \mathbf{q} + \mathbf{q} \circ M^{(v)} \mathbf{p}] \sigma_v$$

and

$$[\mathbf{p} \circ (U - M^{(v)}) \mathbf{q} + \mathbf{q} \circ (U - M^{(v)}) \mathbf{p}] \tau_v.$$

The aggregate allelic numbers at the mating pool in the next generation total

$$\sum_{v=1}^{N} c_v[\mathbf{p} \circ M^{(v)}\mathbf{q} + \mathbf{q} \circ M^{(v)}\mathbf{p}]\sigma_v \text{ for males, and}$$

$$\sum_{v=1}^{N} c_v[\mathbf{p} \circ (U - M^{(v)})\mathbf{q} + \mathbf{q} \circ (U - M^{(v)})\mathbf{p}]\tau_v \text{ for females,}$$

and these normalized to frequency vectors yield the dynamic recursive system

$$\mathbf{q}' = \frac{\displaystyle\sum_{v=1}^{N} c_v[\mathbf{p} \circ M^{(v)}\mathbf{q} + \mathbf{q} \circ M^{(v)}\mathbf{p}]\sigma_v}{2\displaystyle\sum_{v=1}^{N} c_v\langle\mathbf{p},M^{(v)}\mathbf{q}\rangle\sigma_v} \tag{2.17a}$$

and

$$\mathbf{p}' = \frac{\displaystyle\sum_{v=1}^{N} c_v[\mathbf{p} \circ (U - M^{(v)})\mathbf{q} + \mathbf{q} \circ (U - M^{(v)})\mathbf{p}]\tau_v}{2\displaystyle\sum_{v=1}^{N} c_v[1 - \langle\mathbf{p},M^{(v)}\mathbf{q}\rangle]\tau_v}. \tag{2.17b}$$

The normalizing terms

$$m = \sum_v c_v\sigma_v\langle\mathbf{p},M^{(v)}\mathbf{q}\rangle \text{ and } f = \sum_v c_v\tau_v[1 - \langle\mathbf{p},M^{(v)}\mathbf{q}\rangle]$$

can be interpreted as the number of males compared to the number of females at adulthood as a consequence of the sex-determining mechanism and intersex spatial viability differences.

Introducing the matrices

$$M^* = \sum_{v=1}^{N} c_v\sigma_v M^{(v)} \tag{2.18a}$$

and

$$F^* = \sum_{v=1}^{N} c_v\tau_v(U - M^{(v)}), \tag{2.18b}$$

the recursion system (2.17) becomes

$$\mathbf{q}' = \frac{\mathbf{p} \circ M^*\mathbf{q} + \mathbf{q} \circ M^*\mathbf{p}}{2\langle \mathbf{p}, M^*\mathbf{q} \rangle},$$

$$\mathbf{p}' = \frac{\mathbf{p} \circ F^*\mathbf{q} + \mathbf{q} \circ F^*\mathbf{p}}{2\langle \mathbf{p}, F^*\mathbf{q} \rangle} \qquad (2.19)$$

which we recognize as exactly the transformation equations (2.6) for an r-allele, *two-sex viability model* with male and female viability coefficient matrices M^* and F^*, respectively. These matrices, unlike the single-population zygote sex-determination model of (2.4), are generally *not* related such that $F^* = aU + bM^*$ (a,b constant). In the special case where male and female fitnesses are equal and vary with environmental conditions (i.e., with $\sigma_v = \tau_v$ for all v), we have

$$F^* = \left(\sum_{v=1}^{N} c_v \sigma_v \right) U - M^*$$

and the analysis of (2.19) then parallels the discussion and results set forth for the model (2.4). It is noteworthy that even when each separate $M^{(v)}$ reflects dichotomous sex-determination mechanisms (i.e., all the elements of $M^{(v)}$ are either 0 or 1), M^* will *not* be dichotomous unless all $M^{(v)}$ coincide.

Another version of the model is as follows. Mating occurs in a common area as before. However, we now interpret c_v as the fraction of the aggregate mating population contributed by the vth habitat. The frequency-allele vector of males is

$$\frac{\mathbf{p} \circ M^{(v)}\mathbf{q} + \mathbf{q} \circ M^{(v)}\mathbf{p}}{2\langle \mathbf{p}, M^{(v)}\mathbf{q} \rangle} = T_v(\mathbf{p},\mathbf{q})$$

and that of females is

$$\frac{\mathbf{p} \circ (U - M^{(v)})\mathbf{q} + \mathbf{q} \circ (U - M^{(v)})\mathbf{p}}{2[1 - \langle \mathbf{p}, M^{(v)}\mathbf{q} \rangle]} = S_v(\mathbf{p},\mathbf{q}).$$

The combined frequency vectors of males and females participating in mating of the next generation are then

$$\mathbf{q}' = \sum_{v=1}^{N} c_v T_v(\mathbf{p},\mathbf{q})m_v \bigg/ \sum_{v=1}^{N} c_v m_v$$

$$\mathbf{p}' = \sum_{v=1}^{N} c_v S_v(\mathbf{p},\mathbf{q})f_v \bigg/ \sum_{v=1}^{N} c_v f_v. \tag{2.20}$$

where

$$m_v = \frac{\langle \mathbf{p}, M^{(v)}\mathbf{q}\rangle \sigma_v}{\langle \mathbf{p}, M^{(v)}\mathbf{q}\rangle \sigma_v + [1 - \langle \mathbf{p}, M^{(v)}\mathbf{q}\rangle]\tau_v}$$

and

$$f_v = 1 - m_v.$$

These recursions differ markedly from (2.17) and cannot be reduced to the form (2.19) of a two-sex viability model. All these models are analyzed in Chapter 7.

More general models incorporating variable sex-determination coefficients and gender fitnesses over space and time in the framework of general migration and demographic structures can be formulated (cf. Karlin, 1982a).

MODEL V. HAPLODIPLOID MODELS (MIXED PARTHENOGENESIS EXPRESSING ARRHENOTOKY)

Consider a sex determination model controlled at a single autosomal locus involving r alleles $A_1, A_2, \ldots, A_r$. Assume that a fraction $1 - \alpha$ of female gametes independent of the genotype are fertilized and zygotes of genotype $A_i A_j$ with probability $m_{ij}((1 - m_{ij}))$ develop into mature males (females). All unfertilized eggs become haploid males.

Assume that the relative viability of haploid to diploid males is s to 1. In this model viability competition involves haploid versus diploid forms, but is otherwise genotype-independent.

We use the same notation as previously. Under random mating the transformation equations connecting allelic frequencies over two successive generations become

$$\mathbf{q}' = \frac{(1 - \alpha)(\mathbf{p} \circ M\mathbf{q} + \mathbf{q} \circ M\mathbf{p}) + \alpha s \mathbf{p}}{(1 - \alpha)2w + \alpha s} \qquad (2.21a)$$

$$\mathbf{p}' = \frac{[\mathbf{p} \circ (U - M)\mathbf{q} + \mathbf{q} \circ (U - M)\mathbf{p}]}{2(1 - w)} \qquad (2.21b)$$

where $w = \langle M\mathbf{p},\mathbf{q} \rangle$. These equations are based on the assumption that the gametic output from diploid matings is twice that produced by haploid fertilization. An alternative model considers that the gametic output contributed by diploid males versus haploid males are equal. In this perspective, the equations (2.21) are modified, replacing the coefficient of αs from 1 to 2. We could also envision the case where the sex of diploid offspring is under maternal genotypic control.

MODEL VI. SEX ALLOCATION OF RESOURCES TO FEMALE AND MALE FUNCTIONS DETERMINED AT AN AUTOSOMAL LOCUS

Suppose that a typical $A_i A_j$ plant produces $m_{ij}N$ pollen grains and $(1 - m_{ij})n$ seeds compared to a pure male (female) expected to produce N pollen grains (n seeds). This is a case of simultaneous hermaphroditism. The parameters m_{ij} quantify the proportional allocation of fixed resources to male as against female function. The level N relative to n serves only as an intrinsic male versus female viability parameter. The sex determination structure of Model I allowing for intersex viability differences fully encompasses the foregoing sex allocation context without any modifications in the genotype frequency transformation equations (2.4).

MODEL VII. SEX DETERMINATION AT AN X-LINKED LOCUS WITH MULTIPLE ALLELES

Let Y be a dominant male-determining chromosome such that any XY formation necessarily determines a male. Further, suppose that any XX structure carrying $A_i A_j$ at a specific

63

locus produces a male with probability m_{ij} and a female with probability $1 - m_{ij}$. This system can be fully described by the sex-determination matrix

$$\tilde{M} = \left\| \begin{matrix} 1 & \cdots & 1 \\ \vdots & M & \\ 1 & & \end{matrix} \right\|$$

where $M = \|m_{ij}\|_{i,j=1}^{r}$ and the extra "allele" has unit values in the role of the Y chromosome. It is clear that a pure male population cannot occur. In the same way, any dominant male- or female-determining factor can be accomodated by Model I.

MODEL VIII. SEX-BIASED VIABILITIES UNDER MATERNAL CONTROL AT A SEX-LINKED LOCUS

Suppose that females are XX and males XY. A sex-linked gene of r possible alleles $A_1, \ldots, A_r$ is assumed to inflict sex-biased abortions. Let $2p_{ij}$ be the frequency of $A_i A_j$ when $i \neq j$ and p_{ii} when $i = j$ in the female population, and let q_i be the frequency of the allele A_i in the male population. Assume that the expected proportion of males in the progeny of an $A_i A_j$ mother is m_{ij}. Define $z_i = \sum_{k=1}^{r} m_{ik} p_{ik}$ and $w = \sum_{i=1}^{r} z_i$.

In vector notation the following recurrence equations are obtained (cf. (2.14))

$$\mathbf{p}' = \frac{\mathbf{q}}{2} + \frac{\mathbf{p} - \mathbf{z}}{2(1 - w)}, \quad \mathbf{q}' = \frac{\mathbf{z}}{w},$$

$$\mathbf{z}' = \frac{(\mathbf{p} - \mathbf{z}) \circ M\mathbf{q} + \mathbf{q} \circ M(\mathbf{p} - \mathbf{z})}{2(1 - w)} \qquad (2.22)$$

MODEL VIII'. MATERNALLY INHERITED VIABILITY CONTROLLED AT A SEX-LINKED LOCUS

With the interpretation that m_{ij} is the expected number of offspring of an $A_i A_j$ female in the preceding model, with $1:1$ sex ratio at conception, the recurrence equations take the form

$$\mathbf{p}' = \frac{\mathbf{q}}{2} + \frac{\mathbf{z}}{2w}, \mathbf{q}' = \frac{\mathbf{z}}{w}, \mathbf{z}' = \frac{\mathbf{z} \circ M\mathbf{q} + \mathbf{q} \circ M\mathbf{z}}{2w}. \quad (2.23)$$

MODEL IX. MULTIALLELE VIABILITY REGIME ACTIVE IN ONLY ONE SEX

Let m_{ij} be the fitness of an A_iA_j female while there is no differential viability in the male population. We have over successive generations the frequency relations

$$\mathbf{p}' = \frac{\mathbf{p} \circ M\mathbf{q} + \mathbf{q} \circ M\mathbf{p}}{2w}, \mathbf{q}' = \mathbf{p} \quad (2.24)$$

where $w = \langle \mathbf{p}, M\mathbf{q} \rangle$.

APPENDIX A. A GENERAL ONE-LOCUS SELECTION MODEL

The following model is not essential to the discussion of the later chapters but is useful in providing a general perspective as to the nature of natural selection forces acting on populations. More specifically, we formulate a general selection model that incorporates the three common forms of the forces of natural selection: differential viabilities, differential fertilities, and segregation distortion effects. Viability coefficients f_{ij} and m_{ij} for the male and female A_iA_j genotype, $i,j = 1,2, \ldots r$ respectively, reflect the relative viabilities for the alternative genotypes to survive from infancy to maturity. The various fertility intensities indicate the differential mean number of infant offspring produced by a mating type. Thus, associated with each mating type $(A_iA_j) \times (A_kA_l)$ is the measure $\phi(i,j; k,l)$ of the mean number of progeny of the mating. In the most general formulation we can have $\phi(i,j; k,l) \neq \phi(k,l; i,j)$ so that the order of the parental genotypes is important where reciprocal crosses may entail different mean fertility outcomes. Segregation distortion signifies that the proportions of the zygotic possibilities in the cross $(A_iA_j) \times (A_kA_l)$ may differ from an equally likely combination of gametes contributed by each parent.

Under Mendelian segregation the segregation output can be displayed in the form

$$(A_iA_j) \times (A_kA_l) \rightarrow \tfrac{1}{4}(A_iA_k) + \tfrac{1}{4}(A_iA_l) + \tfrac{1}{4}(A_jA_k) + \tfrac{1}{4}(A_jA_l)$$
$$(2.25)$$

(i.e., with chance $1/4$ the zygote A_iA_k is produced, with chance $1/4$ A_iA_l, etc.)

With segregation distortion, we can have the progeny ratios

$$A_iA_j \times A_kA_l \rightarrow \alpha(i,k;j,l)(A_iA_k) + \alpha(i,l;j,k)(A_iA_l)$$
$$+ \alpha(j,k;i,l)(A_jA_k) + \alpha(j,l;i,k)(A_jA_l) \qquad (2.26)$$

where the α's add to 1, but they are not necessarily equal. For the general selection model involving all three forms of selection forces listed, it is *not* true that *random mating* (i.e., random formation of genotype crosses) *is equivalent to random union of gametes*. Conditions where this equivalence holds will be indicated shortly.

Let $\bar{P}_{ij}$ equal the frequency of A_iA_j (in this order) in the infant female population and $P_{ij} = \bar{P}_{ij} + \bar{P}_{ji}$ (=frequency of the unordered A_iA_j when $i \neq j$ and twice the frequency of A_iA_i if $i = j$). The quantities $\bar{Q}_{ij}$ and Q_{ij} are defined analogously with reference to the genotypes in the male population. Differential viability selection acts during the conversion from infants to adults. Let $\tilde{P}_{ij}$ and $\tilde{Q}_{ij}$ denote the frequencies of the associated mature genotypes. In comparison with (2.1) we obtain

$$\tilde{P}_{ij} = \frac{\bar{P}_{ij}f_{ij}}{\sum\limits_{i,j} \bar{P}_{ij}f_{ij}}, \quad \tilde{Q}_{ij} = \frac{\bar{Q}_{ij}m_{ij}}{\sum\limits_{i,j} \bar{Q}_{ij}m_{ij}} \qquad (2.27)$$

Random mating signifies that the cross $A_iA_j \times A_kA_l$ (the ordering is significant where the first genotype corresponds to the female involved and the second to the male) occurs with frequency $\tilde{P}_{ij}\tilde{Q}_{kl}$, the relative number of offspring being $\phi(i,j;k,l)$. We allow general segregation distortion as in (2.26).

66

The number of A_jA_i female infants (the ordering will signify, consistent with the preceding, that the first gamete A_j derives from the female parent and A_i is contributed by the male parent) is

$$\bar{P}'_{ji} = \frac{1}{\mathcal{N}} \left\{ \sum_{\mu,\nu} \tilde{P}_{\mu j}\tilde{Q}_{\nu i}\phi(\mu,j; \nu,i)\alpha(j,i; \mu,\nu) \right.$$

$$+ \sum_{\mu,\nu} \tilde{P}_{\mu j}\tilde{Q}_{i\nu}\phi(\mu,j; i,\nu)\alpha(j,i\ \mu,\nu)$$

$$+ \sum_{\mu,\nu} \tilde{P}_{j\mu}\tilde{Q}_{\nu i}\varphi(j,\mu; \nu,i)\alpha(j,i; \mu,\nu)$$

$$\left. + \sum_{\mu,\nu} \tilde{P}_{j\mu}\tilde{Q}_{i\nu}\phi(j,\mu; i,\nu)\alpha(j,i; \mu,\nu) \right\} \qquad (2.28)$$

where

$$\mathcal{N} = 4 \sum_{i,j,\mu,\nu} \tilde{P}_{j\mu}\tilde{Q}_{i\nu}\phi(j,\mu; i,\nu)\alpha(j,i: \mu,\nu).$$

A similar formula applies to obtain the equation for $\bar{Q}'_{ji}$.

We now concentrate on the special case where the fertility factors are the products

$$\phi(i,j; k,l) = \psi(i,j)\theta(k,l) \qquad (2.29)$$

with a fecundity factor $\psi(i,j)$ attributable to the female parent and $\theta(k,l)$ to the male parent where the effects operate independently. We stipulate the natural symmetry identities $\psi(i,j) = \psi(j,i)$, $\theta(i,j) = \theta(j,i)$. With Mendelian segregation (the factorization $\alpha(i,j; k,l) = \alpha(i)\alpha(j)\alpha(k)\alpha(l)$ would suffice) the combined transformation laws (2.27) and (2.28) can be considerably reduced, yielding

$$\bar{P}'_{ij} = \frac{1}{\mathcal{N}*} \sum_{\mu,\nu} [\bar{P}_{\mu i}\bar{Q}_{\nu j}B_1 + \bar{P}_{\mu i}\bar{Q}_{j\nu}B_2 + \bar{P}_{i\mu}\bar{Q}_{\nu j}B_3$$

$$+ \bar{P}_{i\mu}\bar{Q}_{j\nu}B_4], \qquad (2.30)$$

where

$$B_1 = f_{\mu i}m_{\nu j}\psi(\mu,i)\theta(\nu,j), \quad B_2 = f_{\mu i}m_{j\nu}\psi(\mu,i)\theta(j,\nu),$$

$$B_3 = f_{i\mu}m_{\nu j}\psi(i,\mu)\theta(\nu,j), \quad B_4 = f_{i\mu}m_{j\nu}\psi(i,\mu)\theta(j,\nu),$$

and $\mathcal{N}^*$ is a normalizing constant. But $B_1 = B_2 = B_3 = B_4$ on account of the imposed symmetry identities and (2.30) becomes

$$\bar{P}_{ij} = \frac{1}{\mathcal{N}^*} \sum_{\mu,\nu} (\bar{P}_{\mu i} + \bar{P}_{i\mu})(\bar{Q}_{\nu j} + \bar{Q}_{j\nu})B_1$$

$$= \frac{1}{\mathcal{N}^*} \left(\sum_\mu P_{\mu i} v_{\mu i} \right) \left(\sum_\nu Q_{\nu j} w_{\nu j} \right), \qquad (2.31)$$

where

$$v_{\mu i} = f_{\mu i} \psi(\mu, i), \quad w_{\nu j} = m_{\nu j} \theta(\nu, j),$$

and

$$\mathcal{N}^* = \left(\sum_{\mu,i} P_{\mu i} v_{\mu i} \right) \left(\sum_{\nu,j} Q_{\nu j} w_{\nu j} \right).$$

Defining

$$p'_i = \frac{\sum_\mu P_{\mu i} v_{\mu i}}{\sum_{\mu,i} P_{\mu i} v_{\mu i}}, \qquad q'_j = \frac{\sum_\nu Q_{\nu j} w_{\nu j}}{\sum_{\nu,j} Q_{\nu j} w_{\nu j}}, \qquad (2.32)$$

we obtain

$$\bar{P}_{ij} = p'_i q'_j$$

and

$$P'_{ij} = p'_i q'_j + p'_j q'_i \text{ (and therefore} = Q'_{ij}). \qquad (2.33)$$

In terms of the variables (2.32) we readily compute p''_i for the succeeding generation to be

$$p''_i = \frac{\sum_\mu (p'_\mu q'_i + p'_i q'_\mu) v_{\mu i}}{\sum_{\mu,i} (p'_\mu q'_i + p'_i q'_\mu) v_{\mu i}}$$

$$= \frac{1}{2} \frac{\sum_\mu (p'_\mu q'_i + p'_i q'_\mu) v_{\mu i}}{\sum_{\mu,i} p'_\mu q'_i v_{\mu i}}. \qquad (2.34)$$

The same transformation (with w_{vj} instead of $v_{\mu i}$) connects q_j'' to q_j' in the manner

$$q_j'' = \frac{1}{2} \frac{\sum\limits_{v} (p_v' q_j' + p_j' q_v') w_{vj}}{\sum\limits_{v,j} p_v' q_j' w_{vj}}. \qquad (2.35)$$

Comparing (2.34) and (2.35) to (2.5) we have established that with *product fecundities* (in the sense of (2.29)), *random mating is equivalent to random union of gametes.* In particular if only differential viabilities are manifested, then the notion of panmictic (i.e., random) mating, whether in the guise of random formation of mating types or random union of gametes, is unambiguously defined. However, in the very general situation with no restrictions on the selection parameters, random mating and random union of gametes induce different models.

Some cases of the above result were proposed in various sources, e.g., Kempthorne (1957), Bodmer (1965) Nagylaki and Crow (1974), Karlin (1978).

CHAPTER THREE

Characterizations and Stability Properties of Equilibria in One-Locus Multiallele Sex Determination Models and Related Models

The results for all the sex determination models of Chapter 2 turn out to be quite analogous. Accordingly, we will describe and interpret the principal findings on sex ratio evolution in the context of Model I. We will adhere throughout to the notation of Chapter 2.

An equilibrium for Model I consists of a pair of frequency vectors $\{\mathbf{p}^*, \mathbf{q}^*\}$ having $\mathbf{p}' = \mathbf{p} = \mathbf{p}^*$ and $\mathbf{q}' = \mathbf{q} = \mathbf{q}^*$ in (2.4). Two classes of equilibrium were described in Eshel and Feldman (1982a): the *symmetric equilibria* exhibiting identical allelic frequencies in male and female populations and *even sex-ratio equilibria* $\{\mathbf{p}^*, \mathbf{q}^*\}$ entailing a 1:1 sex ratio, i.e., $w^* = \langle \mathbf{p}^*, M\mathbf{q}^* \rangle = 1/2$, with an equal representation of males and females in the population. This classification follows from the equilibrium relationship $(2w^* - 1)\mathbf{q}^* = (2w^* - 1)\mathbf{p}^*$ obtained by summing (2.4a) and (2.4b) at equilibrium. In generic cases, symmetric equilibria do not carry a one-to-one (1:1) sex ratio.

An important objective of this chapter is to ascertain conditions for the existence and stability of the two types of equilibria. Except in the two-allele case (Eshel, 1975), only the conditions for the initial increase following the introduction of a mutant allele were determined (Eshel and Feldman, 1982a), while the analysis of the system (2.4) was largely unresolved.

In particular, virtually nothing was known about the even sex-ratio equilibria in the multiallelic version of Model I (and for the other models, II–IX as well). The difficulties in the general cases come mainly from the analytic intractability of the multidimensional equations involved. This chapter proposes solutions to these problems.

In Section 3.1, we observe that a symmetric equilibrium state coincides with an equilibrium for a one-sex one-locus multiallele viability model, while even sex-ratio equilibria correspond to a level surface of a spectral radius functional. In order to describe fully the equilibrium structure for the sex determination Model I (Sections 3.3 and 3.4), we introduce a characterization delineating the existence and stability of equilibria in the viability model (2.7) in terms of the same spectral radius functional (Section 3.2). Various optimality properties of even sex-ratio equilibrium realizations are deduced (Section 3.7) and confirmed by numerical simulations (Section 3.6). Section 3.5 focuses on dichotomous and X-linked sex determination models. The corresponding results for the models with mother control are described in Sections 3.8 and 3.9. In Section 3.10, the analysis of haplodiploid models is treated and interpreted separately. Technical proofs appear in the Appendices A–D.

3.1. EXISTENCE CONDITIONS FOR EVEN AND NON-EVEN SEX-RATIO EQUILIBRIA IN THE SEX DETERMINATION MODEL I

The following facts are fundamental to our stability analysis. A symmetric equilibrium $\{\hat{\mathbf{p}},\hat{\mathbf{p}}\}$ (abbreviated by $\hat{\mathbf{p}}$) of (2.4) solves

$$\hat{\mathbf{p}} = \frac{\hat{\mathbf{p}} \circ M\hat{\mathbf{p}}}{\langle \hat{\mathbf{p}}, M\hat{\mathbf{p}} \rangle}. \qquad (3.1)$$

An even sex-ratio equilibrium $\{\tilde{\mathbf{q}},\tilde{\mathbf{p}}\}$ of (2.4), with the tilde inserted to emphasize that $\tilde{\mathbf{q}} \neq \tilde{\mathbf{p}}$, is characterized by the

71

relation

$$\tilde{\mathbf{q}} = \tilde{\mathbf{p}} \circ M\tilde{\mathbf{q}} + \tilde{\mathbf{q}} \circ M\tilde{\mathbf{p}}. \tag{3.2}$$

Note that this relation entails a one-to-one sex ratio. Summing over the coordinates in (3.2), yields $1 = \langle \mathbf{u}, \tilde{\mathbf{q}} \rangle = 2 \langle \tilde{\mathbf{p}}, M\tilde{\mathbf{q}} \rangle$ where $\mathbf{u}$ is the vector of all unit components and therefore $\tilde{w} = \langle \tilde{\mathbf{p}}, M\tilde{\mathbf{q}} \rangle = 1/2$. However, this property alone is not sufficient to ensure equilibrium.

Recall that the transformation

$$T\mathbf{p} = \frac{\mathbf{p} \circ M\mathbf{p}}{\langle \mathbf{p}, M\mathbf{p} \rangle} \tag{3.3}$$

on the simplex Δ of all frequency vectors $\mathbf{p} = (p_1, p_2, \ldots, p_r)$, i.e., $\sum_{i=1}^{r} p_i = 1$ with $p_i \geq 0$ for $i = 1, \ldots, r$, is that of the classical one-locus one-sex viability model with r-alleles (see (2.7)), and Equation (3.1) is simply $T\mathbf{p} = \hat{\mathbf{p}}$.

Equation (3.2) can be viewed in the following manner. For each $\mathbf{p}$ in Δ^0 interior to the allelic frequency simplex Δ, (that is, for a frequency vector $\mathbf{p} = (p_1, \ldots, p_r)$ displaying all positive components), we form the matrix

$$B(\mathbf{p}) = D_{\mathbf{p}}M + D_{M\mathbf{p}} \quad \begin{array}{l} (D_{\mathbf{p}} \text{ is the diagonal matrix with} \\ \text{the components of } \mathbf{p} \text{ down} \\ \text{the main diagonal}). \end{array} \tag{3.4}$$

If M (and therefore $B(\mathbf{p})$) is a positive matrix in the sense that all entries are positive, we know by the Perron-Frobenius theory that the eigenvalue of largest magnitude for $B(\mathbf{p})$ is positive, admitting a unique (apart from a constant multiplicative factor) strictly positive right eigenvector $\mathbf{y} = \mathbf{y}(\mathbf{p})$. We denote by $\rho(\mathbf{p})$ the *spectral radius* of $B(\mathbf{p})$ (which is the eigenvalue of largest magnitude of $B(\mathbf{p})$) and take $\mathbf{y} = (y_1, \ldots, y_r)$ positive and normalized so that $\sum_{i=1}^{r} y_i = 1$ and $y_i > 0$ for $i = 1, \ldots, r$. Then $\mathbf{y} = \mathbf{y}(\mathbf{p})$ and $\rho(\mathbf{p})$ vary continuously (actually analytically) as functions of $\mathbf{p}$ in Δ^0. (See, e.g., Gantmacher, 1959; Lancaster, 1969.)

72

The principal eigenvalue-eigenvector identity is

$$B(\mathbf{p})\mathbf{y}(\mathbf{p}) = \rho(\mathbf{p})\mathbf{y}(\mathbf{p}). \tag{3.5}$$

Comparing to (3.2), we find that an even sex-ratio equilibrium represented by $\{\tilde{\mathbf{q}},\tilde{\mathbf{p}}\}$, both $\tilde{\mathbf{q}}$ and $\tilde{\mathbf{p}}$ interior to Δ, exists if and only if $\rho(\tilde{\mathbf{p}}) = 1$, and then the corresponding $\mathbf{y}(\tilde{\mathbf{p}})$ is $\tilde{\mathbf{q}}$. These conditions can be extended by continuity allowing $\tilde{\mathbf{p}}$ and $\tilde{\mathbf{q}}$ on the same face of Δ, i.e., with the same zero components.

With respect to our study of the existence and stability of the two types of equilibria, of particular interest is the issue whether symmetric and even sex-ratio equilibria can coexist and be stable. To resolve this problem we rely on a new characterization delineating the existence and stability characteristics of symmetric equilibria, i.e., equilibria for the one-sex viability model (3.3) set forth in the following section.

3.2. A CHARACTERIZATION OF EQUILIBRIUM POINTS IN ONE-LOCUS MULTIALLELE VIABILITY SYSTEMS

Consider a one-locus multiallele system with viability matrix M regardless of sex. To avoid unimportant technical details we assume henceforth, unless stated otherwise, that the symmetric matrix M satisfies the following generic conditions.

DEFINITION 3.1. The matrix $M = \|m_{ij}\|$ is said to be generic if:

a) $0 < m_{ij} < 1$ for all i and j;
b) every principal submatrix of M is nonsingular;
c) for any submatrix of indices $i_1 < i_2 < \cdots < i_k$ the solution of

$$\sum_{v=1}^{k} m_{i_\mu, i_v} x_v = 1, \mu = 1, 2, \ldots, k$$

has $\tag{3.6}$

$$\sum_{v=1}^{k} m_{i_0, i_v} x_v \neq 1 \text{ for all } i_0 \neq i_\mu.$$

It is useful to recount the following classical facts (e.g., see Nagylaki, 1977; Ewens, 1979; or Karlin, 1978). A polymorphic equilibrium $\mathbf{p}^*$ (i.e., an internal equilibrium) of the system (3.3) exists if the linear equations

$$\sum_{j=1}^{r} m_{ij}x_j = 1, \, i = 1,2, \ldots, r \text{ admit a positive solution} \quad (3.7)$$

and then $p_i^* = x_i/\sum_{j=1}^{r} x_j$. This polymorphic equilibrium is locally stable if any of the following conditions holds (Kingman, 1961a, b):

i) The eigenvalues of M, $\lambda_1(M) \geq \lambda_2(M) \geq \cdots \geq \lambda_r(M)$, satisfy

$$\lambda_1(M) > 0 > \lambda_2(M) \geq \cdots \geq \lambda_r(M). \quad (3.8)$$

ii) The mean fitness function $\langle M\mathbf{p},\mathbf{p}\rangle = \sum_{i,j=1}^{r} m_{ij}p_ip_j = w(\mathbf{p})$ achieves a strict maximum at $\mathbf{p}^*$ with respect to $\mathbf{p}$ traversing Δ.

iii) The successive principal determinants $D_k = \det\|m_{ij}\|_{i,j=1}^{k}$ strictly alternate in sign.

An important fact in one-locus multiallele selection theory is that a *locally stable polymorphic equilibrium is globally stable* relative to all initial internal states.

With any number of alleles represented, an equilibrium vector $\hat{\mathbf{p}} = (\hat{p}_1, \ldots, \hat{p}_r)$ of (3.3) solves the system of equations

$$(M\hat{\mathbf{p}})_i = \sum_{j=1}^{r} m_{ij}\hat{p}_j = w(\hat{\mathbf{p}}) = \langle \hat{\mathbf{p}},M\hat{\mathbf{p}}\rangle, \, i \in \mathscr{I}(\hat{\mathbf{p}}) \quad (3.9)$$

the equations applying for those indices $\mathscr{I}(\hat{\mathbf{p}})$ where $\hat{p}_i > 0$. The equilibrium $\hat{\mathbf{p}}$ is *internally stable* provided the principal submatrix of M restricted to the rows and columns where $\hat{p}_i > 0$ satisfies (3.8).

External stability of a boundary equilibrium prevails by definition if any new allele introduced in small frequency into the population is ultimately eliminated. External stability is assured if

$$w(\hat{\mathbf{p}}) > (M\hat{\mathbf{p}})_k \text{ applies for all } k \text{ where } \hat{p}_k = 0. \quad (3.10)$$

74

It should be emphasized that, when M is generic in the sense of Definition 3.1, only a finite number of equilibria can exist, and the conditions (3.9) and (3.10) are necessary and sufficient for local stability of $\hat{\mathbf{p}}$, or, equivalently, for strict local maximality of $w(\mathbf{p})$ at $\hat{\mathbf{p}}$. Note that attributes (3.7) and (3.8) can apply together yielding a globally stable polymorphism. Or condition (3.7) can hold without (3.8), presenting an unstable polymorphism. In the case that (3.8) holds without (3.7), a unique globally stable equilibrium (on the boundary of Δ) exists (cf. Karlin, 1978). In general, a unique stable equilibrium (interior or on the boundary of Δ) corresponds to a strict global maximum of $w(\mathbf{p})$ in Δ. Note also that the equilibria of (3.3) are exactly the critical points of $w(\mathbf{p})$ relative to Δ.

Some of the main conclusions of this chapter depend on knowledge of the number of stable equilibria that can coexist in one-locus multiallele viability models. In this respect, the following result is basic (Karlin, 1978; see Appendix D for a proof): if an equilibrium of (3.3) on the boundary of Δ with exactly $r - 1$ positive components (i.e., with exactly one allele not represented) is internally stable, then there exists at most one alternative equilibrium that can be locally stable relative to Δ; if the boundary equilibrium is externally unstable, then there exists one and only one stable equilibrium in Δ.

We now present new characterizations of the equilibrium points of (3.3) in terms of the function $\rho(B(\mathbf{p})) = \rho(\mathbf{p})$ defined as the spectral radius of the matrix $B(\mathbf{p})$ in (3.4).

THEOREM 3.1. *Let $\rho(\mathbf{p})$ be the spectral radius of $B(\mathbf{p}) = D_{M\mathbf{p}} + D_{\mathbf{p}}M$ for $\mathbf{p}$ in Δ where M is a symmetric matrix generic in the sense of Definition 3.1. Let $\mathbf{p}^*$ be a frequency vector interior to Δ, i.e., with all positive components. Then,*

(i) *$\mathbf{p}^*$ is a polymorphic equilibrium for the viability model M (i.e., for the transformation (3.3)) if and only if $\mathbf{p}^*$ is a critical point of the function $\rho(\mathbf{p})$, meaning that the derivative of $\rho(\mathbf{p})$ in every direction at $\mathbf{p}^*$ relative to the simplex Δ is zero.*

(ii) *$\mathbf{p}^*$ is a locally stable polymorphic equilibrium for M if and only if $\rho(\mathbf{p})$ achieves a strict local maximum with respect to Δ at $\mathbf{p}^*$.*

75

With boundary equilibrium points incorporated, Theorem 3.1 is complemented by the following general characterization.

THEOREM 3.2. *With the conditions of Theorem 3.1 in force,*

(i) *An equilibrium $\hat{\mathbf{p}}$ for the viability model M (interior or on the boundary of Δ) is locally stable if and only if $\rho(\hat{\mathbf{p}})$ is a strict local maximum of $\rho(\mathbf{p})$ with respect to Δ.*

(ii) *The local maxima of $\rho(\mathbf{p})$ in Δ are twice the local maxima of $\langle \mathbf{p}, M\mathbf{p} \rangle$ and are achieved at the same points. A local maximum within Δ is actually a global and unique maximum over all Δ.*

3.3. STABILITY NATURE OF THE SYMMETRIC (NON-EVEN SEX RATIO) EQUILIBRIUM STATES FOR THE SEX DETERMINATION MODEL I

We previously emphasized in Chapter 2 that the model of sex determination in a random mating population where an off-spring of genotype A_iA_j is male (female) with probability m_{ij} $((1 - m_{ij}))$ is equivalent to a two-sex viability model with viability matrices $M = \|m_{ij}\|$ and $F = \|1 - m_{ij}\|$ effective for males and females, respectively. In this section we describe the stability conditions for the symmetric (non-even sex ratio) equilibrium states.

A symmetric equilibrium $\{\hat{\mathbf{p}}, \hat{\mathbf{p}}\}$ for the sex determination model (2.4), represented by $\hat{\mathbf{p}}$ for brevity, corresponds to an equilibrium for the viability model M defined by (3.3). To avoid unimportant situations involving special relations among the sex determination parameters $\{m_{ij}\}$, it will be appropriate unless indicated otherwise to impose the following additional (generic-almost always applicable) assumption.

DEFINITION 3.2. *A sex determination matrix M is said to be symmetrically generic if*

(i) *M and $U - M$ are each generic (in the sense of Definition 3.1).*

(ii) *$\langle \hat{\mathbf{p}}, M\hat{\mathbf{p}} \rangle \neq 1/2$ for every symmetric equilibrium $\hat{\mathbf{p}}$ of (2.4). (In particular, $m_{ii} \neq 1/2$ for every i.)*

The requirement (ii) guarantees that a symmetric equilibrium state does not yield an even sex ratio.

The next theorem indicates when a symmetric equilibrium is stable in the sex determination model (2.4).

THEOREM 3.3. *A symmetric interior equilibrium* $\{\mathbf{p}^*,\mathbf{p}^*\}$ *with* $w^* = \langle \mathbf{p}^*,M\mathbf{p}^* \rangle < 1/2$ *is stable in the sex determination model (2.4) if and only if* $\mathbf{p}^*$ *is stable for the viability model of (3.3).*

Analogously, if $w^* > 1/2$, $\{\mathbf{p}^*,\mathbf{p}^*\}$ *is a stable symmetric equilibrium for the sex determination model (2.4) if and only if* $\mathbf{p}^*$ *is stable for the viability model* $U - M$ *(i.e., (3.3) with* $U - M$ *instead of* M).

The stability analysis for a symmetric boundary equilibrium $\{\hat{\mathbf{p}},\hat{\mathbf{p}}\}$ for the system (2.4) entails the internal stability conditions relative to the face of $\hat{\mathbf{p}}$ akin to Theorem 3.3 and the external stability conditions with respect to alleles not represented in the array of $\hat{\mathbf{p}}$ (see (3.9) and (3.10)).

The result is as follows:

THEOREM 3.3'. *A symmetric equilibrium* $\{\hat{\mathbf{p}},\hat{\mathbf{p}}\}$ *with* $\hat{w} = \langle \hat{\mathbf{p}},M\hat{\mathbf{p}} \rangle < 1/2$ *is stable for the sex determination model (2.4) if and only if:*

(i) *The eigenvalues of* $\hat{\mathbf{p}} \circ M/\hat{w} = D_{\mathbf{p}}M/\hat{w}$ *in the face of* $\hat{\mathbf{p}}$, *i.e., with respect to the hyperplane* $\mathscr{L}_{\mathbf{p}} = \left\{ \xi: \sum_{i \in \mathscr{I}(\mathbf{p})} \xi_i = 0 \right\}$ *are all negative;* and

(ii) $\dfrac{(M\hat{\mathbf{p}})_j}{2\hat{w}} + \dfrac{((U - M)\hat{\mathbf{p}})_j}{2(1 - \hat{w})} < 1$, for all $j \notin \mathscr{I}(\hat{\mathbf{p}})$, (3.11a)

or equivalently since $\hat{w} < 1/2$,

$$(M\hat{\mathbf{p}})_j < \hat{w} \text{ for all } j \notin \mathscr{I}(\hat{\mathbf{p}}). \qquad (3.11b)$$

When $\hat{w} > 1/2$, *then the ascertainment of stability for* $\{\hat{\mathbf{p}},\hat{\mathbf{p}}\}$ *in the sex determination model (2.4) corresponds to the stability criteria with respect to the one-locus viability matrix* $U - M$ *instead of* M.

Thus, a symmetric equilibrium (boundary or interior) of sex ratio $\hat{w} < 1/2$ ($\hat{w} > 1/2$) is stable for the sex determination model of (2.4) if and only if it is stable for the one-locus one-sex random mating multiallele system (3.3) with viability matrix $M((U - M)$, respectively). The condition (3.11b) for external stability was pointed out in Eshel and Feldman (1982a). Written in the form (3.11a), this condition is akin to the Shaw-Mohler equation developed by MacArthur (1965) and Charnov

77

(1982), among others, but in the genetic context of a diploid population allowing multiple alleles.

We conclude this section with a direct consequence of Theorems 3.3 and 3.3'.

COROLLARY 3.1. *When* $\hat{\mathbf{p}}$ *is stable for the viability model M of (3.3) and* $\hat{w} = \langle \hat{\mathbf{p}}, M\hat{\mathbf{p}} \rangle > 1/2$, *then* $\{\hat{\mathbf{p}}, \hat{\mathbf{p}}\}$ *is unstable (actually repelling) in the sex determination model (2.4).*

3.4. NATURE AND PROPERTIES OF EVEN SEX-RATIO EQUILIBRIA OF THE SEX DETERMINATION MODEL I

We stipulate the assumption of Definition 3.2. Let $\{\tilde{\mathbf{q}}, \tilde{\mathbf{p}}\}$ be an equilibrium for the sex determination model (2.4) with

$$\langle \tilde{\mathbf{q}}, M\tilde{\mathbf{p}} \rangle = \tfrac{1}{2}. \tag{3.12}$$

Recall from (3.2) that an even sex-ratio equilibrium is completely characterized by the eigenvalue-one equation

$$B(\tilde{\mathbf{p}})\tilde{\mathbf{q}} = \tilde{\mathbf{q}} \tag{3.13}$$

where the matrix $B(\mathbf{p}) = D_{M\mathbf{p}} + D_{\mathbf{p}}M$ is described in (3.4). Under (3.12) and (3.13) then $\tilde{\mathbf{q}} \neq \tilde{\mathbf{p}}$ by Definition 3.2.

An interior or boundary even sex-ratio equilibrium $\{\tilde{\mathbf{q}}, \tilde{\mathbf{p}}\}$ segregating alleles of indices $\mathscr{I}(\tilde{\mathbf{p}})$ satisfies (3.13) with $\tilde{\mathbf{q}}$ interior to the same face as $\tilde{\mathbf{p}}$.

From the discussion of (3.2), we know that an even sex-ratio equilibrium exists if and only if $\rho(\tilde{\mathbf{p}}) = \rho(B(\tilde{\mathbf{p}})) = 1$ and $\tilde{\mathbf{q}}$ is the corresponding unique right eigenvector in Δ for $B(\tilde{\mathbf{p}})$. The basic result dealt with next concerns the existence of even sex-ratio equilibrium states. Such equilibria, when present, comprise a continuum. In the following theorem, it is understood that an even sex-ratio equilibrium point is represented by its corresponding allelic frequencies in the female population (i.e., the vector component $\tilde{\mathbf{p}}$ in the equilibrium pair $\{\tilde{\mathbf{p}}, \tilde{\mathbf{q}}\}$ satisfying (3.13)).

THEOREM 3.4. *Suppose that the sex determination matrix M is symmetrically generic, i.e., the conditions of Definition 3.2 hold. Let any equilibrium* $\{\mathbf{p}^*, \mathbf{q}^*\}$ *of the sex determination model (2.4) be repre-*

sented by the vector component **p***, i.e., the equilibrium allelic frequency
vector in the female population.*

(i) *Even sex-ratio equilibria of the sex determination model (2.4),
when they exist, are part of an equilibrium surface of dimension* $r - 2$
in the simplex of frequency vectors Δ *of dimension* $r - 1$. *(These are
curves of equilibria for the three-allele case. Only in the case of two
alleles are the even sex-ratio equilibrium points isolated.)*

(ii) *A stable symmetric equilibrium (interior or on the boundary of* Δ*)
that is globally stable for the corresponding viability model* (see Theorem 3.3) *cannot coexist with any even sex-ratio equilibrium and any
other stable symmetric equilibrium in* Δ. (If the symmetric equilibrium is only internally stable, then the conclusion holds on the
face of Δ corresponding to the positive components of the equilibrium.)

(iii) *Two symmetric equilibria* $\hat{\mathbf{p}}_\alpha$ *and* $\hat{\mathbf{p}}_\beta$ *with associated sex ratio*
$\hat{w}_\alpha < 1/2$ *and* $\hat{w}_\beta > 1/2$, *respectively, are completely separated by at
least one even sex-ratio equilibrium surface.*

COROLLARY 3.2. *If there exists a stable symmetric polymorphic
equilibrium* **p*** *of noneven sex ratio for the model (2.4), then no even
sex-ratio equilibrium exists and* **p*** *is the unique stable equilibrium for
the sex determination model.*

It must be emphasized (cf. Chapter 4) that stable symmetric
equilibrium states and even sex-ratio equilibria can coexist,
but then the symmetric and unsymmetric equilibrium states
do not segregate the same set of alleles. There even may coexist
one or several surfaces of even sex-ratio equilibria with stable
boundary symmetric equilibria.

Assertion (i) precludes the existence of an isolated even sex-
ratio equilibrium for the case of $r \geq 3$ alleles. It appears, evi-
denced by a variety of numerical runs and analytic studies of
special classes in the sex determination model (2.4) that each
even sex-ratio equilibrium surface is strongly stable to the
extent that, for a slight perturbation off an even sex-ratio equilibrium surface, the iteration of (2.4) returns to the surface
generally geometrically fast. (For exceptions to this last state-
ment, see Table 4.2 of Chapter 4.)

An analytic formula for the surface (or surfaces) of even sex-ratio equilibrium frequencies in the female population is described as the set of all **p** in Δ satisfying

$$\varphi(\mathbf{p}) = \det\left\|\left(\sum_{k=1}^{r} p_k m_{kj}\right)\delta_{ji} + p_i m_{ij} - \delta_{ij}\right\|_{i,j=1}^{r} = 0 \quad (3.14)$$

($\delta_{ij} = 1$ if $i = j$; 0 when $i \neq j$). The determinant $\varphi(\mathbf{p})$ is an algebraic function of degree r in the variables $(p_1, p_2, \ldots, p_r)$. By symmetry a similar formula holds for the even sex-ratio equilibrium frequencies in the male population. The exact pairing is defined by the equilibrium relation (3.13).

In Chapter 4 we highlight a broad spectrum of examples of Theorem 3.4 in the three-allele case.

3.5. STABLE EQUILIBRIUM POSSIBILITIES FOR DICHOTOMOUS AND X-LINKED SEX DETERMINATION

The following general result underscores the central role of the one-to-one sex ratio when the sex phenotype is controlled at a single multiallele locus with exact genotype (zero-one) sex ascertainment. In this section, the generic conditions of Definition 3.2 are *not* assumed.

THEOREM 3.5. *Suppose all genotypes involving r alleles at a single locus divide into two groupings* (all genotypes in $\mathcal{G}_M$ become males unambiguously, while all genotypes in $\mathcal{G}_F$ become females). *Assuming random mating in this dioecious population, the* only *possibly stable equilibria (or equilibrium) entail a* $1:1$ *population sex ratio, i.e., the frequency of a stable equilibrium phenotype representation has* freq($\mathcal{G}_M$) = freq($\mathcal{G}_F$) = $1/2$.

It is interesting to contrast Theorems 3.5 and 3.3. When the genotype-phenotype classes of sex determination are not absolute (that is, $0 < m_{ij} < 1$), then the possibility of a stable *non-even* sex ratio can emerge. Theorem 3.5 asserts that exact genotype sex determination ($m_{ij} = 0$ or 1 for all i, j) creating the sex dimorphism of the two phenotype classes $\mathcal{G}_M$ and $\mathcal{G}_F$ allows only an even sex-ratio outcome as possibly stable for such a randomly mating population.

The corresponding result to Theorem 3.5 prevails even for sex ratio control involving multiple sex modifiers at an X-linked locus if it is assumed that all XY formations develop into males (according to Model VII) and are maintained in the population.

THEOREM 3.6. *Consider a sex determination mechanism partly controlled at an X-linked multiallele locus such that all heterogametic genotypes XY are unambiguously males, while homogametic genotypes XX are subject to probabilistic sex determination controlled at the X-linked locus. Under random mating the only stable sex-ratio equilibrium states maintaining heterogametic genotypes in the population are 1:1.*

More generally, any dominant (male or female) sex determiner in (2.4) can be maintained at equilibrium only with an even sex ratio. Theorems 3.5 and 3.6 can be applied to some one-locus sex reversal systems as studied by Scudo (1964), Bull and Charnov (1977), and Bull (1981b). For example, if a primary chromosomal XX/XY sex determination mechanism applies with all XY immutably males but a constant fraction α (e.g., environmentally determined) of XX females are transformed into male phenotypes, then the proportion p of XY genotypes obeys the recurrence relationship

$$p' = \frac{p}{2[p + \alpha(1 - p)]}$$

and converges to $(1 - 2\alpha)/[2(1 - \alpha)]$ if $\alpha < 1/2$, leading to a one-to-one population sex ratio, or goes to zero if $\alpha > 1/2$.

In contrast to the result of Theorem 3.5, for the sex determination system induced by the two-locus genotype dichotomization

$$\underbrace{\left\{\frac{AB}{ab}, \frac{Ab}{aB}\right\}}_{\mathscr{G}_M} \underbrace{\left\{\frac{AB}{AB}, \frac{AB}{Ab}, \frac{Ab}{Ab}, \frac{AB}{aB}, \frac{Ab}{ab}, \frac{aB}{aB}, \frac{aB}{ab}, \frac{ab}{ab}\right\}}_{\mathscr{G}_F}, \quad (3.15)$$

a stable non-one-to-one sex ratio equilibrium can be established with sufficient recombination (Scudo, 1964). The nature of the sex ratio population realizations under various multilocus sex determination structures will be dealt with in Chapter 5.

3.6. THE RELATIVE OCCURRENCE OF
SYMMETRIC (NON-ONE-TO-ONE) VERSUS
EVEN SEX-RATIO EQUILIBRIA

We pointed out earlier (see also the abundance of examples in Chapter 4) that there can simultaneously exist stable symmetric equilibria and even sex-ratio equilibrium surfaces. But such a symmetric equilibrium necessarily lies on the boundary of Δ with a restricted domain of attraction. When a theoretical model such as (2.4) produces multiple stable equilibria, we may seek an appropriate measure of the "degree of stability" or "strength of the equilibrium state" by which the equilibria may be compared. A number of indices that allow comparisons among stable equilibrium states include: (i) rate of return under small perturbations (rate of asymptotic stability); (ii) the extent (relative coverage) of the domain of attraction of a stable equilibrium; (iii) number of competing stable equilibria; (iv) diversity or heterozygosity measured at an equilibrium; (v) location (centrality versus boundary) and extent of symmetries inherent in the equilibrium frequencies; and (vi) mean fitness value at the equilibrium.

Each proposed measure has some intuitive merit, although the appropriate application may rest on factors exogenous to the population genetic system in question.

We concentrate on the first three criteria. In order to contrast the extent and domain of stability between symmetric versus even sex-ratio equilibrium states, we constructed 200 random sex determination matrices (each entry independently uniformly distributed on [0,1]). In each case we iterated the transformation (2.4) until an equilibrium was achieved. The computer output is recorded in Tables 3.1 and 3.2. (The two-allele case has been treated analytically.)

We observe that as the number of alleles increases, even sex ratio equilibrium surfaces are more likely to occur and their number to increase. Since the existence of such surfaces precludes the presence of an interior stable symmetric equilibrium,

TABLE 3.1. Comparisons between the Occurrences of Symmetric versus Even Sex-Ratio Equilibrium States

Number of alleles	2*	3†	4†	5†	6†
Convergence to 1:1 sex ratio		.69	.83	.90	.93
Matrices leading to:					
i) 1:1 sex ratio only	.39	.55	.72	.74	.85
ii) symmetric equilibria only	.36	.17	.06	.02	.02
iii) either	.25	.28	.22	.24	.13

* Analytical calculations from uniform distributions (see Section 4.1).
† Numerical simulations from 200 random sex-determination matrices with 10 random starting points for each.

TABLE 3.2. Comparisons between the Rates of Convergence

Number of alleles	2	3	4	5	6
Convergence to 1:1 sex ratio	—	134	151	208	215
Matrices leading to:					
i) 1:1 sex ratio only	—	115	145	191	204
ii) symmetric equilibria only	—	413	679	1,028	3,351
iii) either	—	1,114	1,027	1,623	2,161

NOTE: Average numbers of generations for attainment of an equilibrium state. The iterations on (2.4) are terminated when no change in the sixth decimal place occurs for two successive generations.

the domains of attraction of the symmetric equilibria are expected to shrink (and do) with an increase of dimension (with more allelic variants involved in sex determination).

It is worth noting that convergence to a one-to-one sex ratio is in general faster than convergence to a symmetric equilibrium (from three to ten times in our simulations). The disparity increases with more alleles. This phenomenon might be linked to differences in dimensionality (0 compared to $r - 2$) and the nature of the distribution of both kinds of equilibrium.

Theorem 3.7 below further amplifies the predominant endowment of even sex-ratio equilibrium realizations under sex ratio genotypic determination.

3.7. OPTIMALITY PROPERTIES OF EVEN SEX-RATIO EQUILIBRIA

The concept of ESS (evolutionary stable strategy) was introduced in Maynard Smith and Price (1973) to predict what behavioral traits among a given set can be maintained in a population. It was originally defined in a framework of pairwise contests arising from animal conflicts but with the terminology and formalism of game theory. In broad terms, an ESS can be viewed as a strategy (i.e., a behavioral choice) that is more successful than any other when adopted by all members of a population. In the context of sex ratio evolution, an ESS corresponds to an "unbeatable strategy" as defined by Hamilton (1967). The original static concept of ESS based on comparisons of mean fitnesses has been applied in numerous theoretical studies on the evolution of a wide variety of structural and behavioral traits (e.g., altruism, cooperation, dispersal, sexual selection). Notable contributors to the theory and its applications besides J. Maynard Smith and W. D. Hamilton include B. Charlesworth, D. Charlesworth, M. G. Bulmer, P. D. Taylor, E. L. Charnov, S. Parker, I. Eshel, W.G.S. Hines, and E. C. Zeeman, among others. (See Maynard Smith, 1982; Lessard, 1984; and the recent review of Parker, 1984, with references therein.)

For most population biologists, the problem of interest has been to identify the ESS's under varied assumptions on the population structures and behavioral patterns. This problem has usually been solved by examining monomorphic populations in which all members would adopt the same strategy and any deviant strategy would be selected against when rare. In a population genetic perspective, this approach corresponds to finding a fixation state of an allele such that any mutant allele cannot initially increase in frequency. (This defines *the optimality property in the* ESS *sense* referred to throughout). But this does not address the question of the evolution toward an ESS and even less the possibility of a polymorphic population in an evolutionary stable state.

In the framework of the sex determination model (2.4), Eshel and Feldman (1982a) conjectured that even sex-ratio equilibria are optimal (evolutionary genetically stable in their terminology) in the sense that any other equilibrium can be rendered unstable by genetic mutations that bring the population sex ratio closer to one-half. In practice, this optimality property is understood to mean that with the introduction of a new allele from r to $r + 1$ alleles, all possible stable equilibrium states for the extended model cannot attain a sex ratio farther from one-to-one than that existing with the r-allele subsystem. This certainly expresses an evolutionary tendency toward an even sex ratio.

In this framework, consider a symmetric equilibrium represented by $\mathbf{p}^* = (p_1^*, \ldots, p_r^*)$, $p_i^* > 0$, $i = 1, \ldots, r$ for the r-allele model (2.4) with $w^* = \langle M\mathbf{p}^*, \mathbf{p}^* \rangle < 1/2$. For the new allele A_{r+1}, let $m_{i,r+1}$ be the probability for a zygote of genotype $A_i A_{r+1}$ to be a male and $1 - m_{i,r+1}$ to be a female, respectively. The marginal fraction of male progeny carrying allele A_{r+1} at the equilibrium state $\mathbf{p}^*$ is $w_{r+1}^* = \sum_{i=1}^{r} p_i^* m_{i,r+1}$. A symmetric equilibrium $\mathbf{p}^*$, stable for the r-allele sex determination model (2.4) but unstable with the introduction of A_{r+1}, occurs if and only if

$$w_{r+1}^* > w^* \text{ provided } w^* < 1/2. \tag{3.16}$$

The condition for external instability at $\mathbf{p}^*$ is $w_{r+1}^* < w^*$ when $w^* > 1/2$. (See Eshel and Feldman [1982a], (Compare with (3.11) in Theorem 3.3).

These conditions do not require that the marginal sex ratio w_{r+1}^* at $\mathbf{p}^*$ be closer to $1/2$ than w^*. (In fact, w_{r+1}^* close to one is feasible and trivially satisfies (3.16). However, it was conjectured that the augmented allelic system in the long run would attain a sex ratio closer to 1/2 than the previous equilibrium did. We can prove the following fact.

THEOREM 3.7. *Let* $\mathbf{p}^*$ *be a stable symmetric equilibrium for the r-allele model (2.4) with sex ratio* $w^* = \langle \mathbf{p}^*, M\mathbf{p}^* \rangle$ *which becomes unstable following the introduction of a new allele* A_{r+1}. *Then, for the augmented* $(r + 1)$-*allele system in the symmetrically generic case (Definition 3.2), either: (i) there exists a unique stable symmetric equilibrium whose sex-ratio is closer to 1/2 compared to* w^* *and which does not coexist with any even sex-ratio equilibrium, or (ii)* $\mathbf{p}^*$ *is enclosed by an even sex-ratio equilibrium surface containing no stable symmetric equilibria.*

COROLLARY 3.3. *Under the conditions of Theorem 3.7 the only* stable *equilibrium points attainable from* $\mathbf{p}^*$ *entail a sex ratio closer to 1/2 than* w^*.

Our numerical interations (Section 3.6) have persistently shown convergence to either a symmetric equilibrium or an even sex-ratio equilibrium according to the theme of Corollary 3.3. The paths of the sex ratio values over successive generations may exhibit oscillations, but in the long run they always tend (increasingly or decreasingly) toward 1/2.

We interpret Theorem 3.7 as expressing an "optimality" property in favor of *even sex-ratio* realizations. On the basis of this result we surmise that with increasing allelism the population sex ratio is inclined toward an even sex ratio (cf. the numerical runs of Section 3.6). Any perturbations in a sex-ratio equilibrium state, rendering it unstable, ultimately achieves a sex ratio closer to 1/2. Moreover, an even sex-ratio equilibrium persists as stable with the introduction of any additional allele involved in sex determination.

Although the predicted evolutionary sex ratio by virtue of Theorem 3.7 tends toward one-to-one, it is worth stressing that for any finite multiple-allele sex determination model as in (2.4), there can simultaneously coexist a stable symmetric equilibrium with corresponding sex ratio different from one-to-one and a stable surface of even sex-ratio equilibria, as will be exemplified in Chapter 4.

3.8. RESULTS FOR THE MATERNALLY CONTROLLED SEX RATIO DETERMINATION MODELS II AND VIII

As we indicated earlier the characterizations of the equilibria for the transformation equations (2.14) are akin to those described in Sections 3.1–3.7. (The proofs require only more technical elaborations.) We formally state the results for this case.

THEOREM 3.8. i) *The maternal control sex ratio determination model (2.14) with M symmetrically generic* (in the sense of Definition 3.2) *admits as symmetric equilibria the equilibria of the one-locus viability model (3.3) with matrix M, and as even sex-ratio equilibria all triplets* $\{\tilde{\mathbf{p}},\tilde{\mathbf{q}},\tilde{\mathbf{z}}\}$ *with* $\tilde{\mathbf{q}}$ *corresponding to the right eigenvector in* Δ *associated with the eigenvalue* $\rho(B(\tilde{\mathbf{x}})) = 1$ *for the matrix* $B(\mathbf{x}) = D_{M\mathbf{x}} + \mathbf{x} \circ M$ *defined on the simplex of frequency vectors* Δ *while* $\tilde{\mathbf{p}} = (\tilde{\mathbf{q}} + \tilde{\mathbf{x}})/2$ *and* $\tilde{\mathbf{z}} = \tilde{\mathbf{q}}/2$.

ii) *A symmetric equilibrium* $\{\hat{\mathbf{p}},\hat{\mathbf{p}},\hat{\mathbf{z}}\}$, $\hat{\mathbf{z}} = w\hat{\mathbf{p}}$ *with sex ratio* $\hat{w} < 1/2$ $(\hat{w} > 1/2)$ *is stable if and only if* $\hat{\mathbf{p}}$ *is stable for the one-locus viability model (3.3) with viability matrix* M $((U - M),$ *respectively).*

iii) *If an even sex-ratio equilibrium exists, it is embedded in an* $r - 2$ *dimensional surface of even sex-ratio equilibria.*

iv) *Stable symmetric equilibria and even sex-ratio equilibria segregating the same set of alleles cannot coexist.*

v) *The only attainable stable equilibria resulting with the introduction of a mutant allele have a sex ratio representation closer to one-to-one.*

The symmetric and even sex-ratio equilibria share the same properties with zygotic or maternal genotypic sex-determination. The rates of convergence to the stable equilibria generally differ.

The same characterizations hold for the sex ratio determination model (2.22) with maternal control at a sex-linked locus.

3.9. RESULTS FOR ONE-LOCUS MULTIALLELIC MODELS OF MATERNALLY INHERITED VIABILITY

We state the results akin to Theorem 3.3.

THEOREM 3.9. *The equilibria of the maternal inheritance viability model (2.16) correspond to the equilibria of the viability model (3.3) with the same stability or instability properties.*

The same is true for the maternal inheritance viability model (2.23) controlled at a sex-linked locus and the multiallele viability regime (2.24) active only in one sex (see Karlin, 1978, for more details)

3.10. RESULTS FOR THE MIXED PARTHENOGENESIS (HAPLODIPLOID) MODEL V

Again, there are two classes of equilibria: the symmetric equilibria $\{\mathbf{p}^*,\mathbf{p}^*\}$ in which the $\mathbf{p}^*$ coincide with the equilibrium points of the one-locus one-sex viability system (3.3), and unsymmetric equilibrium points associated with a primary male sex ratio in diploid zygotes equal to

$$\tilde{w} = \frac{1 - \alpha - \alpha s}{2(1 - \alpha)} \qquad \text{if } \tilde{w} > 0. \qquad (3.17)$$

Note that $\tilde{w}$ is always smaller than $1/2$ since $s\alpha > 0$. The corresponding *population* sex ratio at maturity is one-to-one (see the discussion following (3.19)).

Again, an unsymmetric equilibrium involving r alleles is part of an $r - 2$ dimensional surface of equilibrium points enjoying the *same properties* as the even sex-ratio equilibrium points for the diploid sex determination model (2.4). Thus, with increased allelism, arrhenotoky with a constant fraction of (male) haploids favors a female-biased sex ratio among diploid zygotes according to the "optimal" ratio of male diploids of (3.17).

THEOREM 3.10. *Consider the haplodiploid sex determination model (2.21) with sex determination matrix M for diploid zygotes symmetrically generic in the sense of Definition 3.2. A symmetric equilibrium*

88

$\langle \mathbf{p}^*, \mathbf{p}^* \rangle$ with $\langle \mathbf{p}^*, M\mathbf{p}^* \rangle = w^*$ *is stable*

> *if* $\mathbf{p}^*$ *is stable for the one-locus viability system* (3.3)
> *corresponding to M when* $w^* < \tilde{w}$,   (3.18)
> *but corresponding to* $U - M$ *when* $w^* > \tilde{w}$.

The conditions (3.18) correspond to $w^* < 1/2$ and $w^* > 1/2$, respectively, of Theorem 3.3. At an unsymmetric equilibrium the relative numbers at maturity are

Haploid males	Diploid males	Diploid females	
αs	$\tilde{w}(1 - \alpha),$	$(1 - \alpha)(1 - \tilde{w})$	(3.19)

This yields a one-to-one sex ratio at maturity if $\tilde{w} > 0$, i.e., $\alpha < 1/(1 + s)$. The alternative model with $2\alpha s$ instead of αs (reflecting equal gametic output contributed by diploid fertilization versus haploid fertilization) leads to an optimal sex ratio $(1 - \alpha)/2(1 - \alpha + \alpha s)$ at maturity if $\alpha < 1/(1 + 2s)$. Otherwise, i.e., if the viability s of haploid males compared to 1 for diploid males is large enough, evolution to a population of all haploid males occurs. The condition $\alpha \geq 1/(1 + 2s)$ for the evolution of pure haplodipoidy is in Hartl and Brown (1970), where the effects of differential male fertility (virility) are also discussed.

Multiallele models leading to the same conditions for the *initial increase* in frequency of haploid males were formulated in Bull (1981c) and Eshel and Feldman (1982b).

The optimality properties of the relative sex proportions (3.19) in haplodiploid populations in analogy with our results in favor of an even sex ratio in diploid populations (Model I) can suggest the direction of evolution.

REMARK. Bull (1979) proposed that haploid males can evolve under mother's control if their fitness s exceeds $1/2$ (compared to 1 for diploid males) without loss of fertility, on the basis that they then transmit more maternal genetic material to the future generations. However, this advantage for haploid males must decrease with a decrease of their fertility. The condition $\alpha \geq 1/(1 + s)$ in the case of a half gametic output compared to $\alpha \geq 1/(1 + 2s)$ in the case of an equal gametic output for the evolution of all haploid males, where α is the fraction of unfertilized eggs, reflects this effect.

APPENDIX A. SOME PRELIMINARIES

The gradient map $T'(\hat{\mathbf{p}})$ at an equilibrium $\hat{\mathbf{p}}$ of the viability model (3.3) (i.e., the local linear approximation to T at $\hat{\mathbf{p}}$) is

$$T'(\hat{\mathbf{p}})\boldsymbol{\xi} = \frac{\boldsymbol{\xi} \circ M\hat{\mathbf{p}}}{w(\hat{\mathbf{p}})} + \frac{\hat{\mathbf{p}} \circ M\boldsymbol{\xi}}{w(\hat{\mathbf{p}})} - \frac{2\langle M\hat{\mathbf{p}},\boldsymbol{\xi}\rangle}{[w(\hat{\mathbf{p}})]^2}\,\hat{\mathbf{p}} \circ M\hat{\mathbf{p}} \quad (A.1)$$

where $\boldsymbol{\xi} = (\xi_1, \ldots, \xi_r)$ satisfies $\sum\limits_{j=1}^{r} \xi_j = 0$ and is such that $\hat{\mathbf{p}} + \varepsilon\boldsymbol{\xi} \in \Delta$ for $\varepsilon > 0$ small enough. For a polymorphic equilibrium $\mathbf{p}^*$, we have $M\mathbf{p}^* = w(\mathbf{p}^*)\mathbf{u}$ where $\mathbf{u}$ is the vector displaying all unit components (cf. (3.9)) and the gradient map $T'(\mathbf{p}^*)$ acting on $\mathscr{L} = \{\boldsymbol{\xi}: \langle\boldsymbol{\xi},\mathbf{u}\rangle = 0\}$ reduces to

$$T'(\mathbf{p}^*) = I + \frac{\mathbf{p}^* \circ M}{w(\mathbf{p}^*)} = \frac{1}{w(\mathbf{p}^*)}\,B(\mathbf{p}^*) \quad (A.2)$$

in agreement with the definition (3.4) for $B(\mathbf{p}^*)$. Actually using the notation

$$\frac{\mathbf{x}}{\mathbf{y}} = \frac{(x_1, \ldots, x_r)}{(y_1, \ldots, y_r)} = \left(\frac{x_1}{y_1}, \ldots, \frac{x_r}{y_r}\right) \quad (A.3)$$

and introducing the matrix

$$M(\mathbf{p}) = \frac{\mathbf{p}}{M\mathbf{p}} \circ M, \quad (A.4)$$

we have, for every $\mathbf{p}$ in Δ,

$$B(\mathbf{p}) = D_{M\mathbf{p}} + \mathbf{p} \circ M = D_{M\mathbf{p}}[I + M(\mathbf{p})]. \quad (A.5)$$

Recall the following facts from matrix theory (see, e.g., Gantmacher, 1959). If M is a symmetric matrix and D is a diagonal matrix with all positive diagonal elements, then DM, MD, and $D^{1/2}MD^{1/2}$ share the same eigenvalues, which are real since $D^{1/2}MD^{1/2}$ is a symmetric matrix. Moreover, the matrix $DM = \mathbf{d} \circ M$, where $\mathbf{d} = (d_1, \ldots, d_r)$ is the vector of the diagonal elements of D, acts as a symmetric matrix with respect to the modified inner product

$$\langle\langle \mathbf{x},\mathbf{y} \rangle\rangle = \left\langle \frac{\mathbf{x}}{\mathbf{d}}, \mathbf{y} \right\rangle = \sum_{i=1}^{r} \frac{x_i y_i}{d_i}. \tag{A.6}$$

Indeed, we have

$$\langle\langle DM\mathbf{x},\mathbf{y} \rangle\rangle = \langle M\mathbf{x},\mathbf{y} \rangle = \langle \mathbf{x},M\mathbf{y} \rangle = \langle\langle \mathbf{x},DM\mathbf{y} \rangle\rangle. \tag{A.7}$$

We say that DM is *symmetrizable*. In such a case, there exists a complete set of eigenvectors orthonormal with respect to the inner product (A.6).

When $\mathbf{p}$ is interior to Δ (i.e., $\mathbf{p} \in \Delta^0$) the matrix $M(\mathbf{p})$ of (A.4) is symmetrizable and positive. It is easy to check that

$$M(\mathbf{p})\mathbf{p} = \mathbf{p} \text{ and } \mathbf{u}(\mathbf{p})M(\mathbf{p}) = \mathbf{u}(\mathbf{p}) \tag{A.8}$$

i.e., 1 is an eigenvalue of $M(\mathbf{p})$ with positive right and left eigenvectors, $\mathbf{p}$ and $\mathbf{u}(\mathbf{p})$, respectively, where $\mathbf{u}(\mathbf{p}) = M\mathbf{p}$. Therefore 1 must be a simple eigenvalue equal to the spectral radius of $M(\mathbf{p})$ owing to the Perron-Frobenius theory for positive matrices. Note that since $\mathbf{u}(\mathbf{p})$ is a left eigenvector, we have

$$\langle M(\mathbf{p})\xi,\mathbf{u}(\mathbf{p}) \rangle = \langle \xi,\mathbf{u}(\mathbf{p})M(\mathbf{p}) \rangle = \langle \xi,\mathbf{u}(\mathbf{p}) \rangle \tag{A.9}$$

and therefore $\langle M(\mathbf{p})\xi,\mathbf{u}(\mathbf{p}) \rangle = 0$ if $\langle \xi,\mathbf{u}(\mathbf{p}) \rangle = 0$. Finally the other eigenvalues of $M(\mathbf{p})$ (apart from 1) are real and strictly less than 1 in magnitude. It follows that for $\mathbf{p} \in \Delta^0$ the matrix $I + M(\mathbf{p})$ is invertible with all positive eigenvalues and so is $D_{M\mathbf{p}}[I + M(\mathbf{p})] = B(\mathbf{p})$ invertible. Let us summarize.

LEMMA A.1. *For $\mathbf{p} \in \Delta^0$, the symmetrizable* (cf. (A.7)) *matrix $M(\mathbf{p}) = (\mathbf{p}/M\mathbf{p}) \circ M$ has a simple eigenvalue 1 with right eigenvector $\mathbf{p}$, which strictly dominates all other eigenvalues in absolute value. Moreover, $M(\mathbf{p})$ transforms the linear manifold $\mathscr{L}(\mathbf{p}) = \{\xi: \langle \xi,M\mathbf{p} \rangle = 0\}$ into itself.*

COROLLARY A.1. *The mappings $I + M(\mathbf{p})$ and $I - M(\mathbf{p})$ for $\mathbf{p} \in \Delta^0$ are invertible on $\mathscr{L}(\mathbf{p})$ onto itself.*

COROLLARY A.2. *If $\mathbf{p} \in \Delta^0$, then the symmetrizable matrix $B(\mathbf{p}) = D_{M\mathbf{p}}[I + M(\mathbf{p})]$ is invertible and all eigenvalues of $B(\mathbf{p})$ are positive.*

Finally, we record the following lemma.

LEMMA A.2. *A frequency vector* $\mathbf{p}^* \in \Delta^0$ *is a polymorphic equilibrium of the viability model (3.3) if and only if*

$$B(\mathbf{p}^*)\mathbf{p}^* = \lambda^*\mathbf{p}^*, \text{ and then } \lambda^* = 2w(\mathbf{p}^*) = 2\langle\mathbf{p}^*,M\mathbf{p}^*\rangle.$$

(A.10)

Moreover, the polymorphic equilibrium $\mathbf{p}^*$ *is locally stable if and only if the eigenvalues of* $B(\mathbf{p}^*)$ *restricted to* $\mathscr{L} = \{\boldsymbol{\xi}: \langle\boldsymbol{\xi},\mathbf{u}\rangle = 0\}$ *are strictly less than* $w(\mathbf{p}^*)$ *(or equivalently by Lemma A.1 the eigenvalues of* $M(\mathbf{p}^*)$ *restricted to* $\mathscr{L}$ *are strictly negative.)*

PROOF of Lemma A.2. For $\mathbf{p}^* \in \Delta^0$, we have

$$B(\mathbf{p}^*)\mathbf{p}^* = D_{M\mathbf{p}^*}[I + M(\mathbf{p}^*)]\mathbf{p}^* = 2\mathbf{p}^* \circ M\mathbf{p}^* \quad (A.11)$$

Therefore $B(\mathbf{p}^*)\mathbf{p}^* = \lambda^*\mathbf{p}^*$ if and only if $M\mathbf{p}^* = (\lambda^*/2)\mathbf{u}$. Then $w(\mathbf{p}^*) = \langle\mathbf{p}^*,M\mathbf{p}^*\rangle = (\lambda^*/2)\langle\mathbf{p}^*,\mathbf{u}\rangle = \lambda^*/2$, i.e., $\mathbf{p}^*$ is a polymorphic equilibrium of (3.3). A necessary and sufficient condition for local stability of $\mathbf{p}^*$ is that the gradient matrix at $\mathbf{p}^*$ relative to $\mathscr{L}$ (cf. (A.2)) exhibits all eigenvalues strictly less than 1 in magnitude as stated in Lemma A.2. The proof is complete.

It will be essential to deal with $B(\mathbf{p})$ where $\mathbf{p}$ is on the boundary of Δ. Recall that $\rho(\mathbf{p}) = \rho(B(\mathbf{p}))$ denotes the spectral radius (the magnitude of the eigenvalue of largest magnitude) of $B(\mathbf{p}) = D_{M\mathbf{p}} + \mathbf{p} \circ M$ for $\mathbf{p} \in \Delta$. When $\mathbf{p} \in \Delta^0$, then $B(\mathbf{p})$ is positive and $\rho(\mathbf{p})$ is defined by (3.5). For $\mathbf{p}$ on the boundary of Δ, let $\mathscr{I}(\mathbf{p})$ be the set of indices corresponding to the positive components of $\mathbf{p}$ and let $B_+(\mathbf{p})$ be the (positive) principal submatrix of $B(\mathbf{p})$ restricted to these indices. The rows of $B(\mathbf{p})$ corresponding to the zero components of $\mathbf{p}$ reduce to the rows of the diagonal matrix $D_{M\mathbf{p}}$. Therefore the eigenvalues of $B(\mathbf{p})$ are the eigenvalues of $B_+(\mathbf{p})$ and the quantities $(M\mathbf{p})_i$ for those i where $p_i = 0$. Denoting the spectral radius of $B_+(\mathbf{p})$ by $\rho_+(\mathbf{p}) = \rho(B_+(\mathbf{p}))$, we have

$$\rho(\mathbf{p}) = \max\{\rho_+(\mathbf{p}),(M\mathbf{p})_i \text{ for all } i \notin \mathscr{I}(\mathbf{p})\}. \quad (A.12)$$

Observe that if

$$\rho_+(\mathbf{p}) > (M\mathbf{p})_i \text{ for all } i \notin \mathscr{I}(\mathbf{p}), \quad (A.13)$$

then the principal right eigenvector $\mathbf{q}$ of $B(\mathbf{p})$ in Δ is such that $\mathscr{I}(\mathbf{q}) = \mathscr{I}(\mathbf{p})$ (i.e., $\mathbf{q}$ and $\mathbf{p}$ are on the same face of Δ) and the positive components of $\mathbf{q}$ form the principal right eigenvector of $B_+(\mathbf{p})$ for the corresponding subsystem. Recall that the even sex-ratio equilibria for the sex determination model (2.4) are characterized by the relation $\rho(\mathbf{p}) = 1$. Therefore with the conditions of Definition 3.1 in force, the condition (A.13) is always satisfied at such equilibria.

APPENDIX B. PROOFS OF THE MAIN RESULTS FOR ONE-LOCUS MULTIALLELE VIABILITY SYSTEMS

We first develop some preliminaries. Let $\mathbf{p}^* \in \Delta^0$. Choose an arbitrary direction $\boldsymbol{\eta} = (\eta_1, \eta_2, \ldots, \eta_r)$ in Δ^0, that is, satisfying $\langle \boldsymbol{\eta}, \mathbf{u} \rangle = \sum_{i=1}^{r} \eta_i = 0$, and s small enough to ensure $\mathbf{p}(s) = \mathbf{p}^* + s\boldsymbol{\eta}$ in Δ^0. The Perron-Frobenius theory for positive matrices affirms that the spectral radius $\rho(s) = \rho(\mathbf{p}(s))$ of $B(\mathbf{p}(s)) = D_{M\mathbf{p}(s)} + \mathbf{p}(s) \circ M$ is a simple eigenvalue of $B(\mathbf{p}(s))$ analytic in s. The components of $\mathbf{q}(s)$, the unique right eigenvector in Δ^0 of $B(\mathbf{p}(s))$ corresponding to $\rho(s)$, are also analytic functions of s. We display these quantities by the relation

$$B(\mathbf{p}(s))\mathbf{q}(s) = \rho(s)\mathbf{q}(s) \text{ where } \rho(s) = \rho(\mathbf{p}(s)). \quad \text{(B.1)}$$

We use hereafter the notation

$$\dot{\mathbf{p}}(s) = \frac{d\mathbf{p}(s)}{ds} = \left(\frac{dp_1(s)}{ds}, \ldots, \frac{dp_r(s)}{ds} \right) \text{ and}$$

$$\dot{\mathbf{q}}(s) = \left(\frac{dq_1(s)}{ds}, \ldots, \frac{dq_r(s)}{ds} \right),$$

$$\ddot{\mathbf{p}}(s) = \frac{d^2\mathbf{p}(s)}{ds^2} \text{ and } \ddot{\mathbf{q}}(s) = \frac{d^2\mathbf{q}(s)}{ds^2}, \quad \text{(B.2)}$$

which we abbreviate to $\dot{\mathbf{p}}$, $\dot{\mathbf{q}}$, $\ddot{\mathbf{p}}$, and $\ddot{\mathbf{q}}$ at $s = 0$.

Since $\langle \mathbf{p}(s), \mathbf{u} \rangle = \langle \mathbf{q}(s), \mathbf{u} \rangle = 1$ where $\mathbf{u}$ is the vector of all unit components, we have

$$\langle \dot{\mathbf{p}}(s), \mathbf{u} \rangle = \langle \ddot{\mathbf{p}}(s), \mathbf{u} \rangle = \langle \dot{\mathbf{q}}(s), \mathbf{u} \rangle = \langle \ddot{\mathbf{q}}(s), \mathbf{u} \rangle = 0. \quad \text{(B.3)}$$

Observe that for $\mathbf{p} \in \Delta^0$ and every vector ξ, we have

$$\xi B(\mathbf{p}) = \xi \circ M\mathbf{p} + [\xi \circ \mathbf{p}]M = [M\mathbf{p}] \circ \xi + M[\mathbf{p} \circ \xi].$$

(B.4)

Choosing $\xi = \mathbf{z} = \mathbf{q}/\mathbf{p}$ where $\mathbf{q}$ is the leading right eigenvector of $B(\mathbf{p})$ in Δ^0, we get

$$\mathbf{z}B(\mathbf{p}) = \rho(\mathbf{p})\mathbf{z}.$$

(B.5)

Indeed, $\mathbf{z}B(\mathbf{p}) = [M\mathbf{p}] \circ \mathbf{q}/\mathbf{p} + M\mathbf{q} = (\mathbf{u}/\mathbf{p}) \circ [\mathbf{q} \circ M\mathbf{p} + \mathbf{p} \circ M\mathbf{q}] = (\mathbf{u}/\mathbf{p}) \circ B(\mathbf{p})\mathbf{q} = \rho(\mathbf{p})\mathbf{z}$. Furthermore, we have

$$\left[\frac{\mathbf{u}}{\mathbf{z}}\right] B(\mathbf{q}) = \rho(\mathbf{p})\mathbf{u}$$

(B.6)

since

$$\left[\frac{\mathbf{u}}{\mathbf{z}}\right] B(\mathbf{q}) = [M\mathbf{q}] \circ \frac{\mathbf{p}}{\mathbf{q}} + M\left[\mathbf{q} \circ \frac{\mathbf{p}}{\mathbf{q}}\right]$$

$$= \frac{\mathbf{u}}{\mathbf{q}} \circ [\mathbf{p} \circ M\mathbf{q} + M\mathbf{p} \circ \mathbf{q}]$$

$$= \frac{\mathbf{u}}{\mathbf{q}} \circ B(\mathbf{p})\mathbf{q} = \rho(\mathbf{p})\mathbf{u}.$$

Differentiating the equation in (B.1) with respect to s produces

$$\dot{B}(\mathbf{p}(s))\mathbf{q}(s) + B(\mathbf{p}(s))\dot{\mathbf{q}}(s) = \dot{\rho}(s)\mathbf{q}(s) + \rho(s)\dot{\mathbf{q}}(s),$$

which reduces to

$$B(\mathbf{q}(s))\dot{\mathbf{p}}(s) + B(\mathbf{p}(s))\dot{\mathbf{q}}(s) = \dot{\rho}(s)\mathbf{q}(s) + \rho(s)\dot{\mathbf{q}}(s). \quad \text{(B.7)}$$

On account of (B.5), the inner product on both sides of (B.7) with the vector $\mathbf{z}(s) = \mathbf{q}(s)/\mathbf{p}(s)$ yields $\langle \mathbf{z}(s)B(\mathbf{q}(s)),\dot{\mathbf{p}}(s)\rangle + \rho(s)\langle \mathbf{z}(s),\dot{\mathbf{q}}(s)\rangle = \dot{\rho}(s)\langle \mathbf{z}(s),\mathbf{q}(s)\rangle + \rho(s)\langle \mathbf{z}(s),\dot{\mathbf{q}}(s)\rangle$, i.e.,

$$\langle \mathbf{z}(s)B(\mathbf{q}(s)),\dot{\mathbf{p}}(s)\rangle = \dot{\rho}(s)\langle \mathbf{z}(s),\mathbf{q}(s)\rangle.$$

(B.8)

Note also that on the basis of (B.6), we have

$$\left\langle \left[\frac{\mathbf{u}}{\mathbf{z}(s)}\right] B(\mathbf{q}(s)),\dot{\mathbf{p}}(s) \right\rangle = \rho(s)\langle \mathbf{u},\dot{\mathbf{p}}(s)\rangle = 0, \quad \text{(B.9)}$$

the equality to 0 resulting from (B.3).

We are now prepared to prove Theorems 3.1 and 3.2.

(i) *Characterization of an internal equilibrium*

Necessity. Let $\mathbf{p}^* = \mathbf{p}(0)$ be an equilibrium of (3.3) in Δ^0. Lemma A.2 ensures that $\mathbf{q}(0)$, the leading right eigenvector of $B(\mathbf{p}^*)$ in Δ^0, is $\mathbf{p}^*$. Therefore $\mathbf{z}(0) = \mathbf{q}(0)/\mathbf{p}(0) = \mathbf{u}$ and (B.8) in view of (B.5) yields

$$\dot{\rho}(0) = \langle \mathbf{u}B(\mathbf{p}^*), \dot{\mathbf{p}}(0) \rangle = \rho(\mathbf{p}^*) \langle \mathbf{u}, \dot{\mathbf{p}}(0) \rangle = 0. \quad (B.10)$$

Sufficiency. If $\mathbf{p}^*$ furnishes a critical point for $\rho(\mathbf{p})$ in Δ^0 and $\mathbf{q}^*$ denotes the leading right eigenvector of $B(\mathbf{p}^*)$ in Δ^0, then the equations (B.8) and (B.9) establish that

$$\left\langle \left[\frac{\mathbf{q}^*}{\mathbf{p}^*} \right] B(\mathbf{q}^*), \boldsymbol{\eta} \right\rangle = 0 \text{ and } \left\langle \left[\frac{\mathbf{p}^*}{\mathbf{q}^*} \right] B(\mathbf{q}^*), \boldsymbol{\eta} \right\rangle = 0 \quad (B.11)$$

for every $\boldsymbol{\eta} \in \mathcal{L}$, i.e., for all $\boldsymbol{\eta}$ satisfying $\langle \boldsymbol{\eta}, \mathbf{u} \rangle = 0$. Since $\mathcal{L}$ is $(n-1)$-dimensional, the equations (B.11) are compatible only if there exists some scalar α such that

$$\left[\frac{\mathbf{q}^*}{\mathbf{p}^*} \right] B(\mathbf{q}^*) = \alpha \left[\frac{\mathbf{p}^*}{\mathbf{q}^*} \right] B(\mathbf{q}^*), \text{ and therefore, } \frac{\mathbf{q}^*}{\mathbf{p}^*} = \alpha \frac{\mathbf{p}^*}{\mathbf{q}^*},$$

$$(B.12)$$

since $B(\mathbf{q}^*)$ is invertible by Corollary A.2. But this is feasible only if $\mathbf{q}^* = \sqrt{\alpha}\mathbf{p}^* = \mathbf{p}^*$. In such a case, Lemma A.2 asserts that $\mathbf{p}^*$ must be a polymorphic equilibrium of (3.3).

(ii) *Stability criterion for an internal equilibrium*

Let $\mathbf{p}^*$ be a polymorphic equilibrium, i.e., a critical point of $\rho(\mathbf{p})$ in Δ^0. Differentiating in (B.7) and evaluating at $s = 0$ produces in the notation of (B.2) (with $\mathbf{q}(0) = \mathbf{p}(0) = \mathbf{p}^*$ and $\rho(\mathbf{p}^*) = 2 \langle \mathbf{p}^*, M\mathbf{p}^* \rangle$ owing to Lemma A.2).

$$B(\mathbf{p}^*)\ddot{\mathbf{p}} + B(\mathbf{p}^*)\ddot{\mathbf{q}} + 2\dot{\mathbf{p}} \circ M\dot{\mathbf{q}} + 2\dot{\mathbf{q}} \circ M\dot{\mathbf{p}}$$
$$= \rho(\mathbf{p}^*)\ddot{\mathbf{q}} + \ddot{\rho}(0)\mathbf{p}^*. \quad (B.13)$$

Observe that $M\mathbf{p}^* = w(\mathbf{p}^*)\mathbf{u}$ with $w(\mathbf{p}^*) = \langle \mathbf{p}^*, M\mathbf{p}^* \rangle$ so that $B(\mathbf{p}^*) = w(\mathbf{p}^*)(I + M(\mathbf{p}^*))$ where we have introduced the notation

$$M(\mathbf{p}^*) = \frac{\mathbf{p}^* \circ M}{w(\mathbf{p}^*)} \quad (\text{cf. (A.4)}). \quad (B.14)$$

With reference to $\mathbf{u}B(\mathbf{p}^*) = 2w(\mathbf{p}^*)\mathbf{u}$ and in view of (B.3), taking the inner product on both sides of (B.13) with $\mathbf{u}$ gives

$$\ddot{\rho}(0) = 4\langle \dot{\mathbf{p}}, M\dot{\mathbf{q}} \rangle. \tag{B.15}$$

In terms of the modified inner product

$$\langle\langle \mathbf{x}, \mathbf{y} \rangle\rangle = w(\mathbf{p}^*)\left\langle \frac{\mathbf{x}}{\mathbf{p}^*}, \mathbf{y} \right\rangle, \tag{B.16}$$

we have

$$\ddot{\rho}(0) = 4\langle\langle \dot{\mathbf{p}}, M(\mathbf{p}^*)\dot{\mathbf{q}} \rangle\rangle. \tag{B.17}$$

We next rewrite (B.7) at $s = 0$ in the form

$$(I - M(\mathbf{p}^*))\dot{\mathbf{q}} = (I + M(\mathbf{p}^*))\dot{\mathbf{p}}. \tag{B.18}$$

Both $I + M(\mathbf{p}^*)$ and $I - M(\mathbf{p}^*)$ are invertible on $\mathscr{L} = \{\xi: \langle \xi, \mathbf{u} \rangle = 0\}$ and leave this linear manifold invariant (Corollary A.1). Moreover, $M(\mathbf{p}^*)$ is symmetric in the inner product (B.16) (cf. (A.7)). Thus, every vector $\mathbf{x}$ in $\mathscr{L}$ can be represented using an orthonormal basis $\{\boldsymbol{\varphi}_i\}$ of eigenvectors of $M(\mathbf{p}^*)$ in $\mathscr{L}$ such that

$$\mathbf{x} = \sum_{i=1}^{r-1} \langle\langle \mathbf{x}, \boldsymbol{\varphi}_i \rangle\rangle\boldsymbol{\varphi}_i. \tag{B.19}$$

In particular,

$$\dot{\mathbf{q}} = \sum_{i=1}^{r-1} \langle\langle \dot{\mathbf{q}}, \boldsymbol{\varphi}_i \rangle\rangle\boldsymbol{\varphi}_i, \; \dot{\mathbf{p}} = \sum_{i=1}^{r-1} \langle\langle \dot{\mathbf{p}}, \boldsymbol{\varphi}_i \rangle\rangle\boldsymbol{\varphi}_i. \tag{B.20}$$

Using $M(\mathbf{p}^*)\boldsymbol{\varphi}_i = \lambda_i\boldsymbol{\varphi}_i$, the equation (B.18) implies

$$(1 - \lambda_i)\langle\langle \dot{\mathbf{q}}, \boldsymbol{\varphi}_i \rangle\rangle = (1 + \lambda_i)\langle\langle \dot{\mathbf{p}}, \boldsymbol{\varphi}_i \rangle\rangle \text{ for } i = 1, \ldots, r-1 \tag{B.21}$$

where λ_i is real and $|\lambda_i| < 1$ (Lemma A.1). Expanding (B.17) via (B.20) and using (B.21) leads to

$$\ddot{\rho}(0) = 4 \sum_{i=1}^{r-1} \lambda_i \langle\langle \dot{\mathbf{p}}, \boldsymbol{\varphi}_i \rangle\rangle^2 \left\{ \frac{1 + \lambda_i}{1 - \lambda_i} \right\}. \tag{B.22}$$

As $\dot{\mathbf{p}} = \boldsymbol{\eta}$ is an arbitrary vector in $\mathscr{L}$, we can conclude that $\ddot{\rho}(0) < 0$ for all directions within Δ if and only if $\lambda_i < 0$, $i =$

$1,2, \ldots, r - 1$, that is, $\mathbf{p}^*$ is locally stable (Lemma A.2) if and only if $\rho(\mathbf{p})$ exhibits a (strict) local maximum at $\mathbf{p}^*$ with respect to Δ.

This completes the proof of Theorem 3.1.

(iii) *Characterization of a Locally Stable Boundary Equilibrium*

Let $\hat{\mathbf{p}}$ be a boundary equilibrium of (3.3) and define $\mathbf{p}(s) = \hat{\mathbf{p}} + s\boldsymbol{\eta}$ such that $\mathbf{p}(s) \in \Delta$ for $s \geq 0$ sufficiently small. Assume

$$\rho(\hat{\mathbf{p}}) > (M\hat{\mathbf{p}})_j \text{ for all } j \notin \mathscr{I}(\hat{\mathbf{p}}), \qquad (B.23)$$

and let $\mathbf{q}(s)$ be the principal right eigenvector of $B(\mathbf{p}(s))$ in Δ. Then according to (A.12) and Lemma A.2, we have $\mathbf{q}(0) = \hat{\mathbf{p}}$ and $\rho(\hat{\mathbf{p}}) = 2\langle\hat{\mathbf{p}}, M\hat{\mathbf{p}}\rangle = 2w(\hat{\mathbf{p}})$, and therefore the equation (B.7) for $s = 0$ becomes

$$B(\hat{\mathbf{p}})\dot{\mathbf{p}} + B(\hat{\mathbf{p}})\dot{\mathbf{q}} = \dot{\rho}(0)\hat{\mathbf{p}} + 2w(\hat{\mathbf{p}})\dot{\mathbf{q}},$$

or equivalently,

$$B(\hat{\mathbf{p}})[\dot{\mathbf{p}} + \dot{\mathbf{q}}] - 2w(\hat{\mathbf{p}})\dot{\mathbf{q}} = \dot{\rho}(0)\hat{\mathbf{p}}. \qquad (B.24)$$

Since $\mathbf{u}B(\hat{\mathbf{p}}) = 2M\hat{\mathbf{p}}$, taking the inner product with $\mathbf{u}$ in (B.24) leads to

$$\dot{\rho}(0) = 2\langle M\hat{\mathbf{p}}, \dot{\mathbf{p}} + \dot{\mathbf{q}}\rangle$$

$$= 2w(\hat{\mathbf{p}}) \sum_{j \in \mathscr{I}(\hat{\mathbf{p}})} [\dot{p}_j + \dot{q}_j] + 2 \sum_{j \notin \mathscr{I}(\hat{\mathbf{p}})} (M\hat{\mathbf{p}})_j [\dot{p}_j + \dot{q}_j]$$

$$= 2 \sum_{j \notin \mathscr{I}(\hat{\mathbf{p}})} [(M\hat{\mathbf{p}})_j - w(\hat{\mathbf{p}})][\dot{p}_j + \dot{q}_j]. \qquad (B.25)$$

But for every $j \notin \mathscr{I}(\hat{\mathbf{p}})$, we have from (B.24)

$$(M\hat{\mathbf{p}})_j[\dot{p}_j + \dot{q}_j] - 2w(\hat{\mathbf{p}})\dot{q}_j = 0,$$

i.e.,

$$\dot{q}_j = \left[\frac{(M\hat{\mathbf{p}})_j}{2w(\hat{\mathbf{p}}) - (M\hat{\mathbf{p}})_j}\right]\dot{p}_j. \qquad (B.26)$$

Hence (B.25) can be expressed as follows:

$$\dot{\rho}(0) = -4w(\hat{\mathbf{p}}) \sum_{j \notin \mathscr{I}(\hat{\mathbf{p}})} \left[\frac{w(\hat{\mathbf{p}}) - (M\hat{\mathbf{p}})_j}{2w(\hat{\mathbf{p}}) - (M\hat{\mathbf{p}})_j}\right]\eta_j. \qquad (B.27)$$

Therefore $\dot{\rho}(0) < 0$ in every direction $\boldsymbol{\eta} = (\eta_1, \ldots, \eta_r)$ within Δ (i.e., $\sum\limits_{j=1}^{r} \eta_j = 0$ and $\eta_j \geq 0$ for all $j \notin \mathscr{I}(\hat{\mathbf{p}})$ with strict inequality somewhere) if and only if $(M\hat{\mathbf{p}})_j < w(\hat{\mathbf{p}})$ for all $j \notin \mathscr{I}(\hat{\mathbf{p}})$ (since equalities are precluded in generic cases), that is, $\rho(\mathbf{p})$ is locally decreasing from $\hat{\mathbf{p}}$ toward the interior of Δ if and only if $\hat{\mathbf{p}}$ is externally stable (see (3.10)). Moreover it was established in Theorem 3.1 that $\rho_+(\mathbf{p})$ is internally maximized at $\hat{\mathbf{p}}$ (i.e., within the subsystem corresponding to the positive components of $\hat{\mathbf{p}}$) if and only if $\hat{\mathbf{p}}$ is internally stable.

It remains to show that a strict local maximum of

$$\rho(\mathbf{p}) = \max\{\rho_+(\mathbf{p}), (M\mathbf{p})_j \text{ for all } j \notin \mathscr{I}(\mathbf{p})\} \quad \text{(B.28)}$$

cannot be achieved (only) through the linear terms $(M\mathbf{p})_j$ for $j \notin \mathscr{I}(\mathbf{p})$. Otherwise this must occur at a vertex, say A_1-fixation, and then (relabeling the alleles if necessary) we must have $m_{12} > 2m_{11}$ and $m_{12} > m_{1k}$ for all $k \geq 3$ (with the conditions (3.6) in force). By continuity, this entails that $\rho(\mathbf{p}) = \rho_+(\mathbf{p})$ within the subsystem $\{A_1, A_2\}$ near A_1-fixation. With $M = \|m_{ij}\|_{i,j=1}^{2}$, and $\mathbf{p} = (p_1, p_2)$, we have

$$B(\mathbf{p}) = D_{\mathbf{p}}M + D_{M\mathbf{p}} = \begin{bmatrix} 2p_1 m_{11} + p_2 m_{12} & p_1 m_{12} \\ p_2 m_{12} & 2p_2 m_{22} + p_1 m_{12} \end{bmatrix},$$
$$\text{(B.29)}$$

and at A_1-fixation, the relation (B.7) becomes

$$B(\hat{\mathbf{p}})\dot{\mathbf{p}} + B(\hat{\mathbf{p}})\dot{\mathbf{q}} = \dot{\rho}(0)\hat{\mathbf{q}} + m_{12}\dot{\mathbf{q}} \quad \text{(B.30)}$$

where $\hat{\mathbf{p}} = (1,0)$ and $B(\hat{\mathbf{p}})\hat{\mathbf{q}} = m_{12}\hat{\mathbf{q}}$, i.e.,

$$\hat{q}_2 = \left[\frac{m_{12} - 2m_{11}}{m_{12}}\right]\hat{q}_1, \hat{q}_1 = 1 - \hat{q}_2. \quad \text{(B.31)}$$

Choosing $\dot{\mathbf{p}} = (-1,1)$, the equality for the second component in (B.30) is

$$-\hat{q}_2 m_{12} + 2\hat{q}_2 m_{22} + \hat{q}_1 m_{12} = \dot{\rho}(0)\hat{q}_2,$$

i.e.,

$$\dot{\rho}(0) = 2m_{22} - m_{12} + \left[\frac{\hat{q}_1}{\hat{q}_2}\right] m_{12}$$

$$= \frac{[2m_{22} - m_{12}][m_{12} - 2m_{11}] + m_{12}^2}{m_{12} - 2m_{11}}$$

$$= \frac{2m_{12}[m_{11} + m_{22}] - 4m_{11}m_{22}}{m_{12} - 2m_{11}} > \frac{4m_{11}^2}{m_{12} - 2m_{11}} > 0.$$

This contradicts the assumption that $\rho(\mathbf{p})$ achieves a local maximum at A_1-fixation and completes the proof of part (i) of Theorem 3.2. Part (ii) is a direct consequence of Lemma A.2 coupled with classical results (ascertaining that the mean fitness function $w(\mathbf{p})$ is locally maximized at the stable equilibria of (3.3) and globally maximized if the stable equilibrium is interior to Δ; see Section 3.2).

APPENDIX C. PROOFS OF THE RESULTS OF CHAPTER 3 FOR THE SEX DETERMINATION MODELS AND RELATED MODELS OF CHAPTER 2

C.1. Proofs of Theorems 3.3 and 3.3′ for the symmetric equilibria of (2.4)

Let $\mathbf{p}^*$ be an interior symmetric equilibrium of the system (2.4) with associated sex ratio $w^* = \langle \mathbf{p}^*, M\mathbf{p}^* \rangle$. We recall the following result (e.g., Karlin (1978)).

The relevant gradient matrix at $\{\mathbf{p}^*, \mathbf{p}^*\}$ is of the form

$$\Gamma = \begin{pmatrix} R & R \\ S & S \end{pmatrix} \tag{C.1}$$

acting on $\mathscr{L} \otimes \mathscr{L} = \{\{\xi, \eta\}: \langle \xi, \mathbf{u} \rangle = \langle \eta, \mathbf{u} \rangle = 0\}$ with

$$R = \frac{1}{2}\left[I + \frac{\mathbf{p}^* \circ M}{w^*}\right], \quad S = \frac{1}{2}\left[I - \frac{\mathbf{p}^* \circ M}{1 - w^*}\right]. \tag{C.2}$$

The relevant eigenvalues of Γ are those of $R + S$ relative to the linear manifold $\mathscr{L}$ plus the eigenvalue 0 which occurs with additional multiplicity r. Note that the matrix

$$R + S = I + \left[\frac{1 - 2w^*}{2(1 - w^*)} \right] \frac{\mathbf{p}^* \circ M}{w^*} \tag{C.3}$$

resembles the gradient matrix at $\mathbf{p}^*$ for the one-locus one-sex viability selection model with associated matrix M except for the constant multiplier $(1 - 2w^*)/2(1 - w^*)$ (compare to (A.2)). It follows that if $\mathbf{p}^* \in \Delta^0$ is a stable equilibrium of (3.3) (such that $\mathbf{p}^* \circ M/w^*$ has only negative eigenvalues of magnitude less than 1 with respect to $\mathscr{L}$) and $w^* < 1/2$, then all eigenvalues of $R + S$ are also of magnitude less than 1 relative to $\mathscr{L}$, and conversely. The case $w^* > 1/2$ can be dealt with analogously by interchanging the sexes and replacing M by $U - M$. The proof of Theorem 3.3 is complete.

The stability analysis for a symmetric boundary equilibrium $\{\hat{\mathbf{p}}, \hat{\mathbf{p}}\}$ entails the stability conditions relative to the face of $\hat{\mathbf{p}}$ akin to Theorem 3.3, and the external stability conditions with respect to the alleles not represented in the array of $\hat{\mathbf{p}}$ (cf. (3.10)).

The local linear analysis near $\{\hat{\mathbf{p}}, \hat{\mathbf{p}}\}$ with respect to external alleles reveals the first-order linear approximation

$$(q_j' + p_j') = (q_j + p_j) \left[\frac{(M\hat{\mathbf{p}})_j}{2\hat{w}} + \frac{((U - M)\hat{\mathbf{p}})_j}{2(1 - \hat{w})} \right] \text{ for } j \notin \mathscr{I}(\hat{\mathbf{p}}) \tag{C.4}$$

where $\hat{w} = \langle \hat{\mathbf{p}}, M\hat{\mathbf{p}} \rangle$. Therefore the relevant eigenvalues governing external stability are

$$\frac{(M\hat{\mathbf{p}})_j}{2\hat{w}} + \frac{((U - M)\hat{\mathbf{p}})_j}{2(1 - \hat{w})}$$

$$= 1 + \frac{(1 - 2\hat{w})}{2(1 - \hat{w})} \left[\frac{(M\hat{\mathbf{p}})_j}{\hat{w}} - 1 \right] \text{ for all } j \notin \mathscr{I}(\hat{\mathbf{p}}). \tag{C.5}$$

This completes the proof of Theorem 3.3'.

C.2. Proof of Theorem 3.4 for the even sex-ratio equilibrium surfaces of (2.4). (i) Let $\tilde{\mathbf{p}}$ be an even sex-ratio equilibrium interior to Δ, i.e., $\tilde{\mathbf{p}} \in \Delta^0$ satisfies $\rho(\tilde{\mathbf{p}}) = 1$. Since M is assumed to be symmetrically generic, $\tilde{\mathbf{p}}$ cannot coincide with a symmetric equilibrium, i.e., a critical point of $\rho(\mathbf{p})$ owing to Theorem 3.1. Therefore there exists some direction within Δ^0 along which the derivative of $\rho(\mathbf{p})$ at $\hat{\mathbf{p}}$ is not zero. Hence $\hat{\mathbf{p}}$ must belong to the common frontier of two connected components of the *open sets* $V_+ = \{\mathbf{p} \in \Delta^0 : \rho(\mathbf{p}) > 1\}$ and $V_- = \{\mathbf{p} \in \Delta^0 : \rho(\mathbf{p}) < 1\}$, respectively. That frontier is an even sex-ratio equilibrium surface of dimension $r - 2$ (and simply connected) in the simplex Δ^0 of dimension $r - 1$ by continuity of $\rho(\mathbf{p})$. Clearly, any curve in Δ connecting a point of V_- with a point of V_+ must cross the surface of $\rho(\mathbf{p}) = 1$.

If $\tilde{\mathbf{p}}$ is on the boundary of Δ, then $\rho(\mathbf{p}) = \rho_+(\mathbf{p})$ in some neighborhood of $\tilde{\mathbf{p}}$ by appeal to (A.12) and (A.13). The preceding arguments can be applied *mutatis mutandis* with respect to the face of $\tilde{\mathbf{p}}$, and then extended to all Δ by continuity. (Note that $\tilde{\mathbf{p}}$ cannot be a vertex on account of the assumption of Definition 3.2.)

(ii) An internal stable symmetric equilibrium $\mathbf{p}^*$ with $w^* = \langle \mathbf{p}^*, M\mathbf{p}^* \rangle < 1/2$ corresponds to an internal stable equilibrium for the viability model M (Theorem 3.3) and therefore to the global (unique) maximum of $\rho(\mathbf{p})$ over all Δ (Theorem 3.2). This precludes the existence of another stable symmetric equilibrium. Moreover, since $\rho(\mathbf{p}^*) = 2\langle \mathbf{p}^*, M\mathbf{p}^* \rangle < 1$ (Lemma A.2), the function $\rho(\mathbf{p})$ cannot achieve the value 1 in Δ and even sex-ratio equilibria cannot be present. If $w^* > 1/2$, we simply replace M by $U - M$.

(iii) If $\hat{\mathbf{p}}_\alpha$ and $\hat{\mathbf{p}}_\beta$ are two symmetric equilibria such that $\langle \hat{\mathbf{p}}_\alpha, M\hat{\mathbf{p}}_\alpha \rangle = \hat{w}_\alpha < 1/2$ and $\langle \hat{\mathbf{p}}_\beta, M\hat{\mathbf{p}}_\beta \rangle = \hat{w}_\beta > 1/2$, then we have $\rho_+(\hat{\mathbf{p}}_\alpha) = 2\hat{w}_\alpha < 1$ and $\rho_+(\hat{\mathbf{p}}_\beta) = 2\hat{w}_\beta > 1$ owing to Lemma A.2 and also

$$\rho(\hat{\mathbf{p}}_\alpha) = \max\{\rho_+(\hat{\mathbf{p}}_\alpha), (M\hat{\mathbf{p}}_\alpha)_i, i \notin \mathscr{I}(\hat{\mathbf{p}}_\alpha)\} < 1,$$
$$\rho(\hat{\mathbf{p}}_\beta) = \max\{\rho_+(\hat{\mathbf{p}}_\beta), (M\hat{\mathbf{p}}_\beta)_i, i \notin \mathscr{I}(\hat{\mathbf{p}}_\beta)\} > 1. \quad \text{(C.6)}$$

By continuity of $\rho(\mathbf{p})$, all curves connecting $\hat{\mathbf{p}}_\alpha$ and $\hat{\mathbf{p}}_\beta$ in Δ must intersect the surface of $\rho(\mathbf{p}) = 1$ and these intersection points generate at least one separating even sex-ratio equilibrium surface.

REMARK. Even sex-ratio equilibrium surfaces cannot enclose two connected components entirely in the interior of Δ. In fact, each such component would contain a critical point $\rho(\mathbf{p})$ (a local maximum or minimum) engendering multiple polymorphic equilibria for the viability model M. Such a situation is precluded under the conditions of Definition 3.1. For the same reason, an even sex-ratio equilibrium surface cannot split within Δ yielding a continuum of symmetric equilibrium with associated sex ratio $1/2$.

C.3. PROOF of Theorem 3.5 for dichotomous sex determination models. Let $M = \|m_{ij}\|_{i,j=1}^{r}$ be a dichotomous sex-determination matrix (i.e., $m_{ij} = 0$ or 1 for every i and j). Suppose first that M is nonsingular and let $\mathbf{p}^{*}$ be a polymorphic symmetric equilibrium of (2.4) with $\langle \mathbf{p}^{*}, M\mathbf{p}^{*} \rangle = w^{*} < 1/2$. We show that $\mathbf{p}^{*}$ cannot be stable. Otherwise, $\mathbf{p}^{*}$ would be stable for the viability model M and $w(\mathbf{p}) = \langle \mathbf{p}, M\mathbf{p} \rangle$ would achieve a global maximum in Δ at $\mathbf{p}^{*}$. Hence, we would have $1/2 > w^{*} > m_{ii}$ compelling $m_{ii} = 0$ for all i. But then $m_{ij} = 1$ for some $i \neq j$ and the equilibrium $\hat{\mathbf{p}}$ with $\hat{p}_{i} = \hat{p}_{j} = 1/2$ would have mean fitness $w(\hat{\mathbf{p}}) = 1/2$ exceeding $w(\mathbf{p}^{*}) = w^{*}$, which is a contradiction.

The case $w^{*} > 1/2$ can be dealt with analogously by replacing M by $U - M$. Note that when a polymorphic symmetric equilibrium $\mathbf{p}^{*}$ exists, M is nonsingular if and only if $U - M$ is nonsingular since then $M\mathbf{p}^{*} = w^{*}\mathbf{u}$ has a unique solution if and only if $(U - M)\mathbf{p}^{*} = (1 - w^{*})\mathbf{u}$ has a unique solution.

Consider next a symmetric equilibrium $\mathbf{p}^{*}$ with $w^{*} \neq 1/2$ and corresponding coefficient matrix M singular. Then there exists a nontrivial vector $\boldsymbol{\xi}$ satisfying $M\boldsymbol{\xi} = \mathbf{0}$. Since $0 = \langle \mathbf{p}^{*}, M\boldsymbol{\xi} \rangle = \langle M\mathbf{p}^{*}, \boldsymbol{\xi} \rangle = w^{*}\langle \mathbf{u}, \boldsymbol{\xi} \rangle$, we must have $\langle \boldsymbol{\xi}, \mathbf{u} \rangle = 0$. The line $\mathbf{p}^{*} + t\boldsymbol{\xi}$ in Δ for t real satisfies $M(\mathbf{p}^{*} + t\boldsymbol{\xi}) = M\mathbf{p}^{*} = w^{*}\mathbf{u}$. This symmetric equilibrium line must intersect the boundary of Δ at some point $\mathbf{p}^{**}$. If the submatrix corresponding to the positive components of $\mathbf{p}^{**}$ is singular, we can reduce again the analysis to a smaller face along an equilibrium line with sex ratio $w^{*} \neq 1/2$. Proceeding in this way, we finally arrive at a boundary symmetric equilibrium whose corresponding submatrix is nonsingular and which cannot be internally stable owing to the previous arguments. We conclude that $\mathbf{p}^*$ belongs to an unstable symmetric equilibrium surface.

C.4. PROOF *of Theorem 3.6 for X-linked sex-determination models.* A dominant male (or female) determining factor (as occurs in standard XX-XY systems) can be accommodated by (2.4) if we set $m_{1j} = 1$ for all j in the sex-determination matrix $M = \|m_{ij}\|_{i,j=1}^r$. Since $(M\mathbf{p})_1 = 1$ for all frequency vectors $\mathbf{p}$, the strong sex-determiner represented by allele A_1 cannot be present at any symmetric equilibrium (otherwise the sex ratio would be one which is not biologically feasible). Moreover, a symmetric (non-even sex ratio) equilibrium is stable if and only if it is stable with respect to the subsystem $\tilde{M} = \|m_{ij}\|_{i,j=2}^r$ and associated with a sex ratio exceeding $1/2$ to ensure external stability against A_1. In all circumstances any dominant sex-determiner in (2.4) can be maintained at equilibrium only with an even sex ratio.

C.5. PROOF *of Theorem 3.7 on the optimality properties of even sex-ratio equilibria.* Suppose that $\mathbf{p}^*$ is a stable polymorphic symmetric equilibrium for the sex determination model (2.4) with $\langle \mathbf{p}^*, M\mathbf{p}^* \rangle = w^* < 1/2$. (The case $w^* > 1/2$ can be dealt with analogously.) Then Theorem 3.3 informs us that $\mathbf{p}^*$ is stable for the viability model (3.1) with matrix $M = \|m_{ij}\|_{i,j=1}^r$. But $\mathbf{p}^*$ is also externally unstable in both models (compare (3.16) with (3.10)) after the introduction of allele A_{r+1}, i.e., with respect to the augmented $(r + 1)$-dimensional matrix $\tilde{M} = \|m_{ij}\|_{i,j=1}^{r+1}$. We make appeal to Theorem D.1 (see Appendix D) on the number of stable equilibria that can coexist in one-locus multiallele viability models to ensure the existence of a unique stable equilibrium $\mathbf{p}^{**}$ (interior or on the boundary of the augmented allelic frequency simplex) for the viability model $\tilde{M}$. Moreover, we must have $\langle \mathbf{p}^{**}, \tilde{M}\mathbf{p}^{**} \rangle = w^{**} > w^*$. In the event $w^{**} < 1/2$, $\mathbf{p}^{**}$ is a stable symmetric equilibrium for the sex determination model corresponding to $\tilde{M}$ and actually the unique stable equilibrium for this model with no even sex-ratio equilibria present owing to part (ii) of Theorem 3.4. On the other hand, if $w^{**} > 1/2$, part (iii) of Theorem 3.4 asserts that $\mathbf{p}^{**}$ is separated from $\mathbf{p}^*$ in the augmented system by an even sex-ratio equilibrium surface.

In the latter event (i.e., $w^* < 1/2 < w^{**}$), any stable symmetric equilibrium $\hat{\mathbf{p}}$ for the sex determination model $\tilde{M}$ must be associated with a sex ratio $\langle \hat{\mathbf{p}}, \tilde{M}\hat{\mathbf{p}} \rangle = \hat{w} > 1/2$ (but $\hat{\mathbf{p}} \neq \mathbf{p}^{**}$ by Corollary 3.1). If such a $\hat{\mathbf{p}}$ exists, it must be separated from $\mathbf{p}^*$ by an even sex-ratio equilibrium surface that cannot intersect the faces of $\mathbf{p}^*$ and $\hat{\mathbf{p}}$ because these equilibria are internally stable (see Theorem 3.4). This is possible only if $\hat{\mathbf{p}}$ corresponds to A_{r+1}-fixation and then

$$m_{i,r+1} > m_{r+1,r+1} > 1/2, i = 1, \ldots, r. \tag{C.7}$$

C.6. PROOF of Theorem 3.8 for the maternal control sex-ratio determination model (2.14). Direct inspection of the recurrence system (2.14) reveals two types of equilibria $\{\tilde{\mathbf{p}}, \tilde{\mathbf{q}}, \tilde{\mathbf{z}}\}$ with corresponding sex ratio $\tilde{w}$: either

$$\tilde{\mathbf{p}} = \tilde{\mathbf{q}} = \tilde{\mathbf{z}}/\tilde{w} = \tilde{\mathbf{p}} \circ M\tilde{\mathbf{p}}/\tilde{w} \tag{C.8}$$

and therefore $\tilde{\mathbf{p}}$ is an equilibrium for the viability model (3.3), or $\tilde{w} = 1/2$, $\tilde{\mathbf{z}} = \tilde{w}\tilde{\mathbf{q}}$ and

$$\tilde{\mathbf{q}} = 2(\tilde{\mathbf{p}} - \tilde{\mathbf{z}}) \circ M\tilde{\mathbf{q}} + 2\tilde{\mathbf{q}} \circ M(\tilde{\mathbf{p}} - \tilde{\mathbf{z}}). \tag{C.9}$$

Taking $\tilde{\mathbf{x}} = 2(\tilde{\mathbf{p}} - \tilde{\mathbf{z}})$, Equation (C.9) means that $\tilde{\mathbf{q}}$ is the leading right eigenvector in Δ of $B(\tilde{\mathbf{x}}) = D_{M\tilde{\mathbf{x}}} + \tilde{\mathbf{x}} \circ M$ for $\tilde{\mathbf{x}}$ in Δ such that the leading eigenvalue is $\rho(B(\tilde{\mathbf{x}})) = 1$. Then $\tilde{\mathbf{p}} = (\tilde{\mathbf{q}} + \tilde{\mathbf{x}})/2$ and $\tilde{\mathbf{z}} = \tilde{\mathbf{q}}/2$.

As for the sex determination model (2.4) with zygotic control the sex ratio determination model (2.14) with maternal control exhibits *symmetric equilibria* satisfying (C.8) and *even sex-ratio equilibrium surfaces* defined by (C.9). Moreover, the equilibrium frequencies in the male population coincide in both models. (Compare with (3.1) and (3.2).) Therefore the stability nature of both types of equilibria will be qualitatively the same in both models (and described by Theorems 3.4–3.7) if we can prove that a symmetric equilibrium $\{\mathbf{p}^*, \mathbf{q}^*, \mathbf{z}^*\}$ of (2.14) with $\mathbf{p}^* = \mathbf{q}^* = \mathbf{z}^*/w^*$ and sex ratio $w^* < 1/2$ ($w^* > 1/2$, respectively) is stable if and only if $\mathbf{p}^*$ is a stable equilibrium for the viability model (3.3) with viability matrix M ($U - M$, respectively) (cf. Theorems 3.3 and 3.3′). (Recall that M is assumed to be symmetrically generic in the sense of Definition 3.2 precluding $w^* = 1/2$.)

Assume first that $\mathbf{p}^*$ is interior to Δ (and therefore satisfies $M\mathbf{p}^* = w^*\mathbf{u}$). Given small perturbations in the form $\mathbf{p}^* + \boldsymbol{\xi}$, $\mathbf{q}^* + \boldsymbol{\eta}$, $\mathbf{z}^* + \boldsymbol{\theta}$, the gradient matrix at hand for the transformation (2.14) applies to the coordinates $\{\boldsymbol{\xi},\boldsymbol{\eta},\boldsymbol{\theta}\}$ satisfying $\langle\boldsymbol{\xi},\mathbf{u}\rangle = \langle\boldsymbol{\eta},\mathbf{u}\rangle = 0$ and takes the block form

$$
G = \begin{bmatrix}
\dfrac{I}{2} & 0 & \dfrac{1}{2w^*}[I - \mathbf{p}^* \circ U] \\[2ex]
\dfrac{I}{2} & \dfrac{I}{2(1-w^*)} & \dfrac{-1}{2(1-w^*)}[I - \mathbf{p}^* \circ U] \\[2ex]
\dfrac{B(\mathbf{p}^*)}{2} & \dfrac{B(\mathbf{p}^*)}{2(1-w^*)} & \dfrac{-1}{2(1-w^*)}[B(\mathbf{p}^*) - 2(\mathbf{p}^* \circ M\mathbf{p}^*) \circ U]
\end{bmatrix}
$$

$$(C.10)$$

(I is the $r \times r$ identity matrix and U is the $r \times r$ matrix with all unit entries.)

There are three eigenvectors of G in the form $\{a\mathbf{p}^*, b\mathbf{p}^*, c\mathbf{p}^*\}$ where a,b,c are coefficients satisfying the third-order matrix equation

$$
\begin{bmatrix}
\dfrac{1}{2} & 0 & 0 \\[2ex]
\dfrac{1}{2} & \dfrac{1}{2(1-w^*)} & 0 \\[2ex]
w^* & \dfrac{w^*}{1-w^*} & 0
\end{bmatrix}
\begin{bmatrix} a \\[1ex] b \\[1ex] c \end{bmatrix}
= \mu \begin{bmatrix} a \\[1ex] b \\[1ex] c \end{bmatrix}
\qquad (C.11)
$$

The corresponding three eigenvalues are $\mu = 0$, $1/2$, $1/2(1 - w^*)$. But these can be ignored since $\mathbf{p}^*$ is not orthogonal to $\mathbf{u}$.

105

Next consider a nontrivial vector $\boldsymbol{\xi}$ satisfying $\mathbf{p}^* \circ M\boldsymbol{\xi} = \gamma w^*\boldsymbol{\xi}$ and $\langle \boldsymbol{\xi}, \mathbf{u} \rangle = 0$ so that $\boldsymbol{\xi}$ is a right eigenvector of $\mathbf{p}^* \circ M$ orthogonal to the principal left eigenvector $\mathbf{u}$ associated with the eigenvalue w^*. There are $r - 1$ linearly independent such $\boldsymbol{\xi}$ and the corresponding γ are real since $\mathbf{p}^* \circ M = D_{\mathbf{p}^*}M$ is a symmetric matrix multiplied by a positive diagonal matrix (see (A.7)). Moreover, $|\gamma| < 1$ since the largest eigenvalue in magnitude of the positive matrix $\mathbf{p}^* \circ M$ *is* w^* with right eigenvector $\mathbf{p}^*$ (Lemma A.1). We try an eigenvector for the matrix G in the form $\{a\boldsymbol{\xi}, b\boldsymbol{\xi}, c\boldsymbol{\xi}\}$. This requires an eigenvalue λ such that

$$
\begin{bmatrix}
\dfrac{1}{2} & 0 & \dfrac{1}{2w^*} \\[2ex]
\dfrac{1}{2} & \dfrac{1}{2(1 - w^*)} & -\dfrac{1}{2(1 - w^*)} \\[2ex]
\dfrac{(1 + \gamma)w^*}{2} & \dfrac{(1 + \gamma)w^*}{2(1 - w^*)} & -\dfrac{(1 + \gamma)w^*}{2(1 - w^*)}
\end{bmatrix}
\begin{bmatrix} a \\[2ex] b \\[2ex] c \end{bmatrix}
= \lambda
\begin{bmatrix} a \\[2ex] b \\[2ex] c \end{bmatrix}
$$

$$(C.12)$$

Again, there exist three admissible values for λ, that is, $\lambda_0 = 0$ and the roots λ_1 and λ_2 of the quadratic polynomial

$$
Q(\lambda) = -4\lambda^2(1 - w^*) + 2\lambda\{2 - 2w^* - w^*\gamma\} + \gamma
$$

Suppose $w^* < 1/2$. Since $Q(1/2) = (1 + \gamma)(1 - w^*)$, while $Q(0) = \gamma$ and $Q(1) = \gamma(1 - 2w^*)$, it follows that $\max\{|\lambda_1|, |\lambda_2|\} < 1$ if and only if $\gamma < 0$. Since there are $r - 1$ linearly independent eigenvectors for $\mathbf{p}^* \circ M$ orthogonal to $\mathbf{u}$, we generate $3r - 3$ relevant eigenvalues from which the stability nature of $\{\mathbf{p}^*, \mathbf{q}^*, \mathbf{z}^*\}$ can be ascertained. Therefore in the case $w^* < 1/2$, we have shown that this equilibrium is stable if and only if the eigenvalues of $\mathbf{p}^* \circ M$ different from w^* are all negative, i.e., $\mathbf{p}^*$ is stable for the viability model M (cf. Lemma A.2). The case $w^* > 1/2$ can be handled similarly relative to $U - M$ instead of M.

It remains to prove the foregoing result when $\mathbf{p}^*$ is on the boundary of Δ. We need only to investigate the external stability conditions. For each index i such that $p_i^* = q_i^* = z_i^* = 0$, a first-order approximation of (2.14) near the symmetric equilibrium $\{\mathbf{p}^*,\mathbf{q}^*,\mathbf{z}^*\}$ yields

$$
\begin{bmatrix} q_i' \\[2mm] p_i' \\[2mm] z_i' \end{bmatrix} = \begin{bmatrix} \dfrac{1}{2} & 0 & \dfrac{1}{2w^*} \\[3mm] \dfrac{1}{2} & \dfrac{1}{2(1-w^*)} & -\dfrac{1}{2(1-w^*)} \\[3mm] \dfrac{(M\mathbf{p}^*)_i}{2} & \dfrac{(M\mathbf{p}^*)_i}{2(1-w^*)} & -\dfrac{(M\mathbf{p}^*)_i}{2(1-w^*)} \end{bmatrix} \begin{bmatrix} q_i \\[2mm] p_i \\[2mm] z_i \end{bmatrix}
$$

$$\text{(C.13)}$$

The matrix in (C.13) has the same structure as the matrix in (C.12) but with $(M\mathbf{p}^*)_i/w^*$ replacing $(1 + \gamma)$. Therefore, in the case $w^* < 1/2$, the eigenvalues are less than 1 in modulus if and only if $(M\mathbf{p}^*)_i < w^*$ (cf. (3.10)). We conclude that $\{\mathbf{p}^*,\mathbf{q}^*,\mathbf{z}^*\}$ with $w^* < 1/2$ is externally stable if and only if $\mathbf{p}^*$ is externally stable for the viability model M. The case $w^* > 1/2$ is analogous with $U - M$ replacing M. This completes the proof of Theorem 3.8.

C.7. Proof of Theorem 3.9 for the one-locus multiallele viability model of maternal inheritance (2.16). An equilibrium $\{\hat{\mathbf{p}},\hat{\mathbf{z}}\}$ of (2.16) satisfies $\hat{\mathbf{z}} = \hat{w}\hat{\mathbf{p}} = \hat{\mathbf{p}} \circ M\hat{\mathbf{p}}$ with $\hat{w} = \langle \hat{\mathbf{p}},M\hat{\mathbf{p}} \rangle$, and therefore $\hat{\mathbf{p}}$ is an equilibrium for the viability model (3.3). The $2r$-th order gradient matrix of the system (2.16) at equilibrium has the block form

$$
H = \begin{bmatrix} \dfrac{1}{2}I & \dfrac{1}{2\hat{w}}[I - \hat{\mathbf{p}} \circ U] \\[4mm] \dfrac{1}{2}B(\hat{\mathbf{p}}) & \dfrac{1}{2\hat{w}}[B(\hat{\mathbf{p}}) - 2(\hat{\mathbf{p}} \circ M\hat{\mathbf{p}}) \circ U] \end{bmatrix} \quad \text{(C.14)}
$$

where U is an $r \times r$ matrix of all unit entries and $B(\hat{\mathbf{p}}) = D_{M\hat{\mathbf{p}}} + \hat{\mathbf{p}} \circ M$. If $\hat{\mathbf{p}}$ is interior to Δ, $M\hat{\mathbf{p}} = \hat{w}\mathbf{u}$ and a complete listing of the eigenvalues and eigenvectors of H are available. These are

eigenvectors	eigenvalue	multiplicity
$\{0, \hat{\mathbf{p}}\}$	0	1
$\left\{\dfrac{\hat{\mathbf{p}}}{2\hat{w}}, \hat{\mathbf{p}}\right\}$	$\dfrac{1}{2}$	1
$\left\{\dfrac{\xi}{\hat{w}}, \xi\right\}$	0	$r-1$
$\left\{\dfrac{\xi}{\hat{w}(1+\gamma)}, \xi\right\}$	$1 + \dfrac{\gamma}{2}$	$r-1$

$$(\text{C.15})$$

where ξ satisfies $(\hat{\mathbf{p}} \circ M)\xi = \gamma\hat{w}\xi$ and $\langle\xi,\mathbf{u}\rangle = 0$. There are $r-1$ independent right eigenvectors ξ of $\hat{\mathbf{p}} \circ M$ orthogonal to $\mathbf{u}$ and these are the relevant ones for discerning the stability nature of the equilibrium at hand. Moreover, from the theory of positive matrices and the fact that $\hat{\mathbf{p}} \circ M\hat{\mathbf{p}} = \hat{w}\hat{\mathbf{p}}$, it follows that $|\gamma| < 1$. Inspection of (C.15) reveals that the equilibrium $\{\hat{\mathbf{p}},\hat{\mathbf{z}}\}$ is stable for (2.16) if and only if $\gamma < 0$ in all possible cases. Comparing with Lemma A.2, this is equivalent to the familiar criterion that $\hat{\mathbf{p}}$ is stable for (3.3) with viability matrix M.

When $\hat{\mathbf{p}}$ is a boundary equilibrium, i.e., $\hat{\mathbf{p}} = (\hat{\mathbf{p}}_1, \ldots, \hat{\mathbf{p}}_k, 0, \ldots, 0)$ for some k (and relabeling of alleles if necessary), the external stability conditions require that the eigenvalues of the system

$$p'_i = \frac{1}{2}p_i + \frac{z_i}{2\hat{w}},$$

$$z'_i = \frac{p_i(M\hat{\mathbf{z}})_i + z_i(M\hat{\mathbf{p}})_i}{2\hat{w}} = \frac{p_i(M\hat{\mathbf{p}})_i}{2} + \frac{z_i(M\hat{\mathbf{p}})_i}{2\hat{w}} \quad (\text{C.16})$$

be smaller than 1 in magnitude for every $i \geq k+1$. The corresponding matrix is

$$\begin{bmatrix} \dfrac{1}{2} & \dfrac{1}{2\hat{w}} \\[2ex] \dfrac{(M\hat{\mathbf{p}})_i}{2} & \dfrac{(M\hat{\mathbf{p}})_i}{2\hat{w}} \end{bmatrix} \quad (\text{C.17})$$

whose eigenvalues are 0 and $(1/2)(1 + (M\hat{\mathbf{p}})_i/\hat{w})$. The latter is smaller than 1 in magnitude if and only if $(M\hat{\mathbf{p}})_i < \hat{w}$. This is equivalent to (3.10) and completes the proof of Theorem 3.9.

C.8. PROOF of Theorem 3.10 for the haplodiploid sex-determination model (2.21). At an equilibrium $\{\mathbf{p}^*,\mathbf{q}^*\}$ of (2.21) with $w^* = \langle\mathbf{p}^*,M\mathbf{q}^*\rangle$ we must have

$$\mathbf{p}^*\{1-\alpha-\alpha s-2(1-\alpha)w^*\}=\mathbf{q}^*\{1-\alpha-\alpha s-2(1-\alpha)w^*\}.$$

(C.18)

Then either $\mathbf{p}^* = \mathbf{q}^* = \mathbf{p}^* \circ M\mathbf{p}^*/w^*$ and $\mathbf{p}^*$ coincides with an equilibrium of (3.3), or $w^* = \tilde{w} = [1 - \alpha - \alpha s]/[2(1 - \alpha)]$ and

$$2(1 - \tilde{w})\mathbf{p}^* = \mathbf{p}^* \circ (U - M)\mathbf{q}^* + \mathbf{q}^* \circ (U - M)\mathbf{p}^*, \quad \text{(C.19)}$$

i.e., $\mathbf{p}^*$ is the right eigenvector in Δ of $B_{U-M}(\mathbf{q}^*) = D_{(U-M)\mathbf{q}^*} + \mathbf{q}^* \circ (U - M)$ for $\mathbf{q}^*$ in Δ, corresponding to the leading eigenvalue $2(1 - \tilde{w})$. Therefore the equilibrium configuration is analogous to that of the basic model (2.4) with $\tilde{w}$ replacing $1/2$. It remains only to show that a symmetric equilibrium $\{\mathbf{p}^*,\mathbf{p}^*\}$ of (2.21) is stable if and only if $\mathbf{p}^*$ is stable for the viability model (3.3) with viability matrix M or $U - M$ accordingly as $w^* < \tilde{w}$ or $w^* > \tilde{w}$.

For $\mathbf{p}^*$ interior to Δ, the relevant gradient matrix of the transformation (2.21) takes the form

$$\begin{bmatrix} \dfrac{(1 - \alpha)B(\mathbf{p}^*)}{2(1 - \alpha)w^* + \alpha s} & \dfrac{(1 - \alpha)B(\mathbf{p}^*) + \alpha sI}{2(1 - \alpha)w^* + \alpha s} \\[3mm] \dfrac{I - B(\mathbf{p}^*)}{2(1 - w^*)} & \dfrac{I - B(\mathbf{p}^*)}{2(1 - w^*)} \end{bmatrix} \quad \text{(C.20)}$$

where $B(\mathbf{p}^*) = D_{M\mathbf{p}^*} + \mathbf{p}^* \circ M$. Looking for eigenvectors in the form $\{a\boldsymbol{\xi},b\boldsymbol{\xi}\}$ with $\mathbf{p}^* \circ M\boldsymbol{\xi} = \gamma w^*\boldsymbol{\xi}$ and $\langle\boldsymbol{\xi},\mathbf{u}\rangle = 0$ leads to the search for the eigenvalues of the second-order matrix

$$\begin{bmatrix} \dfrac{(1 - \alpha)(1 + \gamma)w^*}{2(1 - \alpha)w^* + \alpha s} & \dfrac{(1 - \alpha)(1 + \gamma)w^* + \alpha s}{2(1 - \alpha)w^* + \alpha s} \\[3mm] \dfrac{1 - (1 + \gamma)w^*}{2(1 - w^*)} & \dfrac{1 - (1 + \gamma)w^*}{w^*(1 - w^*)} \end{bmatrix} \quad \text{(C.21)}$$

This is a positive matrix since $|\gamma| < \min\{1,(1 - w^*)/w^*\}$. (Recall that $\mathbf{p}^* \circ M\mathbf{p}^* = w^*\mathbf{p}^*$ and $\mathbf{p}^* \circ (U - M)\mathbf{p}^* = (1 - w^*)\mathbf{p}^*$ and therefore $|\gamma w^*| < w^*$ and $|-\gamma w^*| < (1 - w^*)$ owing to the Perron-Frobenius theory for positive matrices). From the characteristic polynomial of (C.21), the largest eigenvalue in magnitude will be greater than 1 if and only $Q(1) < 0$ where

$$Q(\lambda) = 2(1 - w^*)[2(1 - \alpha)w^* + \alpha s]\lambda^2$$
$$- \{2w^*(1 - w^*)(1 - \alpha)(1 + \gamma)$$
$$+ [1 - (1 + \gamma)w^*][2(1 - \alpha)w^* + \alpha s]\}\lambda$$
$$- \alpha s[1 - (1 + \gamma)w^*]. \tag{C.22}$$

Since

$$Q(1) = 4(1 - \alpha)w^*\gamma\{w^* - \tilde{w}\}, \tag{C.23}$$

we find that $\{\mathbf{p}^*,\mathbf{p}^*\}$ is stable for (2.21) if and only if all possible γ are negative in the case $w^* < \tilde{w}$ but positive in the case $w^* > \tilde{w}$. Comparing with Lemma A.2 completes the proof of Theorem 3.10 for an interior equilibrium. With a boundary equilibrium, it suffices to replace $(1 + \gamma)$ by $(M\mathbf{p}^*)_i/w^*$ when $p_i^* = 0$ in the above treatment. Compare with (3.10).

APPENDIX D. ON THE NUMBER OF STABLE EQUILIBRIA THAT CAN COEXIST IN ONE-LOCUS MULTIALLELE VIABILITY MODELS

THEOREM D.1 (Karlin, 1978). *Consider the one-locus r-allele selection model (3.3) having viability matrix $M = \|m_{ij}\|_{j,j=1}^r$. Assume that M is generic in the sense of Definition 3.1. If there exists an internally stable equilibrium on the boundary of Δ with exactly $r - 1$ positive components (so that exactly one of the alleles is not represented), then there can exist at most one alternative equilibrium that can be stable relative to Δ. (If the boundary equilibrium is externally unstable, then there exists one and only one stable equilibrium in Δ.)*

PROOF of Theorem D.1. Recall that a locally stable equilibrium provides a local maximum in Δ of the fitness (quadratic) function $w(\mathbf{p}) = \sum_{i,j=1}^{r} m_{ij}p_ip_j$. Moreover that local maximum is global if the stable equilibrium is interior to Δ. Now suppose that, in addition to an internally stable boundary equilibrium $\mathbf{p}^*$ involving exactly $r - 1$ alleles, there exist at least two alter-

110

native points, $\mathbf{p}^{**}$ and $\mathbf{p}^{***}$ in Δ, where $w(\mathbf{p})$ displays a strict local maximum with respect to Δ. (We know that $\mathbf{p}^{**}$ and $\mathbf{p}^{***}$ cannot lie on the face containing $\mathbf{p}^*$ since $w(\mathbf{p})$ must achieve a global maximum at $\mathbf{p}^*$ with respect to the face of $\mathbf{p}^*$.) Consider the intersection of the two-dimensional linear variety determined by $\mathbf{p}^*$, $\mathbf{p}^{**}$, and $\mathbf{p}^{***}$ with Δ. Call this section Γ. Clearly, Γ is a convex quadrilateral (or at least a triangle) and $\mathbf{p}^*$ lies on the interior of an edge of Γ. Moreover, $w(\mathbf{p})$ restricted to Γ is a (nondegenerate) quadratic form exhibiting two local maxima at $\mathbf{p}^{**}$ and $\mathbf{p}^{***}$ in addition to a local maximum at $\mathbf{p}^*$ with respect to the edge of $\mathbf{p}^*$. The rest of the proof of Theorem D.1 relies on the following lemma whose nondegeneracy assumptions are clearly verified if M is generic in the sense of Definition 3.1.

LEMMA D.1. *Let $Q(x,y)$ be a quadratic form (definitely not linear) in the two variables x and y (i.e., $Q(x,y) = a_1 x^2 + a_2 xy + a_3 y^2 + a_4 x + a_5 y + a_6$). Consider Q defined over Γ, a convex quadrilateral. We assume that Q does not reduce to a constant on any line intersecting Γ. If Q exhibits an interior local maximum with respect to a boundary edge of Γ, then Q can exhibit at most one local maximum elsewhere with respect to Γ.*

REMARK. The quadrilateral Γ may be degenerate to a triangle.

PROOF of Lemma D.1. Some preliminary statements are useful:

(a) By virtue of the property that linear mappings preserve convex figures and transform a quadratic form into another quadratic, we can assume without loss of generality that Q has one of the following three forms:

$$Q(x,y) = \pm xy, \ Q(x,y) = \pm (x^2 + y^2), \ Q(x,y) = x^2 + L(y)$$

$$(\text{D.1})$$

where L is a linear function. The three types will be examined separately.

(b) The quadratic $Q(x,y)$ on any boundary segment K of Γ can exhibit either a unique interior maximum relative to K or two local maxima located at the vertices of K. This is clear since

Q viewed on K reduces to a quadratic polynomial in one variable.

(c) From any local maximum of Q with respect to Γ achieved on the boundary, the level curves of Q decrease toward the interior of Γ.

We continue with the investigation of the separate cases prescribed in (D.1).

Case (i). $Q(x,y) = x^2 + y^2$. Obviously, any local maximum with respect to a boundary segment of Γ must be achieved at a vertex of Γ. Therefore the hypothesis of Lemma D.1 cannot be fulfilled in the present case. (It is possible to generate up to four local maxima but they all appear at the vertices of Γ; see Figure D.1.)

Case (ii). $Q(x,y) = -(x^2 + y^2)$. There can exist one and only one local maximum with respect to Γ as illustrated in Figure D.2. (That local maximum can be achieved at a vertex or on an edge of Γ. It may also occur in the interior of Γ. In this case, it must be achieved at the origin.)

Case (iii). $Q(x,y) = xy$. The level curves increase away from the origin in the first and third quadrants and toward the origin in the second and fourth quadrants. (See Figure D.3.) There can exist at most two local maxima with respect to Γ. These necessarily include any local maximum with respect to a boundary edge of Γ. Conversely, every local maximum with respect to Γ is located on the boundary.

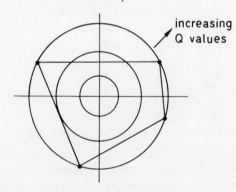

increasing
Q values

Figure D.1

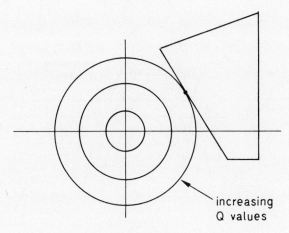

Figure D.2

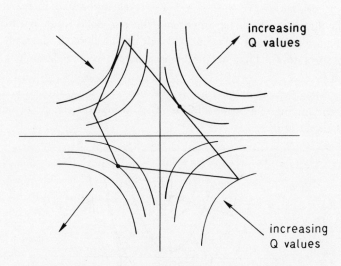

Figure D.3

Case (iv). $Q(x,y) = -xy$. The analysis paraphrases case (iii).

Case (v). $Q(x,y) = x^2 + y$. The values of the level curves increase away from the interior of the parabolas, as drawn in Figure D.4, leading to irrelevant situations for Lemma D.1 since the local maxima on any segment can only occur at the extremity points.

113

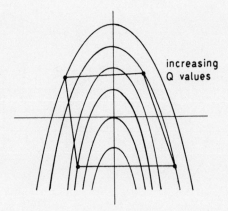

Figure D.4

Case (vi). $Q(x,y) = -x^2 + y$. The level curve values increase in the direction of the interior of the parabolas as shown in Figure D.5. There exists one and only one local maximum with respect to Γ. That maximum is achieved on the boundary of Γ.

The conjunction of cases (i)–(vi) owing to the comment accompanying the generic forms (D.1) completes the proof of Lemma D.1.

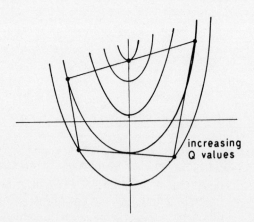

Figure D.5

CHAPTER FOUR

Examples and Special Results for Two- and Three-Allele Sex Determination Models

In this chapter we present more explicit descriptions of the equilibrium configurations for the sex determination model (2.4). Analogous results can be obtained under maternal-control sex determination, haplodiploid genetic systems, and cytoplasmic viability influences.

Our results on models of sex determination mechanisms involving three or more alleles show qualitative features absent in those based on only two alleles. For example, with two alleles a tendency to even sex ratio entails a unique genotypic frequency state whereas with three or more alleles there is a continuum of underlying allelic frequency configurations corresponding to an even sex ratio. At an even sex-ratio equilibrium, the population genotypic frequency composition can change over successive generations, but is constrained, in that the population sex ratio remains at $1:1$. Also, the nature of the evolution of sex ratio in the multiallele model is generally sensitive to initial conditions. A description of all qualitative possibilities in the three-allele case is set forth in Sections 4.2 and 4.3. Documented cases of three-allele sex-determining systems include the platyfish (Kallman, 1973; Orzack et al., 1980), lemmings (Bengtsson, 1977), and other cases.

In Section 4.1, the two-allele case is fully analyzed and the equilibrium configurations are classified with respect to the values of the sex determination coefficients. Results of global convergence are presented in Section 4.5 for the dichotomous models based on three alleles. Sections 4.4 and 4.6 deal with the case where the probability of being male or female depends on being heterozygote or homozygote at a multiallelic locus. In various species with multilocus sex determination, females are homozygous at every relevant locus whereas males are heterozygous at one or more loci; an exemplary case is the wasp *Habrobracon* (see Bull, 1983).

In all models that are fully analyzed, the even sex-ratio equilibrium surfaces are found to be stable when they exist. Situations with stable non-even sex-ratio equilibria are also illustrated.

4.1. A COMPLETE ENUMERATION OF THE STABLE EQUILIBRIUM STATES OF GENOTYPE SEX DETERMINATION INVOLVING TWO ALLELES

The basic model of sex determination of Chapter 2 involving two alleles was studied in Eshel (1975). The parameterization is

$$M = \begin{bmatrix} \alpha & \beta \\ \beta & \gamma \end{bmatrix},$$

α = probability that an A_1A_1 zygote is male,

β = probability that an A_1A_2 zygote is male,

γ = probability that an A_2A_2 zygote is male.

$$(4.1)$$

The transformation equation (2.4) in this case, setting $q_1 = q$ and $p_1 = p$, reduces to

$$q' = \frac{1}{2w} \left[2(\alpha - \beta)pq + \beta(p + q) \right],$$

$$p' = \frac{1}{2(1 - w)} \left[2(\beta - \alpha)pq + (1 - \beta)(p + q) \right]$$

$$(4.2)$$

and $w = (\alpha - 2\beta + \gamma)pq + (\beta - \gamma)(p + q) + \gamma$.

We provide an exhaustive accounting of the equilibrium alternatives for this model in Table 4.1. The validation of global

TABLE 4.1. Stable Equilibrium Alternatives for the Two-Allele Sex Determination Model (4.2) in the case $\beta > 1/2$

Parameter values	Stable equilibria	Domains of attraction
I: $\beta < \gamma < 1, 0 < \alpha < 1/2$	A unique even sex-ratio $\{\tilde{q},\tilde{p}\}$ exists	$\{\tilde{q},\tilde{p}\}$ is globally stable
II: $\beta < \gamma < 1, 1/2 < \alpha < \beta$	Even sex-ratio equilibria do not exist; the pure symmetric state $\{1,1\}$ is uniquely stable	$\{1,1\}$ is globally stable
III: $\beta < \gamma < 1, \beta < \alpha < 1$	$\{p^*,p^*\}$ is uniquely stable	$\{p^*,p^*\}$ is globally stable
IV: $1/2 < \gamma < \beta, 0 < \alpha < 1/2$	$\{0,0\}$ and $\{\tilde{q},\tilde{p}\}$ are stable	$\{p^*,p^*\}$ exists but is unstable; $0 < p^* < \tilde{p}$
V: $1/2 < \alpha < \beta, 1/2 < \gamma < \beta$	Both corners $\{0,0\}$ and $\{1, 1\}$ are stable	$\{p^*,p^*\}$ exists but is unstable
VI: $1/2 < \gamma < \beta, \beta < \alpha < 1$	$\{0, 0\}$ is uniquely stable	$\{0,0\}$ is globally stable
VII: $\gamma < 1/2, \alpha < 1/2, w^* > 1/2$	$\{\tilde{q},\tilde{p}\}$ and $\{\tilde{\tilde{q}},\tilde{\tilde{p}}\}$ are both stable	$\{p^*,p^*\}$ exists and is unstable; $\tilde{p} < p^* < \tilde{\tilde{p}}$
VIII: $0 < \gamma < 1/2, 1/2 < \alpha < \beta$	$\{1,1\}$ and $\{\tilde{q},\tilde{p}\}$ are stable	$\{p^*,p^*\}$ exists and is unstable; $\tilde{p} < p^* < 1$
IX: $0 < \gamma < 1/2, \beta < \alpha < 1$	$\{\tilde{q},\tilde{p}\}$ is uniquely stable	$\{\tilde{q},\tilde{p}\}$ is globally stable
X: $\alpha < 1/2, \gamma < 1/2, w^* < 1/2$	$\{p^*,p^*\}$ is uniquely stable	$\{p^*,p^*\}$ is globally stable

NOTE: The parameter $w^* = (\alpha\gamma - \beta^2)/(\alpha + \gamma - 2\beta)$. A polymorphic symmetric equilibrium is denoted $\{p^*,p^*\}$. An even sex-ratio equilibrium is denoted $\{\tilde{q},\tilde{p}\}$. Two even sex-ratio equilibria $\{\tilde{q},\tilde{p}\}$ and $\{\tilde{\tilde{q}},\tilde{\tilde{p}}\}$ are ordered such that $\tilde{p} < \tilde{\tilde{p}}$.

convergence derives from the monotonicity properties of the transformation equations (4.2) (increasing in each of the variables p and q). (We refer the reader to Chapter 7 for more details). Table 4.1 can be deduced from the graph of the spectral radius functional introduced in (3.5) as we elaborated in Appendix A, Chapter 3. The equilibrium configurations according to the values of the parameters are given explicitly for

117

the case $\beta > 1/2$. (When $\beta < 1/2$ the results are identical with α, β, and γ replaced by $1 - \alpha$, $1 - \beta$, and $1 - \gamma$, respectively.) A polymorphic symmetric equilibrium $\{p^*, q^*\}$ when extant has the formula $q^* = p^* = (\beta - \gamma)/(2\beta - \alpha - \gamma)$. The two possible even sex-ratio equilibria are denoted by $\{\tilde{q}, \tilde{p}\}$ and $\{\tilde{\tilde{q}}, \tilde{\tilde{p}}\}$. Their coordinates are obtained as the solutions $\tilde{p}$ and $\tilde{\tilde{p}}$ of the quadratic equation $x^2[\beta(\alpha + \gamma - 2\beta) - 2(\beta - \gamma)(\alpha - \beta)] + x[(\beta - \gamma) + (\alpha - \beta)(1 - 2\gamma)] + ((1 - \beta)/2)(2\gamma - 1) = 0$ with $\tilde{q} = (\tilde{p}\beta)/(1 - \beta - 2(\alpha - \beta)\tilde{p})$ and similarly for $\tilde{\tilde{q}}$. When two nonsymmetric equilibria exist, we label them by ordering $\tilde{p} < \tilde{\tilde{p}}$. If a single nonsymmetric equilibrium exists, we denote it by $\{\tilde{q}, \tilde{p}\}$.

Case $1/2 < \beta < 1$: The parameter range with respect to α and γ divide into ten regions (I–X) whose characteristics are described in Table 4.1. The curve separating regions X and VII is $w^* = (\alpha\gamma - \beta^2)/(\alpha + \gamma - 2\beta) = 1/2$.

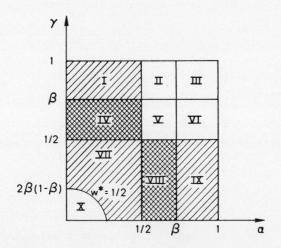

Case $0 < \beta < 1/2$: By symmetry with the previous case, we obtain ten prime regions (I′–X′). The stability characteristics are the same for a prime region as for the corresponding region in Table 4.1.

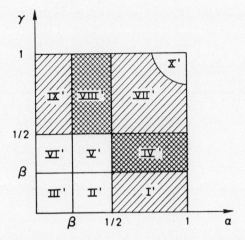

In the accompanying figures, the hatched areas correspond to the regions with stable even sex-ratio only and the cross hatched areas to the regions with stable even sex-ratio equilibria coexisting with stable boundary non-even sex-ratio equilibria. Blank areas are for regions without even sex-ratio equilibria.

It is of interest to assess the parameter domain of (α,β,γ) resulting only in an even sex-ratio equilibrium versus the cases where the stable equilibrium sex ratio is not $1:1$. We evaluate these contingencies under the assumption that α, β, and γ are independently uniformly distributed on $[0,1]$. The probability of only stable even sex-ratio equilibria is

$$\text{Prob}\{\text{I or I}'\} + \text{Prob}\{\text{VII or VII}'\} + \text{Prob}\{\text{IX or IX}'\}$$
$$= \tfrac{1}{8} + \tfrac{5}{36} + \tfrac{1}{8} = \tfrac{14}{36}.$$

The probability of only stable non-even sex-ratio equilibria is

$$\frac{1}{4} + \text{Prob}\{X \text{ or } X'\} = \frac{13}{36} \text{ since Prob}\{X \text{ or } X'\}$$

$$= \frac{1}{4} \int_0^1 [1 - 4\beta(1 - \beta)][\ln\{1 - 4\beta(1 - \beta)\}]\,d\beta$$

$$+ \frac{1}{6} = \frac{1}{9}.$$

119

The probability of coexistence of stable even and non-even sex-ratio equilibria is

$$\text{Prob}\{\text{IV or IV}'\} + \text{Prob}\{\text{VIII or VIII}'\} = \tfrac{1}{8} + \tfrac{1}{8} = \tfrac{1}{4}.$$

Thus, the probability (for the class of models with α, β, and γ uniformly distributed on $[0, 1]$) of an exclusive even sex-ratio equilibrium has probability 14/36 barely in excess to 13/36 of the cases where the only stable sex ratio is different from $1/2$.

4.2. DESCRIPTION OF THREE-ALLELE SEX RATIO OUTCOMES

Let

$$M = \begin{bmatrix} \alpha & \beta & \tilde{\beta} \\ \beta & \gamma & \beta' \\ \tilde{\beta} & \beta' & \alpha' \end{bmatrix} \tag{4.3}$$

be the sex determination coefficient matrix for the model (2.4) with alleles A_1, A_2, and A_3.

For a two-allele sex determination model there can (in generic cases) exist at most two distinct even sex-ratio equilibria (even sex-ratio surfaces involving r alleles are of dimension $r - 2$ according to Theorem 3.4, which reduce to points only in the case $r = 2$).

An exhaustive accounting of the equilibrium outcomes can be carried out in the three-allele model. The determinations for the two-allele case (Table 4.1) are used as a point of departure in conjunction with the results of Chapter 3 to enable us to extrapolate to the three-allele equilibrium configurations.

The diagrams in Figure 4.1 provide a complete schematic representation of all feasible equilibrium configurations involving at least one even sex-ratio equilibrium curve for the three-allele model. An even sex-ratio curve is labeled by S_i. A symmetric equilibrium state is denoted by $\mathbf{p}^*$. Stability is indicated by a domain of attraction with arrows pointing to the relevant equilibrium. Specific parameter ranges for these examples are given in Section 4.3.

The equilibrium configurations described in Figure 4.2 involve only symmetric equilibria. These reflect (by Theorem 3.3)

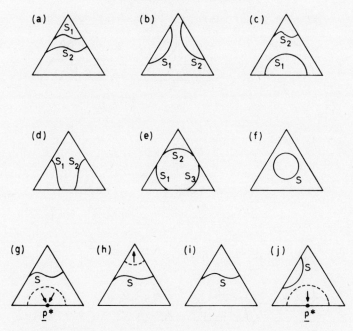

Figure 4.1. Equilibrium configurations involving at least one even sex-ratio equilibrium curve for the three-allele model (4.3)

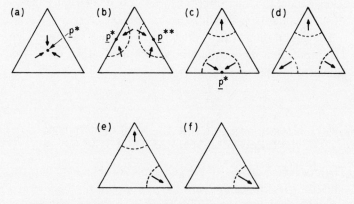

Figure 4.2. Equilibrium configurations of model (4.3) involving no even sex-ratio equilibrium

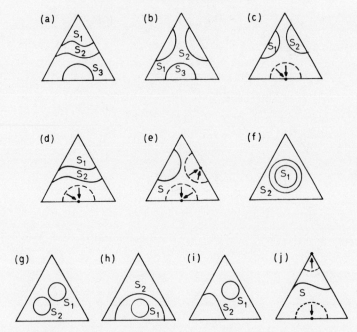

Figure 4.3. Equilibrium configurations precluded for the three-allele model (4.3)

the stable equilibrium alternatives corresponding to a monoecious random mating three-allele viability model. They cover all the qualitative possibilities in this context.

The equilibrium structures of Figure 4.3 are *not* realizable. The proofs that these equilibrium structures cannot occur are presented in Appendix B.

4.3. SOME PARAMETER SPECIFICATIONS FOR THREE-ALLELE SEX RATIO OUTCOMES

The precise conditions on the genotypic sex determination coefficients prescribing the various equilibrium alternatives are accessible (although formidable) but are not too revealing. An

122

exhaustive listing parallel to Table 4.1 of the attainable sex ratio equilibria in terms of the parameters for the three-allele sex determination model will not be given here.

We can amply illustrate the scope of the equilibrium configurations of Figure 4.1 (Figure 4.1j excepted) in terms of the four-parameter sex determination coefficient matrix (symmetric relative to the alleles A_1 and A_3)

$$M = \begin{array}{c} \begin{array}{ccc} A_1 & A_2 & A_3 \end{array} \\ \begin{bmatrix} \alpha & \beta & \tilde{\beta} \\ \beta & \gamma & \beta \\ \tilde{\beta} & \beta & \alpha \end{bmatrix} \end{array}$$

All proofs of the sufficient conditions presented below are detailed in Appendix C.

Conditions for Figure 4.1a

$\alpha < \dfrac{1}{2} < \beta$, $\gamma = 0$ (or γ small), $2\beta(1 - \beta) < \alpha + \gamma - 2\gamma\alpha$,

$\alpha + \tilde{\beta} < 1$, $1 < \beta + \dfrac{\alpha + \tilde{\beta}}{4\beta}$.

Conditions for Figure 4.1b

$\alpha < \dfrac{1}{2} < \beta$, $\gamma = 0$ (or γ small), $2\beta(1 - \beta) < \alpha + \gamma - 2\gamma\alpha$,

$\alpha + \tilde{\beta} < 1$, $1 > \beta + \dfrac{\alpha + \tilde{\beta}}{4\beta}$.

Conditions for Figure 4.1c

$\alpha < \beta < \dfrac{1}{2} < \gamma$, γ close to $\dfrac{1}{2}$, $\tilde{\beta} + \alpha > 1$.

Conditions for Figure 4.1d

$\alpha < \dfrac{1}{2}$, $\alpha + \tilde{\beta} > 1$, β near 1, $\beta < \gamma$.

123

Conditions for Figure 4.1e

$$\alpha, \gamma < \frac{1}{2} < \beta, \ 2\beta(1 - \beta) < \alpha + \gamma - 2\alpha\gamma, \ \alpha + \tilde{\beta} > 1.$$

Conditions for Figure 4.1f are subsumed by the general model of Section 4.4.

Conditions for Figure 4.1g

$$0 < \gamma < \frac{1}{2} < \tilde{\beta} < \alpha < \beta.$$

Conditions for Figure 4.1h

$$0 < \alpha < \frac{1}{2} < \gamma < \beta, \ \tilde{\beta} > \frac{1}{2}, \ \alpha + \tilde{\beta} < 1.$$

Conditions for Figure 4.1i

$$0 < \alpha < \frac{1}{2} < \beta < \gamma, \frac{1}{2} < \tilde{\beta}, \ \alpha + \tilde{\beta} < 1.$$

Conditions for Figure 4.1j.
Given at the end of Appendix C.

Comparisons of Sex Ratio Equilibrium Alternatives

(i) It is noteworthy in both the two- and three-allele sex determination models that it is not possible to have an even sex-ratio equilibrium curve with two or more stable symmetric equilibrium points. Furthermore, the existence of two or more even sex-ratio equilibrium curves excludes the possibility of any stable symmetric equilibrium.

(ii) The coexistence of a single even sex-ratio equilibrium curve with one symmetric stable equilibrium is feasible (cases (g) and (h) of Figure 4.1).

(iii) There is only one feasible configuration with three even sex-ratio equilibrium curves (Figure 4.1e [cf. Figure 4.3]).

(iv) There are no equilibrium arrays allowing four or more even sex-ratio equilibrium curves.

4.4. A SEX DETERMINATION FORM DETERMINED BY THE STATE OF HETEROZYGOSITY VERSUS HOMOZYGOSITY

A sex determination matrix of some biological interest has the form

$$M = (\alpha - \beta)I + \beta U \tag{4.4}$$

where the probability of being a male is either α or β according to whether the genotype is homozygous or heterozygous independently of its allelic composition.

We denote the matrix of type (4.4) with r alleles by $M(r)$. Suppose first that $\alpha < 1/2 < \beta$. It is easily shown that the central frequency vector $\mathbf{p}^*(r) = (1/r, 1/r, \ldots, 1/r)$ is a symmetric equilibrium for $M(r)$ with associated sex-ratio $w^*(r) = \beta + (\alpha - \beta)/r$. Note that $w^*(r)$ increases with r. It is simple to ascertain that $\lambda = (\alpha - \beta) < 0$ is the only non-zero eigenvalue (of multiplicity $r - 1$) for $M(r)$ whose corresponding right eigenvectors $\boldsymbol{\xi}$ satisfy $\langle \boldsymbol{\xi}, \mathbf{u} \rangle = 0$. Therefore $\mathbf{p}^*(r)$ is globally stable (respectively unstable, actually repelling) for the viability model with viability matrix $M(r)$ (respectively $U - M(r)$). It is easy to check that $w^*(r) > 1/2$ if and only if

$$r \geq \hat{r} = \text{int}\left[\frac{\beta - \alpha}{\beta - 1/2} + 1\right].$$

(int$[x]$ stands for the largest integer smaller than or equal to x.) In view of Theorem 3.3, $\mathbf{p}^*(r)$ is uniquely stable for the sex determination model corresponding to $M(r)$ provided $r < \hat{r}$, and unstable when $r \geq \hat{r}$.

In the case at hand

$$B(\mathbf{p}) = D_{M\mathbf{p}} + \mathbf{p} \circ M = \beta I + 2(\alpha - \beta)D_{\mathbf{p}} + \beta D_{\mathbf{p}}U.$$

Therefore, at an even sex-ratio equilibrium of (2.4) with the sex determination matrix (4.4), the relation

$$\tilde{\mathbf{q}} = B(\tilde{\mathbf{p}})\tilde{\mathbf{q}} = \beta\tilde{\mathbf{q}} + 2(\alpha - \beta)\tilde{\mathbf{p}} \circ \tilde{\mathbf{q}} + \beta\tilde{\mathbf{p}}$$

125

holds, or equivalently, the following identity applies to each component:

$$\tilde{q}_i = \frac{\beta \tilde{p}_i}{(1 - \beta) + 2(\beta - \alpha)\tilde{p}_i}$$

This is possible with r alleles if and only if the surface

$$h(\mathbf{p}) = \sum_{i=1}^{r} \frac{\beta p_i}{(1 - \beta) + 2(\beta - \alpha)p_i} = 1$$

with $\mathbf{p} = (p_1, \ldots, p_r)$ intersects the frequency simplex Δ. Note that $h(\mathbf{p})$ spans a concave surface on Δ symmetric in the components of $\mathbf{p}$ whose maximum is

$$h(\mathbf{p}^*(r)) = \frac{\beta}{1 + \beta - 2w^*(r)}.$$

Consistent with Theorem 3.4, $h(\mathbf{p}^*(r)) > 1$ if and only if $w^*(r) > 1/2$. Then even sex-ratio equilibrium surfaces exist, since the value of h at any corner is $\beta/(1 + \beta - 2\alpha) < 1$. A schematic representation of the equilibrium system versus the critical dimension $\hat{r}$ is given in Figure 4.4. This figure depicts the various equilibrium possibilities for the homozygote-heterozygote sex determination model (4.4) with $\alpha < 1/2 < \beta$. The arrows suggest the directions of convergence.

 for $r \le \hat{r} - 1 = \left[\dfrac{\beta - \alpha}{\beta - 1/2}\right]$

 for $r = \hat{r}$

 for $r \ge \hat{r} + 1$

Figure 4.4

126

The case $\alpha > 1/2 > \beta$ is analogous. When $1/2 < \alpha < \beta$ or $\beta < \alpha < 1/2$, only fixation states can be stable. When $1/2 < \beta < \alpha$ or $\alpha < \beta < 1/2$, the polymorphic equilibrium $\mathbf{p}^*(r)$ is globally stable.

4.5. THREE-ALLELE DICHOTOMOUS GENOTYPIC SEX DETERMINATION

There are nine (modulo relabeling of alleles) inequivalent three-allele models of dichotomous sex determination as listed in Table 2.2. The genotypes divide into two phenotypic classes, $\mathscr{G}_M$ and $\mathscr{G}_F$, identified as males and females, and therefore (random) mating types exclusively involve crosses between individuals of $\mathscr{G}_M$ and $\mathscr{G}_F$. We expect, in view of Theorem 3.5, that only even sex-ratio equilibrium curves are stable.

Table 4.2 relists all three-allele sex determination models, and Table 4.3 describes the dynamic properties of these models. Some of these systems were analyzed in Scudo (1964, 1967a), Karlin (1968a), and Bull and Charnov (1977). They bear mainly on the consequences of the introduction of a third sex-determiner into a two-allele sex-determining system.

TABLE 4.2. Dichotomous Sex Partitions with Three Alleles at One Locus

	$\mathscr{G}_M$	$\mathscr{G}_F$
Model I	AA	AB AC BB BC CC
Model II	AB	AA BB BC AC CC
Model III	AA BB	AB AC BC CC
Model IV	AA BC	BB AB AC CC
Model V	AB AC	AA BB CC BC
Model VI	AA AB	AC BC BB CC
Model VII	AA BB AB	AC BC CC
Model VIII	AA BB AC	CC AB BC
Model IX	AA BB CC	AB BC AC

TABLE 4.3. Equilibrium Characteristics for the Dichotomous Sex Determination Models Involving Three Alleles

Model	Even sex-ratio equilibrium curves	Distinct equilibria	Domains of attraction	Rate of convergence
I	S: freq(AA) = 1/2 freq(AC) + freq(AB) = 1/2	none	an equilibrium on S is always attained	one generation
II	S: freq(AA) + freq(BB) = 1/2 freq(AB) = 1/2	central even sex-ratio equilibrium freq(AB) = 1/2 freq(AA) = freq(BB) = 1/4	convergence occurs to the central even sex-ratio equilibrium	algebraic (slow) rate of convergence; allele C is ultimately lost but very slowly
III	S_1: freq(BB) = 1/2 freq(AB) + freq(BC) = 1/2 S_2: freq(AA) = 1/2 freq(AB) + freq(AC) = 1/2 S_3: freq(AA) + freq(BB) = 1/2 freq(AB) = 1/2	central even sex-ratio equilibrium freq(AA) = freq(BB) = 1/4 freq(AB) = 1/2	(i) convergence to S_1 when freq(AA) < freq(BB) and freq(AC) < freq(BC) initially (ii) convergence to S_2 when freq(BB) < freq(AA) and freq(BC) < freq(AC) initially (iii) convergence to the central even sex-ratio equilibrium if freq(AA) compared to freq(BB) alternate in direction over successive generations	geometric rate of convergence for cases (i) and (ii). Case (iii) is attained only for lower-dimensional set of initial frequencies and convergence occurs at an algebraic rate

IV	S_1: freq$(AA) = 1/2$ freq$(AB) +$ freq$(AC) = 1/2$ S_2: freq$(BC) = 1/2$ freq$(BB) +$ freq$(CC) = 1/2$	central non-even sex-ratio equilibrium freq$(AA) =$ freq$(BB) =$ freq$(CC) = 1/9$, freq$(BC) =$ freq$(AB) =$ freq$(AC) = 2/9$. This equilibrium is unstable.	(i) convergence to S_1 if for some generation freq$(BC) < 2$ freq(AA) and freq$(AB) +$ freq$(AC) > 2[$freq$(BB) +$ freq$(CC)]$ (ii) convergence to S_2 occurs if for some generation, freq$(BC) > 2$ freq(AA) and $2[$freq$(BB) +$ freq$(CC)] >$ freq$(AB) +$ freq(AC),	rate of approach is algebraic
V		a one-parameter curve S in terms of freq$(AA) = x^* \leq 1/4$: $v^* =$ freq$(AC) = 1/2 - u^*$ $y^* =$ freq$(BB) = 2(u^*)^2(1 - 2x^*)$ $z^* =$ freq$(CC) = (1/2)(1 - 2u^*)^2(1 - 2x^*)$ $w^* =$ freq$(BC) = 2u^*(1 - 2u^*)(1 - 2x^*)$ $u^* = \dfrac{1}{1 - 4x^*}(u^{(1)} + 2y^{(1)} + w^{(1)}) =$ freq(AB) where $u^{(1)}, y^{(1)}, w^{(1)}$ are the corresponding frequencies at the first generation.	convergence occurs to a point on S, the limiting state depending on the initial conditions	
VI		a unique equilibrium curve S parameterized by say $v^* =$ freq(BC) has freq$(AA) = \dfrac{v^*}{2(1 - v^*)}$, freq$(AB) = \dfrac{1 - 2v^*}{2(1 - v^*)}$ freq$(BB) = \dfrac{(1 - 2v^*)^2}{2}$, freq$(BC) = v^*$ freq$(AC) = v^*(1 - 2v^*)$, freq$(CC) = 0$	convergence appears to occur to a point of S, the limiting equilibrium depending on initial conditions	rate of convergence is unclear

TABLE 4.3 (Continued)

Model	Even sex-ratio equilibrium curves	Distinct equilibria	Domains of attraction	Rate of convergence
VII	a unique equilibrium curve S with parameter a, $0 < a < 1/2$, with freq$(AA) = 2a^2$, freq$(BB) = 2(1/2 - a)^2$, freq$(AB) = 4a(1/2 - a)$, and freq$(AC) = a$, freq$(BC) = (1/2 - a)$, freq$(CC) = 0$	an equilibrium on S is always attained	two generations	
VIII	two classes of even sex-ratio equilibrium curves: (i) freq(AA) + freq(BB) = 1/2, freq(AB) = 1/2 (ii) freq(BB) = 0, the equilibrium curve is the same as in Model VI with the roles of allele C and allele B interchanged.	the domains of attractions are unknown		

IX	S_1: freq(AA) = 1/2
	  freq(AB) + freq(AC) = 1/2
	
	S_2: freq(BB) = 1/2
	  freq(AB) + freq(BC) = 1/2
	
	S_3: freq(CC) = 1/2
	  freq(AC) + freq(BC) = 1/2
	
	S_4: freq(AB) = 1/2
	  freq(AA) + freq(BB) = 1/2
	
	S_5: freq(AC) = 1/2
	  freq(AA) + freq(CC) = 1/2
	
	S_6: freq(BC) = 1/2
	  freq(BB) + freq(CC) = 1/2
		an unstable central non-even sex-ratio equilibrium as in Model IV	see Section 4.6 for the domains of attraction	geometric rate of convergence to S_1, S_2, S_3; algebraic rate for S_4, S_5, S_6

Complete proofs for all models are elaborated in Appendix D. Global convergence to an even sex ratio is the rule.

4.6. CONVERGENCE PROPERTIES FOR THE SEX DETERMINATION MODEL WHERE ALL HOMOZYGOTES DETERMINE FEMALES AND HETEROZYGOTES DETERMINE MALES WITH MULTIPLE ALLELES

(Compare to Section 4.4.) Consider the following genotype dichotomization for sex expression

$$\mathscr{G}_F = \{A_i A_i, \ i = 1,2, \ldots, r\}, \ \mathscr{G}_M = \{A_i A_j, \ i \neq j\}.$$

Assuming random mating, two classes of even sex-ratio equilibrium configurations emerge:

(a) $\text{freq}(A_k A_k) = \dfrac{1}{2}$

while $\displaystyle\sum_{\substack{l=1 \\ l \neq k}}^{r} \text{freq}(A_k A_l) = \dfrac{1}{2}$ for some k,

(b) $\text{freq}(A_i A_i) + \text{freq}(A_j A_j) = \dfrac{1}{2}$, $\text{freq}(A_i A_j) = \dfrac{1}{2}$ for some $i \neq j$.

There are unstable (totally repelling) symmetric equilibria having allele frequencies $\mathbf{p}^*(r) = (1/r, 1/r, \ldots, 1/r)$.

There is global convergence from any initial state. For example, if the allelic frequencies in the initial generation of males and females, respectively, verify the orderings

$$q_1^{(0)} > q_2^{(0)} > \cdots > q_r^{(0)} \text{ and } p_1^{(0)} > p_2^{(0)} > \cdots > p_r^{(0)}$$

then either

$$\lim_{n \to \infty} q_1^{(n)} = \lim_{n \to \infty} q_2^{(n)} = \dfrac{1}{2} \text{ and } \lim_{n \to \infty} (p_1^{(n)} + p_2^{(n)}) = 1$$

or

$$\lim_{n \to \infty} p_1^{(n)} = 1 \text{ while } \lim_{n \to \infty} q_1^{(n)} = \dfrac{1}{2} \text{ and } \lim_{n \to \infty} \sum_{k=2}^{r} q_k^{(n)} = \dfrac{1}{2}$$

(the rate of convergence being algebraic in the second case).

131

Generally, if in the initial generation

$$q_1^{(0)} > q_2^{(0)} > \cdots > q_r^{(0)} \text{ and } p_{k_1}^{(0)} > p_{k_2}^{(0)} > \cdots > p_{k_r}^{(0)},$$

where $\{k_i\}$ is a permutation of $1, 2, \ldots, r$, it can again be proved that convergence occurs to one of the curves exhibiting an even sex ratio. See Appendix E.

APPENDIX A. ANALYSIS OF THE GENERAL TWO-ALLELE SEX-DETERMINATION MODEL

Consider the sex-determination model (2.4) in the case of two alleles, A_1 and A_2, with sex determination matrix

$$M = \begin{bmatrix} \alpha & \beta \\ \beta & \gamma \end{bmatrix} \tag{A.1}$$

An interior symmetric equilibrium exists if and only if $\alpha, \gamma < \beta$ or $\alpha, \gamma > \beta$, and then the equilibrium frequency of A_1 is

$$p^* = \frac{\gamma - \beta}{\alpha + \gamma - 2\beta} \tag{A.2}$$

and the equilibrium sex ratio is

$$w^* = \frac{\alpha\gamma - \beta^2}{\alpha + \gamma - 2\beta}. \tag{A.3}$$

To ascertain the existence of even sex-ratio equilibria and completely describe the equilibrium structure, it suffices to study the spectral radius functional $\rho(p) = \rho(B(\mathbf{p}))$ where $B(\mathbf{p}) = D_{M\mathbf{p}} + \mathbf{p} \circ M$ with $\mathbf{p} = (p, 1 - p)$ owing to Theorems 3.1–3.4. In the case at hand, the values of $\rho(p)$ at both fixation states and at the polymorphic symmetric equilibrium p^* (the only possible critical point) when extant are decisive. With reference to Appendix A of Chapter 3, these values are $\rho(0) = \max\{2\gamma, \beta\}$, $\rho(1) = \max\{2\alpha, \beta\}$, and $\rho(p^*) = 2w^*$.

Without loss of generality, we may assume $\beta > 1/2$ (otherwise, take $U - M$ instead of M) and $\alpha < \gamma$ (relabeling the alleles if necessary). Figure A.1 illustrates all equilibrium possibilities. These are also summarized in Table 4.1. Global convergence follows from monotonicity properties (see Appendix A of Chapter 7).

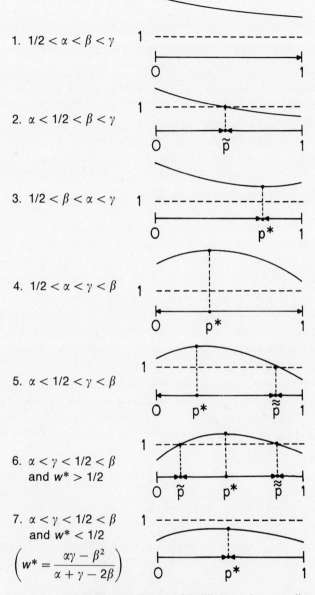

1. $1/2 < \alpha < \beta < \gamma$

2. $\alpha < 1/2 < \beta < \gamma$

3. $1/2 < \beta < \alpha < \gamma$

4. $1/2 < \alpha < \gamma < \beta$

5. $\alpha < 1/2 < \gamma < \beta$

6. $\alpha < \gamma < 1/2 < \beta$
   and $w^* > 1/2$

7. $\alpha < \gamma < 1/2 < \beta$
   and $w^* < 1/2$

$$\left(w^* = \frac{\alpha\gamma - \beta^2}{\alpha + \gamma - 2\beta}\right)$$

Figure A.1. All possible arrangements of equilibrium points according to the graph of $\rho(p)$ for the two-allele sex determination model (4.2) in the case $\beta > 1/2$ and $\alpha < \gamma$

NOTE: $\mathbf{p}^*$ designates an interior symmetric equilibrium while $\tilde{\mathbf{p}}$ and $\tilde{\tilde{\mathbf{p}}}$ denote even sex-ratio equilibria. The arrows indicate the directions of convergence.

APPENDIX B. PROOF THAT THE EQUILIBRIUM
STRUCTURES IN FIGURE 4.3 CANNOT OCCUR

Consider the equilibrium configuration drawn in Figure B.1. for three alleles A_1, A_2, A_3 at one locus controlling sex determination. The curves S_i defined on the frequency simplex Δ are determined by the level where the spectral radius function $\rho(\mathbf{p}) = \rho(B(\mathbf{p}))$ (see equation (3.5)) has the value of 1. Without loss of generality we can take $\rho(\mathbf{p}) > 1$ in regions $\mathcal{R}_1$ and $\mathcal{R}_2$; otherwise take the matrix $U - M$ rather than M. The maximum of $\rho(\mathbf{p})$ in $\mathcal{R}_2$ is achieved at a stable equilibrium $\mathbf{p}^{***}$ (interior or on the boundary of Δ) for the selection model of viability matrix M (Theorem 3.1). Similarly, for the same viability selection model there exists a stable equilibrium $\hat{\mathbf{p}}$ in $\mathcal{R}_1$ characterized as the point in $\mathcal{R}_1$ where $\rho(\mathbf{p})$ is maximized. The point $\hat{\mathbf{p}}$ cannot be a polymorphism interior to Δ for then $\mathbf{p}^{***}$ would not be stable. So $\hat{\mathbf{p}}$ must be on the boundary and then necessarily coincident with either $\mathbf{p}^*$ or $\mathbf{p}^{**}$, the two possible boundary equilibria in $\mathcal{R}_1$. Suppose $\hat{\mathbf{p}}$ equal to $\mathbf{p}^*$ on the edge $A_1 A_2$.

It is clear that $\rho(\mathbf{p})$ along the edge $\overline{A_2 A_3}$ achieves a maximum at $\mathbf{p}^{**}$ because $\rho(\mathbf{p}) = 1$ at the points where the curves S_1 and S_2 cut $\overline{A_2 A_3}$ and $\rho(\mathbf{p}^{**}) > 1$. This means that $\mathbf{p}^{**}$ is a stable equilibrium for the viability model M with respect to the edge

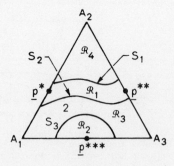

Figure B.1

$\overline{A_2 A_3}$. By virtue of Theorem D.1 (Appendix D of Chapter 3) there can exist at most one alternative stable equilibrium for the viability matrix M with respect to the frequency simplex Δ. But $\mathbf{p}^*$ and $\mathbf{p}^{***}$ are stable, yielding a contradiction. To avert this contradiction we must preclude the equilibrium structure of Figure 4.3a.

The arguments for exclusion of the putative equilibrium configurations described in Figures 4.3b–e paraphrase the foregoing case. For the contingencies of Figure 4.3f–i we can deduce the coexistence of a stable polymorphism for the viability model M(or $U - M$) with a second stable equilibrium. This is impossible (see Section 3.2).

We finally prove that the equilibrium configuration of Figure 4.3j is not possible. Consider a general three allele sex determination coefficient matrix

$$
M = \begin{array}{c} \\ \end{array} \begin{array}{ccc} A_1 & A_2 & A_3 \end{array} \\ \begin{bmatrix} \alpha_1 & \beta_1 & \tilde{\beta} \\ \beta_1 & \gamma & \beta_2 \\ \tilde{\beta} & \beta_2 & \alpha_2 \end{bmatrix}
$$

Without loss of generality we take $\alpha_1 < 1/2$. In order to have an even sex-ratio equilibrium and the A_2 allele fixation stable with respect to the edge $\overline{A_1 A_2}$ only case IV of Table 4.1 is tenable. Then

$$
\tfrac{1}{2} < \gamma < \beta_1, 0 < \alpha_1 < \tfrac{1}{2}.
$$

The corresponding requirements with respect to the edge $\overline{A_1 A_3}$ since $\gamma > 1/2$ requires $1/2 < \gamma < \beta_2$ and $0 < \alpha_2 < 1/2$. A unique stable symmetric equilibrium $\mathbf{p}^*$ segregating alleles A_1 and A_3 requires

$$
\tilde{\beta} > \tfrac{1}{2} \text{ and } 2\tilde{\beta}(1 - \tilde{\beta}) > \alpha_1 + \alpha_2 - 2\alpha_1\alpha_2. \tag{B.1}
$$

At $\mathbf{p}^+ = (0,1,0)$ we have $\rho(\mathbf{p}^+) = 2\gamma > 1$. Therefore, stipulating the configuration of Figure 4.3j implies $\rho(\mathbf{p}^*) < 1/2$. The condition of external stability of $\mathbf{p}^*$ for the viability model M requires

$$
0 > (\beta_1 - \alpha_1)(\tilde{\beta} - \alpha_2) + (\beta_2 - \tilde{\beta})(\tilde{\beta} - \alpha_1).
$$

135

This inequality implies since both β_1, β_2 exceed $1/2$ that

$$0 > (1/2 - \alpha_1)(\tilde{\beta} - \alpha_2) + (1/2 - \tilde{\beta})(\tilde{\beta} - \alpha_1)$$

$$= \tilde{\beta} - \tilde{\beta}^2 - \frac{\alpha_1}{2} - \frac{\alpha_2}{2} + \alpha_1\alpha_2.$$

But the last expression is positive on account of (B.1). This contradiction can only be averted if $\mathbf{p}^*$ is unstable. This analysis establishes that the equilibrium configuration Figure 4.3j is impossible.

APPENDIX C. CONDITIONS VALIDATING THE EQUILIBRIUM CONFIGURATIONS DESCRIBED IN FIGURE 4.1

Consider the three-allele sex determination coefficient matrix

$$M = \begin{array}{c} \begin{array}{ccc} A_1 & A_2 & A_3 \end{array} \\ \begin{bmatrix} \alpha & \beta & \tilde{\beta} \\ \beta & \gamma & \beta \\ \tilde{\beta} & \beta & \alpha \end{bmatrix} \end{array}$$

with $\alpha,\gamma < 1/2 < \beta$, $2\beta(1 - \beta) < \alpha + \gamma - 2\alpha\gamma$. Examining the possibilities with two alleles (Table 4.1) there exist two even sex-ratio equilibria $\tilde{\mathbf{p}}$ and $\tilde{\tilde{\mathbf{p}}}$ segregating alleles A_1 and A_2 and two even sex-ratio equilibria $\mathbf{p}^+$ and $\mathbf{p}^{++}$ involving only alleles A_2 and A_3. (See Figure C.1.) Observe that for $\tilde{\beta} > 1/2$ a sym-

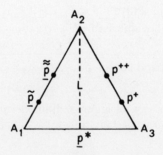

Figure C.1

metric equilibrium $\mathbf{p}^*$ exists on the edge $\overline{A_1 A_3}$. We have two even sex-ratio equilibria $\mathbf{p}'$ and $\mathbf{p}''$ segregating A_1 and A_3 if and only if $\alpha + \tilde{\beta} > 1$.

We calculate now $\rho(\mathbf{p}) = \rho(B(\mathbf{p}))$ for $\mathbf{p} = (u, 1 - 2u, u)$, $0 < u < 1/2$, traversing the line L. Note that

$$M\mathbf{p} = \begin{bmatrix} (\alpha + \tilde{\beta})u + \beta(1 - 2u) \\ 2\beta u + \gamma(1 - 2u) \\ (\alpha + \tilde{\beta})u + \beta(1 - 2u) \end{bmatrix}$$

so that

$$B(\mathbf{p}) = D_{M\mathbf{p}} + \mathbf{p} \circ M =$$

$$\begin{bmatrix} (2\alpha + \tilde{\beta})u + \beta(1 - 2u) & \beta u & \tilde{\beta} u \\ \beta(1 - 2u) & 2\beta u + 2\gamma(1 - 2u) & \beta(1 - 2u) \\ \tilde{\beta} u & \beta u & (2\alpha + \tilde{\beta})u + \beta(1 - 2u) \end{bmatrix}$$

On account of the symmetry in the first and the third variables, the principal eigenvector of the principal eigenvalue (the spectral radius) of the positive matrix $B(\mathbf{p})$ necessarily has the form (ξ, η, ξ). Thus the relevant matrix by which to ascertain $\rho(\mathbf{p})$ is that of the 2×2 matrix

$$L(u) = \begin{bmatrix} (2\alpha + 2\tilde{\beta})u + \beta(1 - 2u) & \beta u \\ 2\beta(1 - 2u) & 2\beta u + 2\gamma(1 - 2u) \end{bmatrix}. \quad \text{(C.1)}$$

Since the above matrix is positive the spectral radius is less than 1 if and only if its characteristic polynomial $Q(\lambda)$ satisfies $Q(1) > 0$ with $Q'(1) > 0$ where

$$Q(\lambda) = \lambda^2 - \lambda[2(\alpha + \tilde{\beta})u + \beta + 2\gamma(1 - 2u)] + 4(\alpha + \tilde{\beta})\beta u^2 + 4\gamma(\alpha + \tilde{\beta})u(1 - 2u) + 2\beta\gamma(1 - 2u)^2.$$

For $\alpha + \tilde{\beta} < 1$ the fact that $Q'(1) = 2 - 2(\alpha + \tilde{\beta})u - \beta - 2\gamma(1 - 2u) > 0$ is confirmed by direct verification for $u = 0$ and $u = 1/2$ and then also for $0 < u < 1/2$ since $Q'(1)$ is linear in u. Set

$$R(u) = Q(1) = 1 - \beta - 2(\alpha + \tilde{\beta})u - 2\gamma(1 - 2u) \quad \text{(C.2)}$$
$$+ 4(\alpha + \tilde{\beta})\beta u^2 + 4\gamma(\alpha + \tilde{\beta})u(1 - 2u)$$
$$+ 2\beta\gamma(1 - 2u)^2.$$

For $u = 0$,

$$Q(1) = (1 - 2\gamma)(1 - \beta) > 0 \quad \text{since} \quad \gamma < 1/2, \beta < 1.$$

For $u = 1/2$,

$$Q(1) = (1 - \alpha - \tilde{\beta})(1 - \beta)$$

which is negative for $\alpha + \tilde{\beta} > 1$ and positive for $\alpha + \tilde{\beta} < 1$.

For $\alpha + \tilde{\beta} < 1$ then $\rho(\mathbf{p}) < 1$ at the boundary points $u = 0$ and $u = 1/2$. To ease the calculations we take $\gamma = 0$ (the results apply for small γ), leading to

$$R(u) = 1 - \beta - 2(\alpha + \tilde{\beta})u + 4(\alpha + \tilde{\beta})\beta u^2.$$

The minimum of $R(u)$ is achieved at $u_0 = 1/4\beta < 1/2$ because $\beta > 1/2$ yielding

$$\min_{0 \leq u \leq 1/2} R(u) = 1 - \beta - \frac{\alpha + \tilde{\beta}}{4\beta}.$$

If $\min_{0 \leq u \leq 1/2} R(u) < 0$, then the equilibrium configuration necessarily takes the form of Figure C.2, where $\rho(\mathbf{p}) > 1$ in $\mathscr{R}^+$ and $\rho(\mathbf{p}) < 1$ in $\mathscr{R}_0$ and $\mathscr{R}_{00}$.

For $\alpha + \tilde{\beta} < 1$, $\beta > 1/2$, and $1 - \beta - (\alpha + \tilde{\beta})/2\beta > 0$, we necessarily have $\rho(\mathbf{p}) < 1$, along all of L and the equilibrium configuration is necessarily that of Figure C.3, where $\rho > 1$ in $\mathscr{R}_1 \cup \mathscr{R}_2$ and $\rho < 1$ in the complementary open set.

Thus the proof of the conditions of Section 4.3 guaranteeing Figures 4.1a and 4.1b is complete.

Retaining the conditions $\alpha, \gamma < 1/2 < \beta$, $2\beta(1 - \beta) < \alpha + \gamma - 2\alpha\gamma$, but now with $\alpha + \tilde{\beta} > 1$, we have a symmetric equi-

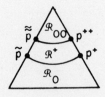

Figure C.2

138

Figure C.3

librium with two unsymmetric (even sex-ratio) equilibria on each edge. Proceeding as previously, we infer (along L) that $\rho < 1$ near the edge $\overline{A_1 A_3}$ while $\gamma < 1$ near A_2-fixation because $\gamma < 1/2$. Since the even sex-ratio equilibria traverse analytic curves, the boundary equilibria can be connected (cf. Theorem 3.4), admitting only the three contingencies of Figure C.4. The occurrence of (a) and (b) were precluded in Appendix B. Therefore for $\alpha + \tilde{\beta} > 1$ the even sex-ratio equilibrium curves necessarily conform to the structure (c) and the case of Figure 4.1e results.

We prepare now to establish the equilibrium configurations of Figures 4.1c and d. To this end consider the matrix

$$M = \begin{bmatrix} \alpha & \beta & \tilde{\beta} \\ \beta & \gamma & \beta \\ \tilde{\beta} & \beta & \alpha \end{bmatrix} \text{ where } \alpha < \frac{1}{2} < \beta < \gamma \qquad (\text{C.3})$$

with $\tilde{\beta}$ satisfying $\alpha + \tilde{\beta} > 1$. Consulting Table 4.1, cases I and X, we ascertain two even sex-ratio equilibrium points on the edge $\overline{A_1 A_3}$ and one on each of the edges $\overline{A_1 A_2}$ and $\overline{A_2 A_3}$.

As before we determine that $\rho(\mathbf{p})$ for $\mathbf{p}$ on L corresponds to the spectral radius of the 2×2 matrix $L(u)$ for $0 \leq u \leq 1/2$ defined in (C.1). For $u = 0$, $\rho = \max(\beta, 2\gamma) > 1$, and for $u = 1/2$,

(a) 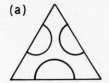 (b) 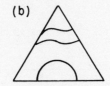 (c)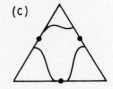

Figure C.4

$\rho = \max(\alpha + \tilde{\beta}, \beta) > 1$. Accordingly, there occur zero or two sign changes for $\rho(\mathbf{p})$ as u runs from 0 to $1/2$.

Under the conditions of (C.3) with $\alpha + \tilde{\beta} > 1$ by choosing β near 1, it is easy to argue that $\rho(\mathbf{p}) > 1$ along all of L (because $Q'(1) < 0$ for all $0 \leq u \leq 1/2$), and then the outcome of Fig. 4.1d necessarily prevails.

Consider next the case where

$$\alpha < \beta < \tfrac{1}{2} < \gamma, \gamma \text{ close to } \tfrac{1}{2}, \tilde{\beta} + \alpha > 1. \qquad \text{(C.4)}$$

Consulting Table 4.1 for these parameter specifications, we deduce that the boundary even sex-ratio equilibria $\tilde{\mathbf{p}}, \mathbf{p}^+, \mathbf{p}'$, and $\mathbf{p}''$ exist as indicated in Figure C.5.

Referring to the polynomial $R(u) = Q(1)$ of (C.2) we have

$$R(0) = (1 - 2\gamma)(1 - \beta) < 0, R(1/2) = (1 - \alpha - \tilde{\beta})(1 - \beta) < 0.$$

But for $u = 1/4$, we obtain

$$R(1/4) = (1 - \gamma - \beta) - (1 - \gamma - \beta/2)(\alpha + \tilde{\beta})/2 + \beta\gamma/2$$
$$> (1 - \gamma - \beta/2)[1 - (\alpha + \tilde{\beta})/2] + \beta\gamma.$$

Note $R(1/4) > 0$ if γ is close to $1/2$. Then $[\rho(\mathbf{p}) - 1]$ changes sign twice as $\mathbf{p}$ traverses L. It follows that the equilibrium configuration conforms to that described in Figure 4.1c.

The conditions $0 < \alpha < 1/2 < \gamma < \beta$ (see Table 4.1, case IV), imply the existence of a single even sex-ratio equilibrium on each of the edges $\overline{A_1 A_2}$ and $\overline{A_2 A_3}$ where the vertex equilibrium of only A_2 is stable. For $\tilde{\beta} > 1/2$ satisfying $\alpha + \tilde{\beta} < 1$, we are in the preserve of Figure C.6 and the central symmetric equilibrium $\mathbf{p}^*$ segregating alleles A_1 and A_3 equally has $\rho(\mathbf{p}^*) < 1$. Since $\alpha + \tilde{\beta} < 2\beta$ holds, the equilibrium $\mathbf{p}^*$ is externally un-

Figure C.5

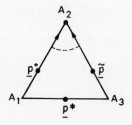

Figure C.6

stable in the associated sex ratio model (Theorem 3.3). In this circumstance, the equilibrium structure of Figure 4.1h ensues. With the same conditions but $1/2 < \beta < \gamma$, A_2-fixation is also unstable and we get Figure 4.1i.

A similar analysis confirms that the conditions $0 < \gamma < 1/2 < \tilde{\beta} < \alpha < \beta$ lead to the equilibrium configuration of Figure 4.1g, since $\gamma < 1/2 < \beta$ ensures that A_2-fixation is unstable while $1/2 < \tilde{\beta} < \alpha$ and $\alpha + \tilde{\beta} < 2\beta$ compels internal and external stability of the central equilibrium on the edge $\overline{A_1 A_3}$.

We conclude this appendix indicating parameter specifications leading to Figure 4.1j. Consider the sex determination coefficient matrix

$$M = \begin{bmatrix} \alpha & \beta & \tilde{\beta} \\ \beta & \gamma & \beta^* \\ \tilde{\beta} & \beta^* & \alpha \end{bmatrix}$$

obeying the relations

$$\alpha, \gamma < 1/2 < \beta, \ 2\beta(1 - \beta) < \alpha + \gamma - 2\alpha\gamma \qquad \text{(C.5a)}$$

$$1/2 < \tilde{\beta}, \ \alpha + \tilde{\beta} < 1, \ \beta + \beta^* < \alpha + \tilde{\beta}. \qquad \text{(C.5b)}$$

The condition (C.5a) implies the existence of two even sex-ratio equilibria on the edge $\overline{A_1 A_2}$. The inequality $\alpha + \tilde{\beta} < 1$ guarantees that the central symmetric equilibrium $\mathbf{p}^*$ on $\overline{A_1 A_3}$ has $\rho(\mathbf{p}^*) < 1$. The inequalities $\beta + \beta^* < \alpha + \tilde{\beta}$ and $\tilde{\beta} > \alpha$ entail that $\mathbf{p}^*$ is stable in the sex ratio model. Reference to Figure 4.3e and Appendix D, Chapter 3 implies that no other stable equilibria are possible. These facts together compel the equilibrium structure of Figure 4.1j.

APPENDIX D. GLOBAL CONVERGENCE BEHAVIOR FOR SOME THREE-ALLELE DICHOTOMOUS SEX DETERMINATION MODELS

Model I

Consider the sex dichotomization

	$\mathscr{G}_M$	$\mathscr{G}_F$			
	AA	AB	AC	BB	CC	BC
initial frequencies	$x^{(0)}$	$y^{(0)}$	$z^{(0)}$	$u^{(0)}$	$x^{(0)}$	$w^{(0)}$

It is easy to verify after one generation that $x = 1/2$ and then $y = y^{(1)}/2(y^{(1)} + z^{(1)})$, $z = z^{(1)}/2(y^{(1)} + z^{(1)})$.

Model II

	$\mathscr{G}_M$	$\mathscr{G}_F$			
	AB	AA	BB	BC	AC	CC
frequencies at generation n	v_n	u_n	w_n	x_n	y_n	z_n

Recursion relations for the six variables are easily obtained. Inspection of the equations reveals after one generation $z_n = 0$, $x_n = y_n$, $u_n + w_n = v_n$, and $x_n + v_n = 1/2$. Moreover, $u_n/x_n = u_{n-1}/x_{n-1} + 1/2 = m/2 + u_0/x_0$, $w_n/y_n = n/2 + w_0/y_0$, and $(2u_n - v_n) = (2u_{n-1} - v_{n-1})/2(2x_{n-1} + v_{n-1}) = (2u_{n-1} - v_{n-1})/(1 + 2x_{n-1})$ for $n \geq 1$. It follows immediately that $u_n \to 1/4$, $w_n \to 1/4$, $v_n \to 1/2$, $x_n \to 0$, $y_n \to 0$, and $z_n = 0$, $n \geq 1$, independent of the initial constitution of the population with the rate of approach to the equilibrium being algebraic. *The allele C is lost at an algebraic rate.*

Model III

	$\mathscr{G}_M$		$\mathscr{G}_F$			
	AA	BB	AB	AC	BC	CC
frequencies at generation n	u_n	w_n	v_n	y_n	x_n	z_n

The recursion equations over two successive generations are

(i) $T_{n-1}u_n = u_{n-1}(v_{n-1} + y_{n-1})$,

(ii) $T_{n-1}w_n = w_{n-1}(v_{n-1} + x_{n-1})$,

(iii) $T_{n-1}v_n = u_{n-1}(v_{n-1} + x_{n-1}) + w_{n-1}(v_{n-1} + y_{n-1})$,

(iv) $T_{n-1}y_n = u_{n-1}(y_{n-1} + x_{n-1})$,

(v) $T_{n-1}x_n = w_{n-1}(y_{n-1} + x_{n-1})$,

(vi) $z_n = 0, n \geq 1$,

$$(D.1)$$

where $T_{n-1} = 2(u_{n-1} + w_{n-1})(v_{n-1} + x_{n-1} + y_{n-1})$ is a normalizing factor.

From the above relations it is easy to establish three equilibrium curves:

$$S_1: w_e = \tfrac{1}{2}, v_e + x_e = \tfrac{1}{2},$$
$$S_2: u_e = \tfrac{1}{2}, v_e + y_e = \tfrac{1}{2},$$
$$S_3: v_e = \tfrac{1}{2}, u_e + w_e = \tfrac{1}{2}.$$

We sketch a proof of the convergence of the vector $(u_n, w_n, v_n, x_n, y_n)$ to one of S_1, S_2, or S_3 and characterize the domains of convergence to the respective curves. Consider first the case $u_0/w_0 \leq 1, y_0/x_0 < 1$. Examination of (D.1) reveals that $\{u_n/w_n\}$ is a strictly decreasing sequence. It therefore converges to some $\alpha, 0 \leq \alpha < 1$. Since $y_n/x_n = u_{n-1}/w_{n-1}$ also $\{y_n/x_n\}$ decreases to α. We prove by contradiction that $\alpha = 0$. Suppose $\alpha > 0$. Then

$$\frac{v_n}{x_n + y_n} = \frac{v_{n-1}}{x_{n-1} + y_{n-1}} + \frac{u_{n-1}/w_{n-1} + y_{n-1}/x_{n-1}}{(1 + u_{n-1}/w_{n-1})(1 + y_{n-1}/x_{n-1})}$$

$$(D.2)$$

The last term on the right is bounded away from zero, since $\alpha > 0$. Hence there exists positive constants C and D such that $(x_n + y_n)/v_n \leq (nC + D)^{-1}$. It follows that $x_n + y_n \to 0$. From the fact that

$$\left| \frac{u_n + w_n - v_n}{y_n - x_n} \right| = \left| \frac{y_{n-1} - x_{n-1}}{y_{n-1} + x_{n-1}} \right| \leq 1, \qquad (D.3)$$

we infer that $u_n + w_n \to 1/2$, $v_n \to 1/2$. To obtain the contradiction, note that (D.2) implies $(x_n + y_n)/v_n \geq 1/(nE + F)$ for some positive constants E and F. Now, $v_n \to 1/2$, so that $\sum_{n=1}^{\infty} (x_n + y_n) = \infty$. Because $(x_n - y_n)/(x_n + y_n) \to (1 - \alpha)/(1 + \alpha) > 0$, we have $\sum_{n=1}^{\infty} (x_n - y_n) = +\infty$. We may therefore write

$$\frac{u_n}{w_n} = \frac{u_0}{w_0} \prod_{k=1}^{n} \left(1 - \frac{x_k - y_k}{v_k + x_k}\right) \tag{D.4}$$

from which we deduce that the infinite product converges to zero. Consequently, $u_n/w_n \to 0$, giving the required contradiction. Since $y_n \to 0$ and $u_n \to 0$, we can deduce $w_n - v_n - x_n \to 0$, i.e., $w_n \to 1/2$, $v_n + x_n \to 1/2$.

Now, operating on (D.2) leads to

$$c \left[2 \sum_{k=1}^{n-1} \frac{u_{k-1}}{w_{k-1}} + \frac{y_0}{x_0} \right] + \frac{v_0}{x_0 + y_0}$$

$$\leq \frac{v_n}{x_n + y_n} \leq \frac{v_0}{x_0 + y_0}$$

$$+ 2 \sum_{k=1}^{n} \frac{u_{k-1}}{w_{k-1}} + \frac{y_0}{x_0}, \tag{D.5}$$

where $c = 1/(1 + u_0/w_0)(1 + y_0/x_0)$. To prove $\lim_{n \to \infty} x_n$ exists and is *positive*, it suffices to establish $\sum u_k/w_k < \infty$. Indeed, suppose that $\sum u_n/w_n$ is convergent and therefore also $\sum y_n/x_n < \infty$. If $A_n = u_n/w_n$, then $y_n/x_n = u_{n-1}/w_{n-1} = A_{n-1}$. From (D.2) we deduce

$$\frac{v_n}{x_n + y_n} - \frac{v_r}{x_r + y_r} = \sum_{r=k}^{n} \left(\frac{A_r + A_{r-1}}{(1 + A_r)(1 + A_{r-1})}\right).$$

Since the right-hand side converges, it follows that

$$\lim_{n \to \infty} v_n/(x_n + y_n) = K$$

exists finitely. Cognizance of (D.5) further reveals that K is bounded away from zero. As $y_n \to 0$ and $v_n + x_n \to 1/2$ we deduce that $\lim x_n$ exists.

144

It remains to prove $\sum u_k/w_k < \infty$. Assume to the contrary that $\sum u_k/w_k = \infty$. From (D.1)

$$\frac{u_n}{w_n} - \frac{u_{n-1}}{w_{n-1}} = \frac{(u_{n-1}/w_{n-1})\{(u_{n-2}/w_{n-2}) - 1\}}{(v_{n-1}/x_{n-1}) + 1} \qquad \text{(D.6)}$$

Now, since u_n/w_n tends to 0, for every $\varepsilon > 0$ there is an $\mathcal{N}(\varepsilon)$ such that $n \geq \mathcal{N}$ implies

$$\frac{u_n}{w_n} - \frac{u_{n-1}}{w_{n-1}} \leq (\varepsilon - 1)\left\{\frac{u_{n-1}/w_{n-1}}{(v_{n-1}/x_{n-1}) + 1}\right\}. \qquad \text{(D.7)}$$

From (D.5), since $\sum u_k/w_k$ diverges, we see that v_n/x_n is of the order of $\sum_{k=1}^{n} u_k/w_k$. Now sum both sides of (D.7) over n; the left-hand side is a telescoping series and remains bounded while the right-hand side (by virtue of the familiar theorem on divergent series to the effect that if $\sum a_n = \infty$, $\sum a_n \geq 0$, and $A_n = \sum_{k=1}^{n} a_k$ then $\sum a_n/A_n = \infty$) diverges to $-\infty$. This is the required contradiction.

Thus we have shown that if $u_0/w_0 \leq 1$, $y_0/x_0 \leq 1$, $(u_0 y_0)/(w_0 x_0) < 1$, there is convergence to a member of the family $w_e = 1/2$, $v_e + x_e = 1/2$, $v_e > 0$, $x_e > 0$. The rate of convergence is geometric. Symmetrical arguments prove that if $w_0/u_0 \leq 1$, $x_0/y_0 \leq 1$, $w_0 x_0/u_0 y_0 < 1$ the limit equilibrium has the form $u_e = 1/2$, $v_e + y_e = 1/2$, $v_e > 0$, $y_e > 0$. A somewhat simpler argument shows that if $\{(u_{n-1}/w_{n-1}) - 1\}$ and $\{(u_n/w_n) - 1\}$ are of continually alternating signs as $n \to \infty$, then $u_n \to 1/4$, $w_n \to 1/4$, $v_n \to 1/2$, convergence occurring at an algebraic rate. These results imply convergence in all cases.

Model IV

	$\mathscr{G}_M$			$\mathscr{G}_F$		
	AA	BC	BB	AB	AC	CC
frequencies	x	y	u	z	w	v

The sex determination matrix of the three alleles $\{A, B, C\}$ for this case is

$$M = \begin{bmatrix} 1 & 0 & 0 \\ 0 & 0 & 1 \\ 0 & 1 & 0 \end{bmatrix}$$

The recursion equations for gene frequencies over two successive generations are

$$Tx' = x\left(\frac{z+w}{2}\right)$$

$$Ty' = \frac{y}{2}\left(u + v + \frac{z+w}{2}\right)$$

$$Tu' = \frac{y}{2}\left(u + \frac{z}{2}\right)$$

$$Tv' = \frac{y}{2}\left(v + \frac{w}{2}\right) \qquad \text{(D.8)}$$

$$Tw' = x\left(v + \frac{w}{2}\right) + \frac{y}{2}\left(\frac{z+w}{2}\right)$$

$$Tz' = x\left(u + \frac{z}{2}\right) + \frac{y}{2}\left(\frac{z+w}{2}\right)$$

with $T = (x + y)(u + v + w + z)$.

In this model there are three classes of fixed points, two of which are boundary curves $S_1 = \{x = 1/2,\ w + z = 1/2\}$, $S_2 = \{y = 1/2,\ u + v = 1/2\}$, and an interior symmetric equilibrium that is unstable (by Theorem 3.4)

$$x = u = v = \tfrac{1}{9}, y = z = w = \tfrac{2}{9} \qquad \text{(D.9)}$$

corresponding to allele frequencies $\text{freq}(A) = \text{freq}(B) = \text{freq}(C) = 1/3$. We prove the following global convergence results.

(a) If for some n, $y^{(n)} < 2x^{(n)}$ and $(z^{(n)} + w^{(n)})/2 > u^{(n)} + v^{(n)}$, the same relations persist for all n thereafter and the system D.8 converges to S_1.

(b) If for some n, $y^{(n)} > 2x^{(n)}$ and $(z^{(n)} + w^{(n)})/2 < w^{(n)} + v^{(n)}$, then convergence occurs to S_2.

(c) If for each n, $2x^{(n)} > y^{(n)}$ and $u^{(n)} + v^{(n)} > (z^{(n)} + w^{(n)})/2$, or $2x^{(n)} < y^{(n)}$ and $(z^{(n)} + w^{(n)})/2 > u^{(n)} + w^{(n)}$, then convergence to the isolated equilibrium (D.9) transpires. Since this isolated equilibrium is unstable the contingencies of (c) are relegated to special initial conditions of dimension span ≤ 1.

Case (a). Direct analysis shows that $2x > y$ and $(z + w)/2 > u + v$ implies the same relations for the next generation, that is, $2x' > y'$ and $z' + w' > 2(u' + v')$. Next, observe that $2x'/y' = (2x/y)(2(z + w)/(2(u + v) + z + w) > 2x/y$. Accordingly, $2x^{(n)}/y^{(n)}$ is increasing. If $\lim_{n \to \infty} 2x^{(n)}/y^{(n)} = \alpha < \infty$ $(\alpha > 1)$, then

$$\frac{z^{(n)} + w^{(n)}}{u^{(n)} + v^{(n)} + \dfrac{z^{(n)} + w^{(n)}}{2}} \to 1. \tag{D.10}$$

We claim that $z^{(n)} + w^{(n)}$ is bounded away from zero or otherwise $z^{(n)} + w^{(n)} + u^{(n)} + v^{(n)} \to 0$, which is impossible. Since $T((z' + w')/2 - (u' + v')) > (x - y/2)((u + v + z + w)/2)$ we deduce that $\lim_{n \to \infty} (z^{(n)} + w^{(n)})/2(u^{(n)} + v^{(n)}) > 1$, an outcome incompatible with (D.10).

To avert this contradiction we must have $\lim_{n \to \infty} 2x^{(n)}/y^{(n)} = \infty$ and therefore $y^{(n)} \to 0$. But $y^{(n)} = u^{(n)} + v^{(n)}$ for all $n \geq 1$ implying that $u^{(n)}$ and $v^{(n)}$ converge to zero. Also, $T^{(n)}(z^{(n+1)} + w^{(n+1)} - x^{(n+1)}) \leq u^{(n)} + v^{(n)} + y^{(n)} \to 0$, and $T^{(n)}$ is necessarily bounded away from zero. It follows that $z^{(n)} + w^{(n)} \to 1/2$ and $x^{(n)} \to 1/2$.

It is easy to check that

$$u^{(n)} + v^{(n)} + w^{(n)} + z^{(n)} \geq x^{(n)} + y^{(n)} \text{ for all } n \geq 1. \tag{D.11}$$

In view of the convergence properties established we ascertain that $y^{(n)}$, $u^{(n)}$, and $v^{(n)}$ go to zero at least geometrically fast.

We next prove that if some subsequence of $z^{(n)} \to 0$ then $z^{(n)} \to 0$. In fact, from the recursion relations we have

$$z^{(n+1)} = \frac{x^{(n)} z^{(n)}}{2(x^{(n)} + y^{(n)})(1 - x^{(n)} - y^{(n)})} + \varepsilon_n$$

where $\sum \varepsilon_n < \infty$. We see from (D.11) that $x^{(n)} + y^{(n)} < 1/2$ and therefore $x^{(n)}/2(x^{(n)} + y^{(n)})(1 - x^{(n)} - y^{(n)}) < 1$ implying $z^{(n+1)} < z^{(n)} + \varepsilon_n$.

Where $z^{(n_k)} \to 0$ coupled to the fact $\sum \varepsilon_n < \infty$ we easily deduce that $\lim z^{(n)} = 0$. Similarly, if $w^{(n_k)} \to 0$ along a subsequence we have that $w^{(n)} \to 0$.

Consider next the case where $z^{(n)}$ is bounded away from zero. Direct iteration in view of $\sum y^{(k)} < \infty$, $\sum u^{(k)} < \infty$, $\sum v^{(k)} < \infty$ produces

$$\frac{w^{(n)}}{z^{(n)}} = \frac{w^{(0)}}{z^{(0)}} \prod_{k=1}^{n} \left(\frac{1 + \varepsilon_k}{1 + \delta_k} \right)$$

where $\sum_k \varepsilon_k$ and $\sum_k \delta_k$ are positive convergent series. This implies that $\lim w^{(n)}/z^{(n)}$ exists. Since also $w^{(n)} + z^{(n)} \to 1/2$, we deduce that $\lim w^{(n)}$ and $\lim z^{(n)}$ each converge.

Case (b). The analysis paraphrases that of Case (a) where now we find that $y^{(n)} \to 1/2$ and $u^{(n)} + v^{(n)} \to 1/2$, $\lim u^{(n)}$ and $\lim v^{(n)}$ exist.

Case (c). If $(z^{(n)} + w^{(n)})/2 < u^{(n)} + v^{(n)}$ and $2x^{(n)}/y^{(n)} > 1$ for all n, we deduce easily that $2x^{(n)}/y^{(n)}$ decreases to 1. A similar argument applies if $(z^{(n)} + w^{(n)})/2 > u^{(n)} + v^{(n)}$ with $2x^{(n)}/y^{(n)} < 1$ where this sequence then increases to 1. Once this is proved it is easy to ascertain convergence to the symmetric equilibrium.

In the case where $2x^{(n)}/y^{(n)} - 1$ alternates in sign we determine that $\left| 2x^{(n+1)}/y^{(n+1)} - 1 \right| < \left| 2x^{(n)}/y^{(n)} - 1 \right|$ and again $2x^{(n)}/y^{(n)} \to 1$. In fact, suppose

$$2x > y, \; u + v > \frac{z + w}{2} \text{ and } 2x' < y', \; u' + v' < \frac{z' + w'}{2}$$

We claim that $1 - 2x'/y' < 2x/y - 1$. In fact, $u' + v' < (z' + w')/2$ implies $(x/y)[1 + (z + w)/(u + v + (z + w)/2)] > 1$. But $x'/y' = (x/y)(z + w)/(u + v + (z + w)/2)$ yielding $x/y + x'/y' > 1$ or $1 - 2x'/y' < 2x/y - 1$. In a similar way in the reverse order we get $2x'/y' - 1 < 1 - 2x/y$.

The central equilibrium (D.9) is not totally repelling, i.e., the local approximation matrix (gradient matrix) at (D.9) has relevant eigenvalues greater than and less than 1 in magnitude. Therefore the domain of convergence to (D.9) is likely a curve of dimension 1, which separates the domain of attraction to the even sex-ratio equilibrium curves S_1 and S_2.

Model V

	$\mathscr{G}_M$		$\mathscr{G}_F$			
	AB	AC	AA	BB	CC	BC
frequencies	u	v	x	y	z	w

The recursion equations are

$$Tu' = \frac{u+v}{2}\left(y + \frac{w}{2}\right) + \frac{u}{2}x$$

$$Tv' = \frac{u+v}{2}\left(z + \frac{w}{2}\right) + \frac{v}{2}x$$

$$Tx' = \frac{(u+v)}{2}x$$

$$Ty' = \frac{u}{2}\left(y + \frac{w}{2}\right) \tag{D.12}$$

$$Tz' = \frac{v}{2}\left(z + \frac{w}{2}\right)$$

$$Tw' = \frac{u}{2}\left(z + \frac{w}{2}\right) + \frac{v}{2}\left(y + \frac{w}{2}\right)$$

$$T = (u+v)(x+y+z+w).$$

We immediately have $u^{(n)} + v^{(n)} = 1/2$, $n \geq 1$, and therefore $T^{(n)} = 1/4$ for all $n \geq 1$. Then $x^{(n)} = x^*$ is invariant for $n \geq 1$. We denote $\mathcal{Z} = z + w/2$, $\Upsilon = y + w/2$, which are the C and B allele frequencies, respectively, in the second mating group. The equations (D.12) reduce to

$$\Upsilon' = \tfrac{1}{2}\Upsilon + u(\tfrac{1}{2} - x^*)$$
$$u' = \Upsilon + 2ux^*.$$

This is a linear system with eigenvalues $\lambda_1 = 1$, and $\lambda_2 = 2x^* - \tfrac{1}{2}$. The iterates of this linear transformation converge yielding

$$\lim_{n \to \infty} \Upsilon^{(n)} = \frac{1 - 2x^*}{3 - 4x^*}\left(u^{(1)} + 2\Upsilon^{(1)}\right)$$

149

$$\lim_{n \to \infty} u^{(n)} = \frac{1}{3 - 4x^*} \left(u^{(1)} + 2\Upsilon^{(1)} \right).$$

By symmetry considerations we further deduce

$$\lim v^{(n)} \to \frac{1}{3 - 4x^*} \left(v^{(1)} + 2\mathcal{Z}^{(1)} \right)$$

$$\lim \mathcal{Z}^{(n)} \to \frac{1 - 2x^*}{3 - 4x^*} \left(v^{(1)} + 2\mathcal{Z}^{(1)} \right)$$

Since $T^{(n)}$, $n \geq 1$ is constant we secure global convergence. The limiting equilibrium curve can be described by a one-parameter family of even sex-ratio equilibrium points

$$w^{(\infty)} = 2u^*(1 - 2u^*)(1 - 2x^*)$$
$$v^{(\infty)} = \tfrac{1}{2} - u^*$$
$$z^{(\infty)} = (1 - 2u^*)^2(1 - 2x^*)/2$$
$$y^{(\infty)} = 2u^{*2}(1 - 2x^*)$$

where $u^* + x^* + w^{(\infty)} + v^{(\infty)} + z^{(\infty)} + y^{(\infty)} = 1$.

Model VI

	$\mathcal{G}_M$			$\mathcal{G}_F$		
	AA	AB	AC	BC	BB	CC
frequencies	x	y	v	w	u	z

The recursion equations are (clearly $z^{(n)} = 0$, $n \geq 1$)

$$Tx' = \left(x + \frac{y}{2} \right) \frac{v}{2}$$

$$Ty' = \left(x + \frac{y}{2} \right) \left(u + \frac{w}{2} \right) + \frac{vy}{4}$$

$$Tv' = \left(x + \frac{y}{2} \right) \left(\frac{v + w}{2} \right)$$

$$Tw' = \frac{y}{2} \left(\frac{v + w}{2} \right)$$

$$Tu' = \frac{y}{2} \left(u + \frac{w}{2} \right)$$

where $T = (x + y)(u + v + w)$. Observe that $T(v' + w') = (x + y)((v + w)/2)$. Therefore at equilibrium $\hat{u} + \hat{v} + \hat{w} = 1/2 = \hat{x} + \hat{y}$ and $\hat{T} = 1/4$.

The equilibria consist of a single even sex-ratio curve, which parameterized in terms of v has the form

$$\hat{x} = \frac{v}{2(1 - v)}, \quad \hat{y} = \frac{1 - 2v}{2(1 - v)},$$

$$\hat{u} = \frac{(1 - 2v)^2}{2}, \quad \hat{w} = v(1 - 2v)$$

for any $0 < v < 1/2$.

Numerical iteration suggests that global convergence occurs to this curve at a geometric rate. A formal proof seems difficult.

Model VII

	$\mathcal{G}_M$			$\mathcal{G}_F$		
	AA	BB	AB	AC	BC	CC
frequencies	x	y	z	u	v	w

Upon writing the recursion relations, we find after two generations an equal sex ratio. The changes in genotype frequencies then reduce to a two-variable problem and global convergence results to the equilibrium curve

$$\{2a^2, 2(\tfrac{1}{2} - a)^2, 4a(\tfrac{1}{2} - a)\}, \{a, \tfrac{1}{2} - a, 0\}.$$

with $0 < a < 1/2$.

Model VIII

	$\mathcal{G}_M$			$\mathcal{G}_F$		
	AA	BB	AC	CC	AB	BC
frequencies	x	y	z	u	v	w

Let $T = (x + y + z)(u + v + w)$. The recursion relations are

$$Tx' = \left(x + \frac{z}{2} \right) \frac{v}{2} \tag{D.13a}$$

$$Ty' = y \left(\frac{v + w}{2} \right) \tag{D.13b}$$

$$T z' = \left(x + \frac{z}{2} \right)\left(u + \frac{w}{2} \right) + \frac{zv}{4} \qquad \text{(D.13c)}$$

$$T u' = \left(u + \frac{w}{2} \right)\frac{z}{2} \qquad \text{(D.13d)}$$

$$T v' = \left(x + \frac{z}{2} \right)\left(\frac{v + w}{2} \right) + \frac{yv}{2} \qquad \text{(D.13e)}$$

$$T w' = y\left(u + \frac{w}{2} \right) + \frac{z}{2}\left(\frac{v + w}{2} \right). \qquad \text{(D.13f)}$$

We will show that there exists no equilibrium segregating all genotypes.

Consider the equations (D.13) at equilibrium. When $y > 0$, then from equation (D.13b) we deduce

$$T = \frac{v + w}{2}. \qquad \text{(D.14)}$$

Using this relation in (D.13a) implies

$$wx = \frac{vz}{2}. \qquad \text{(D.15)}$$

Combining (D.13e) and (D.13f) yields

$$2T^2 = T(x + y + z) + yu \qquad \text{(D.16)}$$

Adding the first three equations of (D.13), taking account of (D.14), leads to

$$\frac{zu}{2} + xu = \frac{zw}{4}. \qquad \text{(D.17)}$$

Multiplying by w and using (D.15) gives

$$z(2uw + 2vu) = zw^2$$

So under the assumption $z > 0$ we have

$$w^2 = 2u(v + w) = 4uT. \qquad \text{(D.18)}$$

In equation (D.13d) substituting for wz from (D.17) gives

$$Tu = u(z + x).$$

152

So if $u > 0$ also

$$T = z + x. \tag{D.19}$$

From (D.16) and (D.19) we find $T^2 = y(T + u)$ or

$$y = \frac{T^2}{T + u}. \tag{D.20}$$

Substituting from (D.19) and (D.20) gives

$$x + z + y = \frac{T(2T + u)}{T + u}.$$

But $T = (x + z + y)(u + v + w)$ so that

$$1 = (u + v + w) \frac{(2T + u)}{T + u} = \frac{(u + 2T)^2}{T + u},$$

which is the equation

$$(u + 2T)^2 - (T + u) = 0. \tag{D.21}$$

Thus u is determined from (D.21) in terms of T and because of $w^2 = 4Tu$ from (D.18). Now by (D.19 and D.14), $w(z + x) = wT = w(v + w)/2$. Using (D.15), this simplifies to $wz + vz/2 = w(v + w)/2$ or

$$z = \frac{w\left(\dfrac{v + w}{2}\right)}{\dfrac{w + v}{2} + \dfrac{w}{2}} = \frac{wT}{T + \dfrac{w}{2}}. \tag{D.22}$$

But $x = T - z = T(T - w/2)/(T + w/2)$ and $y = T^2/(T + u)$. Thus all the variables x, y, z, u, v, w are determined in terms of u and T, which are related as in (D.21). Moreover, it is easy to verify that $x + y + z + u + v + w = 1$ provided u and T satisfy (D.21). Substituting into equation (D.13e) and simplifying we obtain

$$T(2T - w) = \frac{T^3}{T + \dfrac{w}{2}} + \frac{\left(T - \dfrac{w}{2}\right)T^3}{T^2 + \dfrac{w^2}{4}} = \frac{2T^5}{\left(T + \dfrac{w}{2}\right)\left(T^2 + \dfrac{w^2}{4}\right)},$$

153

which reduces to

$$2\left(T^4 - \frac{w^4}{16}\right) = 2T^4 \tag{D.23}$$

and therefore $w = 0$.

The conditions $w = 0$ and $y > 0$ require either $v = 0$ and/or $z = 0$ in view of (D.18).

Case 1: $y > 0$, $z = 0$. Then the solution involves

AA	BB	CC	AB
x	y	u	v

and the necessary equilibrium satisfy $x + y = 1/2$, $v = 1/2$.

Case 2: $y > 0$, $z > 0$, $v = 0$, $w = 0$. This is impossible.

Case 3: $y = 0$. The model then reduces to Model VI with allele B and C interchanged. A one-parameter curve of equilibrium points exists described in the form given in Table 4.1.

To sum up: In Model VIII there are two classes of equilibria, one is a curve parameterized as in Model VI where the frequency of genotype BB equals zero and allele C plays the role of allele B and vice versa. The second class of equilibria has the frequency of AB equal to $1/2$.

APPENDIX E. GLOBAL CONVERGENCE FOR THE SEX DETERMINATION MODEL WHERE ALL HOMOZYGOTES DETERMINE FEMALES AND HETEROZYGOTES DETERMINE MALES

Compare with Model IX of Section 4.5. Consider an r-allele system with sex dichotomization

$$\text{females} \qquad\qquad \text{males}$$
$$\mathcal{G}_F = \{A_iA_i, i = 1, \ldots, r\} \quad \mathcal{G}_M = \{A_iA_j, i \neq j\} \tag{E.1}$$

Thus the sex determination matrix in this case is $M = U - I$. Letting

$$p_i = \text{freq}(A_i) \text{ in the female population},$$

$$q_i = \text{freq}(A_i) \text{ in the male population},$$

the transformation equations (2.4) become

$$p_i' = \frac{p_i q_i}{\sum_{k=1}^{r} p_k q_k} \text{ and } q_i' = \frac{p_i + q_i - 2p_i q_i}{2(1 - \sum_{k=1}^{r} p_k q_k)}, \, i = 1, 2, \ldots, r$$

$$(E.2)$$

with sex ratio $\sum_{k=1}^{r} p_k q_k$ in the current generation. Let $p_i^{(n)}$ and $q_i^{(n)}$ denote the frequencies of alleles A_i $(i = 1, \ldots, r)$ in the female and male populations, respectively, at generation n.

Case (1). Assume first in the initial generation

$$p_1^{(0)} > p_2^{(0)} > \cdots > p_r^{(0)} \text{ and } q_1^{(0)} > q_2^{(0)} > \cdots > q_r^{(0)}. \quad (E.3)$$

We prove:

LEMMA E.1. *For all succeeding generations and $i < j$, $p_i^{(n)} > p_j^{(n)}$ and $q_i^{(n)} > q_j^{(n)}$.*

PROOF. We merely need to prove these inequalities for $n = 1$. For $i < j$, we have

$$\frac{p_i'}{p_j'} = \frac{p_i}{p_j} \frac{q_i}{q_j} > \frac{p_i}{p_j} > 1. \quad (E.4)$$

Consider next

$$\frac{q_i'}{q_j'} = \frac{p_i + q_i - 2p_i q_i}{p_j + q_j - 2p_j q_j} \quad (E.5)$$

For $q_i \leq 1/2$, and because $q_j < q_i$ and necessarily $2p_j \leq 1, j \geq 2$, the right-hand fraction is diminished if q_j is increased to q_i. The resulting ratio exceeds 1 if and only if $(p_i - p_j)(1 - 2q_i) \geq 0$ which is correct when $q_i \leq 1/2$.

Suppose next that $q_i > 1/2$. Then q_j varies between 0 and $1 - q_i$. The condition that the right-hand side in (E.5) exceeds 1 is

$$p_i - p_j + q_i - q_j > 2p_i q_i - 2p_j q_j. \quad (E.6)$$

This is linear in the variable q_j. For $q_j = 0$, (E.6) becomes $p_i - p_j + q_i > 2p_i q_i$, which is linear in q_i. The last inequality is correct for $q_i = 1/2$ and for $q_i = 1$ (with equalities permitted) because $p_j + p_i \leq 1$. Therefore (E.6) holds for $q_j = 0$. When $q_j = 1 - q_i$, the inequality reduces to $p_i + p_j + 2q_i - 1 > 2q_i(p_i + p_j)$ which holds for $q_i = 1/2$ and $q_i = 1$ as before, and therefore for all $1/2 < q_i < 1$.

We have thus established (E.6) in all cases implying $q'_i > q'_j$ and the proof of Lemma E.1 is complete.

We next prove:

Lemma E.2. *(i)* $p_k^{(n)} \to 0$ *as* $n \to \infty$ *for* $k = 3, \ldots, r$. *(ii) Either* $p_2^{(n)} \to 0$ *and then* $q_1^{(n)} \to 1/2$ *and* $q_2^{(n)} + \cdots + q_r^{(n)} \to 1/2$, *or* $\lim_{n \to \infty} p_2^{(n)} > 0$ *and* $\lim_{n \to \infty} q_1^{(n)} = \lim_{n \to \infty} q_2^{(n)} = 1/2$.

Thus, either $A_1 A_1$ is exclusively established in $\mathscr{G}_F$ and $\mathscr{G}_M = \{A_1 A_k, k = 2, \ldots, r\}$, or the limit composition consists of $\mathscr{G}_F = \{A_1 A_1, A_2 A_2\}$ and $\mathscr{G}_M = \{A_1 A_2\}$ with $\text{freq}(\mathscr{G}_F) = \text{freq}(\mathscr{G}_M) = 1/2$ in both cases.

PROOF. If $p_1^{(n)}/p_{k+1}^{(n)} \uparrow \infty$, then $p_{k+1}^{(n)}, p_{k+1}^{(n)}, \ldots, p_r^{(n)} \to 0$. Moreover, since $p_1^{(n)}/p_i^{(n)}$ increases in all cases by (E.4) and $\sum_i p_i^{(n)} = 1$, we deduce that $\lim_{n \to \infty} p_i^{(n)} = p_i^*$ exists.

Suppose $\lim_{n \to \infty} p_1^{(n)}/p_k^{(n)}$ is finite. Then $p_i^{(n)}/p_{i+1}^{(n)}$ increases to a finite limit for each $i \leq k - 1$. Because of the identity in (E.4), this is only possible if $q_1^{(n)}/q_k^{(n)} \to 1$. We show that this convergence is impossible for $k = 3$. Otherwise $p_2^{(n)}/p_3^{(n)}$ increases to a finite limit and necessarily $q_2^{(n)}/q_3^{(n)} \to 1$. But,

$$\frac{q_2^{(n+1)}}{q_3^{(n+1)}} = \frac{p_2^{(n)} + q_2^{(n)} - 2p_2^{(n)} q_2^{(n)}}{p_3^{(n)} + q_3^{(n)} - 2p_3^{(n)} q_3^{(n)}}, \qquad (E.7)$$

which converges (at least for a subset of indices n) to some quantity

$$\frac{p_2^* + q_2^* - 2p_2^* q_2^*}{p_3^* + q_3^* - 2p_3^* q_3^*}, \qquad (E.8)$$

which strictly exceeds 1 because $q_3^* = q_2^* < 1/2$, while $p_2^* > p_3^*$. To avoid the contradiction, we must have $p_1^{(n)}/p_3^{(n)} \uparrow \infty$.

Consider next the contingency $p_1^{(n)}/p_2^{(n)} \uparrow \alpha < \infty$. It follows then that $p_1^{(n)} \to p_1^*$, $p_2^{(n)} \to p_2^*$ ($p_1^* + p_2^* = 1$ with $p_1^* \geq p_2^* >$

0) and $q_1^{(n)}/q_2^{(n)} \to 1$. But

$$\frac{q_1^{(n+1)}}{q_2^{(n+1)}} = \frac{p_1^{(n)} + q_1^{(n)} - 2p_1^{(n)}q_1^{(n)}}{p_2^{(n)} + q_2^{(n)} - 2p_2^{(n)}q_2^{(n)}} \to 1, \qquad (E.9)$$

which is possible only if $q_1^{(n)}$ and $q_2^{(n)}$ converge and the limits are $q_1^* = q_2^* = 1/2$.

In the case that $p_1^{(n)} \to 1$, we find that $q_k^{(n+1)} \approx q_k^{(n)}/(2(1-q_1^{(n)}))$, $k = 2, \ldots, r$, $q_1^{(n+1)} \approx (1-q_1^{(n)})/2(1-q_1^{(n)}) = 1/2$ as $n \to \infty$. So $q_1^{(n)} \to 1/2$ and $\sum_{k=2}^{r} q_k^{(n)} \to 1/2$ the convergence rate being algebraic. We can even prove that each $q_k^{(n)}$ converges as $n \to \infty$. This completes the analysis of the initial conditions (E.3).

Case (2). Suppose next

$$p_1^{(0)} > p_2^{(0)} > \cdots > p_r^{(0)} \quad \text{(this is no restriction and reflects merely a labeling of the alleles)}$$

but now

$$q_2^{(0)} > q_1^{(0)} > q_3^{(0)} > \cdots > q_r^{(0)}. \qquad (E.10)$$

We deduce, parallel to case (1), that $p_i^{(n)}/p_j^{(n)} \uparrow \infty$ and $q_i^{(n)}/q_j^{(n)} > 1$ for all $j \geq 3$, $i < j$.

Suppose now $p_1^{(n)} > p_2^{(n)}$ and $q_2^{(n)} > q_1^{(n)}$ hold for all $n \geq 1$. It follows that $p_1^{(n)}/p_2^{(n)} \downarrow \alpha \geq 1$ and consequently $q_1^{(n)}/q_2^{(n)} \to 1$.

Adapting the arguments of case (1) we deduce that $\lim_{n \to \infty} q_1^{(n)} = \lim_{n \to \infty} q_2^{(n)} = 1/2$ if $\alpha > 1$. If $\alpha = 1$, then $[q_1^{(n+1)} + q_2^{(n+1)}] \simeq 1/[2 - q_1^{(n)} - q_2^{(n)}]$ as $n \to \infty$ and therefore $q_1^{(n)} + q_2^{(n)} \to 1$. In every case the resulting limiting genotype composition is

$$\begin{array}{cc} \mathscr{G}_F & \mathscr{G}_M \\ \{A_1A_1, A_2A_2\} & \{A_1A_2\} \end{array} \qquad (E.11)$$

If $p_2^{(n)} > p_1^{(n)}$ and $q_1^{(n)} > q_2^{(n)}$ hold for all $n \geq 1$, the same ultimate outcome as (E.11) is realized.

Finally, if for some n, the relationships

$$p_1^{(n)} > p_2^{(n)} \text{ and } q_1^{(n)} > q_2^{(n)} \qquad (E.12)$$

or

$$p_2^{(n)} > p_1^{(n)} \text{ and } q_2^{(n)} > q_1^{(n)} \qquad (E.13)$$

take place then the results of case (1) apply, noting that (E.13) emerges from (E.3) because we are again in case (1) recognizing that the alleles A_2 and A_1 have labels interchanged. The property of global convergence holds, *mutatis mutandis*.

Case (3). Suppose

$$p_1^{(0)} > p_2^{(0)} > \cdots > p_r^{(0)}$$

but now

$$q_2^{(0)} > q_3^{(0)} > q_1^{(0)} > q_4^{(0)} > \cdots > q_r^{(0)}. \qquad (E.14)$$

We deduce as in case (1), $p_j^{(n)} \to 0$ for $j \geq 4$, $p_2^{(n)}/p_3^{(n)} \uparrow$ and $q_2^{(n)}/q_3^{(n)} > 1$. If $\lim_{n \to \infty} p_2^{(n)}/p_3^{(n)}$ is finite, then $q_2^{(n)}/q_3^{(n)} \to 1$. These results are consistent only if $\lim_{n \to \infty} q_2^{(n)} = \lim_{n \to \infty} q_3^{(n)} = 1/2$ and therefore $q_1^{(n)} \to 0$ which compels $p_1^{(n)} \to 0$. But then the ordering of (E.14) cannot be maintained for all n.

Another possibility has $p_2^{(n)}/p_3^{(n)} \uparrow \infty$ or $p_3^{(n)} \to 0$. If also $p_1^{(n)} > p_2^{(n)}$ and $q_2^{(n)} > q_3^{(n)} > q_1^{(n)}$ are maintained for all n, then $p_1^{(n)}/p_2^{(n)} \downarrow \alpha \geq 1$ and $q_1^{(n)}/q_2^{(n)} \to 1$, implying as in case (2) that $\lim_{n \to \infty} q_2^{(n)} = \lim_{n \to \infty} q_1^{(n)} = 1/2$, and therefore $q_3^{(n)} \to 0$, which is a contradiction.

The foregoing arguments establish that

$$p_1^{(n)} > p_2^{(n)} > p_3^{(n)} \text{ and } q_2^{(n)} > q_3^{(n)} > q_1^{(n)}$$

cannot prevail for all n.

The switch to

$$p_2^{(n)} > p_1^{(n)} > p_3^{(n)} \text{ and } q_2^{(n)} > q_3^{(n)} > q_1^{(n)} \qquad (E.15)$$

or

$$p_2^{(n)} > p_3^{(n)} > p_1^{(n)} \text{ and } q_1^{(n)} > q_2^{(n)} > q_3^{(n)}, \qquad (E.16)$$

if maintained, would compel $p_2^{(n)} \to 1$ and $q_2^{(n)} \to 1/2$ while the possibilities

$$p_1^{(n)} > p_2^{(n)} > p_3^{(n)} \text{ and } \min(q_2^{(n)}, q_1^{(n)}) > q_3^{(n)} \qquad (E.17)$$

for some n reduce directly to case (1) or (2) by proper relabeling if necessary.

158

APPENDIX E

We next examine the possibility of

$$p_1^{(n)} > p_2^{(n)} > p_3^{(n)} \text{ and } q_3^{(n)} > q_2^{(n)} > q_1^{(n)}. \tag{E.18}$$

If this persists for all n, we can argue that

$$\lim_{n\to\infty} p_1^{(n)} = \lim_{n\to\infty} p_2^{(n)} = \lim_{n\to\infty} p_3^{(n)} = \frac{1}{3}$$

and

$$\lim_{n\to\infty} q_3^{(n)} = \lim_{n\to\infty} q_2^{(n)} = \lim_{n\to\infty} q_1^{(n)} = \frac{1}{3}. \tag{E.19}$$

But the central equilibrium representing equally alleles A_1, A_2, and A_3 is totally repelling (see Section 4.4) and therefore the convergence (E.19) cannot occur. Therefore the possibilities reduce to those of cases (1) or (2) modulo a relabeling of alleles.

Case (4). The general case

$$p_1^{(0)} > p_2^{(0)} > \cdots > p_r^{(0)}$$

and

$$q_{k_1}^{(0)} > q_{k_2}^{(0)} > \cdots > q_{k_r}^{(0)}$$

where $\{k_i\}$ is a permutation of $\{1, \ldots, r\}$, is analyzed by the same methods involving more arduous details.

For example, in the extreme case, if

$$p_1^{(n)} > \cdots > p_r^{(n)}$$

and

$$q_r^{(n)} > q_{r-1}^{(n)} > \cdots > q_1^{(n)} \tag{E.20}$$

holds for all $n \geq 0$ we easily deduce that

$$\lim_{n\to\infty} p_i^{(n)} = \frac{1}{r}, i = 1,2,\ldots,r$$

and

$$\lim_{n\to\infty} q_i^{(n)} = \frac{1}{r}, i = 1,2,\ldots,r. \tag{E.21}$$

159

But the central symmetric equilibrium point is totally repelling so that the outcome (E.21) is precluded. Thus the reverse ordering (E.20) *cannot* be maintained and a switch of order must occur.

We ultimately secure either convergence to

$$\mathscr{G}_F = \{A_k A_k\} \text{ for some } k,$$

while

$$\mathscr{G}_M = \{A_k A_l, l = 1, \ldots k - 1, k + 1, \ldots r\},$$

or for some pair $i \neq j$,

$$\mathscr{G}_F = \{A_i A_i, A_j A_j\} \text{ and } \mathscr{G}_M = \{A_i A_j\}$$

with frequency $(\mathscr{G}_F) = $ frequency $(\mathscr{G}_M) = 1/2$ in both cases.

CHAPTER FIVE

Some Multilocus Sex Determination Systems

A theoretical model of two-locus sex determination that suggests a means for passing from male heterogamety to female heterogamety, or conversely, starts with the classification

$$\overset{\text{\Large ♀}}{\{XX,Mm\}} \quad \overset{\text{\Large ♂}}{\{XY,mm\}}$$

with the double homozygote $\{XX,mm\}$ being male or female and the double heterozygote $\{XY,Mm\}$ lethal. When the first locus dominates sex expression, male heterogamety is in force, whereas if the second locus is determining, then female heterogamety is clearly operative.

Different forms of heteromorphism may develop different systems through multilocus intermediaries. Moreover, several modifier genes can act to switch the place of heterogamety (see Bull and Charnov, 1977; Bull, 1983, p. 240). In this perspective, multilocus arrays may be regarded as early stages of sex chromosomes that essentially originated from autosomes, which gradually accumulated differences in tight linkage.

Examples of multilocus sex determination systems are relatively sparse, partly due to the inherent difficulties in ascertaining the basis of sex expression when sex chromosomes do not exist or are subject to strong autosomal modification, as seems to occur in many invertebrates. Some cases in Diptera and houseflies suggest a sex determination mechanism controlled at two loci under the dichotomization

$$\overset{\text{\Large ♀}}{\{XX,bb\}} \quad \overset{\text{\Large ♂}}{\{Aa,bb\}} \quad \text{or} \quad \{aa,Bb\}$$

This scheme can obviously be generalized to allow any number of loci, females being homozygous at every locus and males being heterozygous at least at one of the loci. Bull (1983) reviews other cases of two-, three-, and multilocus sex-determining mechanisms that have been observed in natural populations.

Pamilo (1982) implemented a series of numerical simulations for sex ratio realizations in eusocial haplodiploid populations incoporating multilocus genetic systems and behavioral influences. Convergence to a 1:1 sex ratio occurs in all the cases with total queen control and those with caste-specific loci. In the former case, there is maintenance of gametic polymorphism while in the latter a balance between all female and all male progenies is established. In general, partial or total worker control leads to a female-biased sex ratio with an increase in the variance of progeny sex ratio.

These observations for multilocus sex ratio determination models may depend upon the specific assumptions used for the simulations (e.g., additivity, dominance/recessivity, free recombination).

In this chapter, we focus on cases where a non-even sex ratio can be maintained by recombination in multilocus sex determination systems. This occurs for models based on the state of heterozygosity at two or several loci under sufficient recombination. Such models contrast with additive multilocus effect models whose dynamical properties are generally independent of recombination.

The models of this chapter, where the sex phenotype is determined from the genotypic composition of several genes (loci), are analyzed by using the general theory of sex-differentiated selection effects and allowing for general recombination schemes. It is interesting that for two or more unlinked genes controlling *exact* sex determination, the stable population sex ratios are generally *not* 1:1. This contrasts sharply with the results of Chapters 2–4 attesting to the tendency for an even sex-ratio equilibrium when sex expression is controlled at a single multiallelic locus. In view of the predominant even sex-

162

ratio manifestation for higher organisms, it appears doubtful that sex ratio determination entails loosely linked multiple loci. However, for lower organisms, which often exhibit biased sex ratios, the involvement of several loci contributing to sex determination may be a feasible mechanism.

5.1. TWO-LOCUS SEX DICHOTOMIZATION WITH MALE DOUBLE HETEROZYGOTES

We start with two specific examples introduced in Scudo (1964) of dichotomous sex determination based on a partition of genotypes from two loci into two (sex) phenotypes. The sex ratio outcomes for this mechanism contrast with the general results of Chapters 2–4.

Model I. Consider the case of asymmetric sex determination where the coupling double heterozygote behaves as one sex phenotype (labeled the male), whereas all other genotypes perform as the second sex phenotype. Specifically, we consider the sex phenotypic partition

$$\mathcal{G}_M \qquad\qquad\qquad \mathcal{G}_F$$

$$\left\{\frac{AB}{ab}\right\} \left\{\frac{AB}{AB} \quad \frac{AB}{Ab} \quad \frac{AB}{aB} \quad \frac{Ab}{Ab} \quad \frac{Ab}{aB} \quad \frac{Ab}{ab} \quad \frac{aB}{aB} \quad \frac{aB}{ab} \quad \frac{ab}{ab}\right\}$$

with corresponding frequencies

$$x, u_1, u_2, u_3, u_4, u_5, u_6, u_7, u_8, u_9.$$

Let r be the recombination rate between the two contributing loci. Under random mating the genotypic frequency recursion equations reduce to the form (omitting the equation for x' since $x + u_1 + u_2 + \cdots + u_9 = 1$)

$$Tu_1' = \frac{1-r}{2} u_1 + \frac{1-r}{4} u_2 + \frac{1-r}{4} u_3 + \frac{r(1-r)}{4} u_5$$

$$Tu_2' = \frac{r}{2}u_1 + \frac{1}{4}u_2 + \frac{r}{4}u_3 + \frac{1-r}{2}u_4$$

$$+ \left[\frac{(1-r)^2 + r^2}{4}\right]u_5 + \frac{1-r}{4}u_6$$

$$Tu_3' = \frac{r}{2}u_1 + \frac{r}{4}u_2 + \frac{1}{4}u_3 + \left[\frac{(1-r)^2 + r^2}{4}\right]u_5$$

$$+ \frac{1-r}{2}u_7 + \frac{1-r}{4}u_8$$

$$Tu_4' = \frac{r}{4}u_2 + \frac{r}{2}u_4 + \frac{r(1-r)}{4}u_5 + \frac{r}{4}u_6$$

$$Tu_5' = \frac{r}{4}u_2 + \frac{r}{4}u_3 + \frac{r}{2}u_4 + \frac{r(1-r)}{2}u_5 + \frac{r}{4}u_6 + \frac{r}{2}u_7 + \frac{r}{4}u_8$$

$$Tu_6' = \frac{1-r}{4}u_2 + \frac{1-r}{2}u_4 + \left[\frac{(1-r)^2 + r^2}{4}\right]u_5 + \frac{1}{4}u_6$$

$$+ \frac{r}{4}u_8 + \frac{r}{2}u_9$$

$$Tu_7' = \frac{r}{4}u_3 + \frac{r(1-r)}{4}u_5 + \frac{r}{2}u_7 + \frac{r}{4}u_8$$

$$Tu_8' = \frac{1-r}{4}u_3 + \left[\frac{(1-r)^2 + r^2}{4}\right]u_5 + \frac{r}{4}u_6 + \frac{1-r}{2}u_7$$

$$+ \frac{1}{4}u_8 + \frac{r}{2}u_9$$

$$Tu_9' = \frac{r(1-r)}{4}u_5 + \frac{1-r}{4}u_6 + \frac{1-r}{4}u_8 + \frac{1-r}{2}u_9 \qquad (5.1)$$

where T is the constant $1 - x$. After we convert to frequency vectors by multiplying with a constant, the transformation equations (5.1) can be compactly written in the form

$$\mathbf{v}' = \frac{A\mathbf{v}}{\langle \mathbf{e}, A\mathbf{v} \rangle} \qquad (5.2)$$

164

where A is the coefficient matrix on the right in (5.1), $\mathbf{v} = (u_1, \ldots, u_9)/\sum_{i=1}^{9} u_i$ and $\mathbf{e} = (1, \ldots, 1)$ is a vector of all unit components. Iteration produces the nth generation frequency vector $\mathbf{v}^{(n)} = A^n \mathbf{v}^{(0)}/\langle \mathbf{e}, A^n \mathbf{v}^{(0)} \rangle$, where $\mathbf{v}^{(0)}$ is the frequency vector for the initial generation. Since A is an irreducible aperiodic non-negative matrix (i.e., some power of A is a positive matrix) the spectral radius of A coincides with the principal eigenvalue λ_1 of A with corresponding positive right eigenvector $\boldsymbol{\varphi}$. It follows that

$$\mathbf{v}^{(n)} = \frac{A^n \mathbf{v}^{(0)}}{\langle \mathbf{e}, A^n \mathbf{v}^0 \rangle} \to \frac{\boldsymbol{\varphi}}{\langle \mathbf{e}, \boldsymbol{\varphi} \rangle} = \mathbf{v}^{(\infty)} \text{ as } n \to \infty,$$

establishing global convergence. We next identify λ_1 and $\boldsymbol{\varphi} = (\varphi_1, \ldots, \varphi_9)$. In cognizance of the symmetry of the model and the uniqueness of $\boldsymbol{\varphi}$ we must have a principal eigenvector of the form

$$\varphi_1 = \varphi_9 \ (=y_1), \varphi_2 = \varphi_3 = \varphi_6 = \varphi_8 \ (=y_2)$$
$$\varphi_4 = \varphi_7 \ (=y_3), \varphi_5 (=y_4).$$

The matrix A acting on (y_1, y_2, y_3, y_4) coalesces to

$$\frac{1}{2} \begin{bmatrix} 1-r & 1-r & 0 & \dfrac{r(1-r)}{2} \\[2ex] r & 1 & 1-r & \dfrac{(1-r)^2 + r^2}{2} \\[2ex] 0 & r & r & \dfrac{r(1-r)}{2} \\[2ex] 0 & 2r & 2r & r(1-r) \end{bmatrix}.$$

It has two zero eigenvalues and the eigenvalues

$$\lambda_{\pm} = \tfrac{1}{4}(2 + r - r^2 \pm \sqrt{r^4 + 2r^3 - 3r^2 + 4r})$$

so that

$$\lambda_1 = \lambda_+ = \tfrac{1}{4}(2 + r - r^2 + \sqrt{r^4 + 2r^3 - 3r^2 + 4r}).$$

For $r = 0$, $\lambda_1 = 1/2$ and for $r = 1/2$, $\lambda_1 = 7/8$.

165

Let $\boldsymbol{\varphi}(r)$ be the eigenvector corresponding to $\lambda_1(r)$ (displaying its dependence on r) so that

$$A\boldsymbol{\varphi}(r) = \lambda_1(r)\boldsymbol{\varphi}(r).$$

Comparing to (5.1) at equilibrium we see that $1 - x^{(\infty)} = \lambda_1(r)$. Thus the globally stable equilibrium for $r = 0$ has $x^{(\infty)} = 1/2$ (an even sex-ratio equilibrium), while for all r, $0 < r < 1/2$,

$$1 - x^{(\infty)}(r) = \frac{y(r)}{2}$$

where $y(r)$ is the maximal positive root of the quadratic equation

$$\xi^2 - (2 + r - r^2)\xi + (1 - r^3) = 0.$$

When $r = 1/2$, we find the following limit values for the original variables

$$x^{(\infty)}(\tfrac{1}{2}) = \tfrac{1}{8} \text{ and } \mathbf{u}^{(\infty)}(\tfrac{1}{2}) = (\tfrac{1}{16}, \tfrac{1}{8}, \tfrac{1}{8}, \tfrac{1}{16}, \tfrac{1}{8}, \tfrac{1}{8}, \tfrac{1}{16}, \tfrac{1}{8}, \tfrac{1}{16}). \quad (5.3)$$

For $0 < r < \tfrac{1}{2}$ we easily deduce $dy(r)/dr > 0$ i.e., $dx^{(\infty)}(r)/dr < 0$. Thus, with positive recombination the realized sex ratio is globally stable and *not* $1:1$. This model of exact sex determination controlled at two-loci with positive recombination contrasts sharply with the case of one-locus multiallele dichotomous sex determination (Theorem 3.5) where only an even sex-ratio equilibrium is realizable.

Model II. In this model the sex expression depends on the level of heterozygosity corresponding to the phenotypic classes

$$\overbrace{\left\{\begin{matrix} AB & Ab \\ \hline ab & aB \end{matrix}\right\}}^{\mathscr{G}_M} \overbrace{\left\{\begin{matrix} AB & AB & AB & Ab & Ab & aB & aB & ab \\ \hline AB & Ab & aB & Ab & ab & aB & ab & ab \end{matrix}\right\}}^{\mathscr{G}_F}$$

Let the associated frequencies be

$$x, y, u_1, u_2, u_3, u_4, u_5, u_6, u_7, u_8.$$

Set $T = (x + y)(1 - x - y)$. The successive genotypic frequency recursion equations under random mating and Mendelian

166

segregation are

$$Tu'_1 = \frac{(1-r)x + ry}{2}\left[u_1 + \frac{u_2}{2} + \frac{u_3}{2}\right]$$

$$Tu'_8 = \frac{(1-r)x + ry}{2}\left[u_8 + \frac{u_7}{2} + \frac{u_5}{2}\right]$$

$$Tu'_2 = \frac{(1-r)x + ry}{2}\left[\frac{u_2}{2} + u_4 + \frac{u_5}{2}\right]$$
$$+ \frac{rx + (1-r)y}{2}\left[u_1 + \frac{u_2}{2} + \frac{u_3}{2}\right]$$

$$Tu'_3 = \frac{(1-r)x + ry}{2}\left[\frac{u_3}{2} + u_6 + \frac{u_7}{2}\right]$$
$$+ \frac{rx + (1-r)y}{2}\left[u_1 + \frac{u_2}{2} + \frac{u_3}{2}\right]$$

$$Tu'_5 = \frac{(1-r)x + ry}{2}\left[\frac{u_2}{2} + u_4 + \frac{u_5}{2}\right]$$
$$+ \frac{rx + (1-r)y}{2}\left[\frac{u_5}{2} + \frac{u_7}{2} + u_8\right]$$

$$Tu'_7 = \frac{(1-r)x + ry}{2}\left[\frac{u_3}{2} + u_6 + \frac{u_7}{2}\right]$$
$$+ \frac{rx + (1-r)y}{2}\left[\frac{u_5}{2} + \frac{u_7}{2} + u_8\right]$$

$$Tu'_4 = \frac{rx + (1-r)y}{2}\left[\frac{u_2}{2} + u_4 + \frac{u_5}{2}\right]$$

$$Tu'_6 = \frac{rx + (1-r)y}{2}\left[\frac{u_3}{2} + u_6 + \frac{u_7}{2}\right]$$

$$Tx' = \frac{(1-r)x + ry}{2}\left[u_1 + u_8 + \frac{u_2}{2} + \frac{u_3}{2} + \frac{u_5}{2} + \frac{u_7}{2}\right]$$

$$Ty' = \frac{rx + (1-r)y}{2}\left[u_4 + u_6 + \frac{u_2}{2} + \frac{u_3}{2} + \frac{u_5}{2} + \frac{u_7}{2}\right]. \quad (5.4)$$

By introducing the new variables (exploiting the invariance under allelic substitution, cf. Karlin and Avni, 1981)

$$y_1 = u_1 + u_8, y_2 = u_2 + u_3 + u_5 + u_7, y_3 = u_4 + u_6$$
$$y_4 = (1 - r)x + ry, y_5 = rx + (1 - r)y \tag{5.5}$$

the recursion system (5.4) reduces to

$$Ty_1' = \frac{1}{2}y_4\left[y_1 + \frac{y_2}{2}\right]$$

$$Ty_2' = y_4\left[y_3 + \frac{y_2}{2}\right] + y_5\left[y_1 + \frac{y_2}{2}\right]$$

$$Ty_3' = \frac{1}{2}y_5\left[y_3 + \frac{y_2}{2}\right]$$

$$Ty_4' = \frac{1 - r}{2}y_4\left[y_1 + \frac{y_2}{2}\right] + \frac{r}{2}y_5\left[y_3 + \frac{y_2}{2}\right]$$

$$Ty_5' = \frac{1 - r}{2}y_5\left[y_3 + \frac{y_2}{2}\right] + \frac{r}{2}y_4\left[y_1 + \frac{y_2}{2}\right]. \tag{5.6}$$

After some manipulations in terms of the variables

$$u = \frac{y_1 + \dfrac{y_2}{2}}{y_3 + \dfrac{y_2}{2}}, \quad v = \frac{y_4}{y_5}$$

at a polymorphic equilibrium we deduce the relation

$$v = u^2 \tag{5.7}$$

where u satisfies the equation

$$ru^5 - (1 - r)u^3 + (1 - r)u^2 - r = 0. \tag{5.8}$$

This equation allows at most three positive solutions. The solution $u = 1 = v$ by further examination of (5.6) unravels the evaluations $y_4 = y_5 = 1/8$ and $y_1 = y_3 = 1/8$, $y_2 = 1/2$, $T = 3/16$. To be consistent, the original frequencies must be

$$x^* = y^* = \tfrac{1}{8}, u_1^* = u_8^* = u_4^* = u_6^* = \tfrac{1}{16},$$
$$u_2^* = u_3^* = u_5^* = u_7^* = \tfrac{1}{8}, \tag{5.9}$$

which we refer to as the *central equilibrium*.

A stability analysis of the equilibrium (5.9) reveals that the central equilibrium is stable only for sufficiently loose linkage, precisely for

$$r > \tfrac{1}{6}. \tag{5.10}$$

The *equilibrium sex ratio* at this point is

$$x^* + y^* = \tfrac{1}{4}. \tag{5.11}$$

Numerical studies suggest that this equilibrium is globally stable for $r > 1/6$.

Again, for this two-locus sex determination model a biased sex ratio remains feasible even when $0 < r < 1/6$, while the central equilibrium is no longer stable. Setting $\delta = (1 - 2r)/r$, more extensive algebra reveals the equilibria

$$\tilde{u} = \tfrac{1}{4}(\sqrt{9 + 4\delta} - 1) + \tfrac{1}{2}\sqrt{\delta - \tfrac{1}{3}(3 + \sqrt{9 + 4\delta})}$$
$$\tilde{\tilde{u}} = \tfrac{1}{4}(\sqrt{9 + 4\delta} - 1) - \tfrac{1}{2}\sqrt{\delta - \tfrac{1}{3}(3 + \sqrt{9 + 4\delta})}$$

with corresponding $\tilde{v} = \tilde{u}^2$ and $\tilde{\tilde{v}} = \tilde{\tilde{u}}^2$. These generate proper equilibria provided $r < 1/6$, i.e., the corresponding y_i and u_i are frequencies. Explicitly, we obtain for $r < 1/6$

$$y_5 = \frac{1}{\left[1 + v + (u + 1)\left\{ \dfrac{1 + u + v}{1 - r + ru^3} \right\} \right]} \text{ and } y_4 = vy_5$$

with a proportion of males $y_4 + y_5 < 1/2$.

5.2. A TWO-LOCUS MULTIPLICATIVE SEX DETERMINATION MODEL

We consider the following general two-locus multiplicative sex determination coefficient matrix

$$M = \begin{array}{c} \begin{array}{cccc} AB & Ab & aB & ab \end{array} \\ \begin{bmatrix} \alpha\beta & \alpha\gamma & \beta & \gamma \\ \alpha\gamma & \alpha\beta & \gamma & \beta \\ \beta & \gamma & \alpha\beta & \alpha\gamma \\ \gamma & \beta & \alpha\gamma & \alpha\beta \end{bmatrix} \end{array} = \begin{array}{c} \begin{array}{cc} A & a \end{array} \\ \begin{bmatrix} \alpha & 1 \\ 1 & \alpha \end{bmatrix} \end{array} \otimes \begin{array}{c} \begin{array}{cc} B & b \end{array} \\ \begin{bmatrix} \beta & \gamma \\ \gamma & \beta \end{bmatrix} \end{array} \tag{5.12}$$

with $\alpha < 1$, $\beta < \gamma \leq 1$, where the symbol $\otimes$ indicates the Kronecker product of the two attendant second-order matrices. Note that the sex determination coefficients in (5.12) depend on the state of heterozygosity at only two specific loci.

The array (5.12) represents a multiplicative two-locus viability matrix encompassed in the formulation of generalized nonepistatic selection regimes in bisexual populations (Karlin and Liberman, 1979a, b, c).

As with one locus, the sex expression is considered determined by the zygote genotype. The population male and female gamete frequency vectors $\mathbf{x} = (x_1, x_2, x_3, x_4)$ and $\mathbf{y} = (y_1, y_2, y_3, y_4)$, respectively, change under random mating as for a *two-locus* trait with male and female viability matrices M and $F = U - M$ (cf. (2.4)) where U is the fourth-order matrix of all unit entries. The recurrence equations for the general two-sex multilocus viability systems are formalized in (5.26) below. For the case at hand, the central equilibrium of gamete frequencies $\mathbf{x}^* = \mathbf{y}^* = (1/4, 1/4, 1/4, 1/4) = \mathbf{c}^*$ exists for any recombination rate. Its stability conditions are set forth in Karlin and Liberman (1979a). Note that M of (5.12) and $U - M = F$ have the generalized nonepistatic structure based on the marginal viability matrices

$$\begin{bmatrix} \alpha & 1 \\ 1 & \alpha \end{bmatrix} \text{ at locus 1}$$

$$\begin{bmatrix} \beta & \gamma \\ \gamma & \beta \end{bmatrix} \text{ at locus 2.}$$

According to Karlin and Liberman (1979b, c), the stability conditions of the central equilibrium are equivalent to that of a *one-sex* generalized nonepistatic selection regime with viability matrix

$$w_f^* M + w_m^* (U - M) = (w_f^* - w_m^*) M + w_m^* U, \quad (5.13)$$

where w^* is the mean fitness at $\mathbf{c}^*$ for the matrix $M = \|m_{ij}\|$, and w_f^* is the corresponding quantity for the viability matrix

$F = \|f_{ij}\| = U - M$. In the case at hand

$$w_m^* = \left(\frac{1+\alpha}{2}\right)\left(\frac{\gamma+\beta}{2}\right)$$

$$w_f^* = 1 - \left(\frac{1+\alpha}{2}\right)\left(\frac{\gamma+\beta}{2}\right)$$

(5.14)

and therefore

$$w_f^* - w_m^* = 1 - \tfrac{1}{2}(1+\alpha)(\gamma+\beta).$$

Note that the sex ratio at $\mathbf{c}^*$ provided by w_m^* and w_f^* is generally *not* one-to-one.

With reference to Karlin and Liberman (1979a), we find that $\mathbf{c}^*$ is stable subject to the existence of sufficient recombination between loci 1 and 2, as follows.

Case 1. $2 > (1+\alpha)(\gamma+\beta)$. $\mathbf{c}^*$ is stable provided

$$r > r_0 = \frac{\left[1 - \dfrac{1}{2}(1+\alpha)(\gamma+\beta)\right]\dfrac{(\alpha-1)(\beta-\gamma)}{4}}{\left[1 - \dfrac{1}{2}(1+\alpha)(\gamma+\beta)\right]\gamma + \dfrac{(1+\alpha)(\gamma+\beta)}{4}}.$$

(5.15a)

It is easy to check that $r_0 < 1/2$. For $\alpha = \beta = 0$, $\gamma = 1$, we recover Model II of Section 5.1 and $r_0 = 1/6$ as indicated in (5.10).

Case 2. $2 < (1+\alpha)(\gamma+\beta)$. The stability condition becomes

$$r > r_0 = \max \left\{ \frac{\left[1 - \dfrac{1}{2}(1+\alpha)(\gamma+\beta)\right]\dfrac{(\alpha-1)(\beta+\gamma)}{4}}{D}, \right.$$

$$\left. \frac{\left[1 - \dfrac{1}{2}(1+\alpha)(\gamma+\beta)\right]\dfrac{(1+\alpha)(\beta-\gamma)}{4}}{D} \right.$$

(5.15b)

where D is the same denominator as in (5.15a).

5.3. SEX DETERMINATION AT SEVERAL LOCI: A GENERAL SYMMETRIC HETEROZYGOSITY REGIME

In this section, we consider sex determination involving n-loci in the form of a generalized symmetric heterozygosity selection regime such that the probability of a genotype being male depends only on the degree of heterozygosity at the constituent loci.

Let $A_1^{(k)}, A_2^{(k)}, \ldots, A_{m_k}^{(k)}$ designate the possible alleles at the kth locus $(k = 1, 2, \ldots, n)$. The associated *gamete* types are described by n-tuples,

$$\mathbf{i}_0 = (i_0^{(1)}, i_0^{(2)}, \ldots, i_0^{(n)}),$$

where $i_0^{(k)}$ is one of $A_1^{(k)}, \ldots, A_{m_k}^{(k)}$. A typical *genotype* composed of two gametes is displayed in the form

$$\begin{pmatrix} \mathbf{i}_0 \\ \mathbf{i}_1 \end{pmatrix} = \begin{pmatrix} i_0^{(1)}, i_0^{(2)}, \ldots, i_0^{(n)} \\ i_1^{(1)}, i_1^{(2)}, \ldots, i_1^{(n)} \end{pmatrix} \tag{5.16}$$

signifying that the allelic composition at locus k consists of alleles $i_0^{(k)}$ and $i_1^{(k)}$. The probability of this genotype being male is

$$w\begin{pmatrix} \mathbf{i}_0 \\ \mathbf{i}_1 \end{pmatrix} = w\begin{pmatrix} i_0^{(1)}, \ldots, i_0^{(n)} \\ i_1^{(1)}, \ldots, i_1^{(n)} \end{pmatrix} = w(\mathbf{i}_0, \mathbf{i}_1) \tag{5.17}$$

and that of being female is $1 - w(\mathbf{i}_0, \mathbf{i}_1)$. The array of these values over all gamete pairs generates a sex determination matrix M of order $\mathcal{N} \times \mathcal{N}$, where $\mathcal{N} = \prod_{k=1}^{n} m_k$.

The *recombination-segregation frequencies* are summarized by the array of non-negative quantities

$$R(\boldsymbol{\varepsilon}) = R(\varepsilon_1, \varepsilon_2, \ldots, \varepsilon_n), \tag{5.18}$$

where the n-tuples $\boldsymbol{\varepsilon} = (\varepsilon_1, \varepsilon_2, \ldots, \varepsilon_n)$ satisfy $\varepsilon_i = 0$ or 1, $i = 1, 2, \ldots, n$. These numbers are to be interpreted as follows. The recombination event associated with $\boldsymbol{\varepsilon} = (\varepsilon_1, \varepsilon_2, \ldots, \varepsilon_n)$ means that at the positions (loci) where $\varepsilon_\nu = 1$, an interchange of genetic material occurs and subsequent Mendelian segregation produces the recombinant gametes

$$(i^{(1)}_{\varepsilon_1}, i^{(2)}_{\varepsilon_2}, \ldots, i^{(n)}_{\varepsilon_n}) \text{ and } (i^{(1)}_{1-\varepsilon_1}, i^{(2)}_{1-\varepsilon_2}, \ldots, i^{(n)}_{1-\varepsilon_n}).$$

Thus, the gamete output from segregation comprises equally likely the gametes $(i^{(1)}_{\varepsilon_1}, i^{(2)}_{\varepsilon_2}, \ldots, i^{(n)}_{\varepsilon_n})$ or $(i^{(1)}_{1-\varepsilon_1}, i^{(2)}_{1-\varepsilon_2}, \ldots, i^{(n)}_{1-\varepsilon_n})$ with frequency $R(\varepsilon) = R(\varepsilon_1, \varepsilon_2, \ldots, \varepsilon_n)$. The recombination frequencies obey the relations $R(\varepsilon) = R(\varepsilon_1, \varepsilon_2, \ldots, \varepsilon_n) = R(1 - \varepsilon_1, 1 - \varepsilon_2, \ldots, 1 - \varepsilon_n) = R(1 - \varepsilon)$, expressing the fact that the two parental gametes contribute in a symmetrical manner to the segregation process; also $\sum_{\varepsilon} R(\varepsilon) = 1$ holds where the sum extends over all $\varepsilon = (\varepsilon_1, \varepsilon_2, \ldots, \varepsilon_n)$, $\varepsilon_i = 0$ or 1, $i = 1, 2, \ldots, n$. We indicate some examples.

Absolute Linkage (No Recombination) $\mathscr{R}^{(0)}$:

$$R(\mathbf{0}) = R(\mathbf{1}) = \tfrac{1}{2}, R(\varepsilon) = 0 \text{ for } \varepsilon \neq \mathbf{0} \text{ or } \mathbf{1}, \qquad (5.19)$$

where $\mathbf{0} = (0, 0, \ldots, 0)$ reflects no exchange of genetic material and $\mathbf{1} = (1, 1, \ldots, 1)$ signifies total exchange which is effectively equivalent to no exchange.

Free Recombination $\mathscr{R}^{(f)}$: Here

$$R(\varepsilon) = \frac{1}{2^n} \text{ independent of } \varepsilon. \qquad (5.20)$$

When all loci segregate independently, then (5.20) applies.

Recombination Arrays Reflecting Specific Characteristics of Loci: Suppose that the events of breaks between successive positions are independent (the no-interference postulate). Let r_i be the probability of a crossover event between loci i and $i + 1$. Then

$$R(\varepsilon) = \frac{1}{2} \prod_{i=1}^{n-1} r_i^{|\varepsilon_i - \varepsilon_{i+1}|} (1 - r_i)^{1 - |\varepsilon_i - \varepsilon_{i+1}|}. \qquad (5.21)$$

It is possible to generate a hierarchy of recombination distributions that take account of the natural physical ordering

173

of the loci through notions of renewal processes, order statistics, countermechanisms, and count-location chiasma formation processes; consult Karlin and Liberman (1979d) and Karlin (1984) for their descriptions, recent developments, and references.

Although sex-dependent recombination distributions are not uncommon, this assumption is not considered for the sex determination model at hand.

The *general symmetric heterozygosity regime* (SH-model) involving n loci entails that *the male sex determination probabilities of the genotypes depend only on the number and positions of the heterozygous loci.* Accordingly, the value of (5.17) may be parameterized by the array

$$\gamma(\delta_1, \ldots, \delta_n) = \gamma(\boldsymbol{\delta}), \tag{5.22}$$

where $\boldsymbol{\delta} = (\delta_1, \delta_2, \ldots, \delta_n)$ is an n-tuple of 0 or 1 components indicating the homozygous and heterozygous loci, respectively, such that

$$w\left(\frac{i_0^{(1)}, \ldots, i_0^{(n)}}{i_1^{(1)}, \ldots, i_1^{(n)}}\right) = \gamma(\boldsymbol{\delta}), \tag{5.23}$$

such that $i_0^{(k)}$ and $i_1^{(k)}$ represent different alleles exactly for those loci k where $\delta_k = 1$. Thus, the maleness probability value of a genotype is a function only of the gene positions where the genotype is heterozygous but not of the constituent alleles appearing at these loci.

The particular case where the values depend only on the *number* of heterozygous loci (and *not* their positions) is termed *the complete symmetric aggregate heterozygosity model.* In this latter situation there are $n + 1$ quantities γ_k, $k = 0, 1, \ldots, n$, such that

$$\gamma_k = \textit{the probability of a genotype being male}$$
$$\textit{if } k \textit{ among its loci are heterozygous.} \tag{5.24}$$

We know that genotypic sex ratio evolution models behave as special cases of two-sex viability models (see Chapter 2).

Let

$$w_m\begin{pmatrix}\mathbf{i}_0\\\mathbf{i}_1\end{pmatrix} = w_m\begin{pmatrix}i_0^{(1)},i_0^{(2)},\ldots,i_0^{(n)}\\i_1^{(1)},i_1^{(2)},\ldots,i_1^{(n)}\end{pmatrix},$$

$$w_f\begin{pmatrix}\mathbf{i}_0\\\mathbf{i}_1\end{pmatrix} = w_f\begin{pmatrix}i_0^{(1)},i_0^{(2)},\ldots,i_0^{(n)}\\i_1^{(1)},i_1^{(2)},\ldots,i_1^{(n)}\end{pmatrix} = 1 - w_m\begin{pmatrix}\mathbf{i}_0\\\mathbf{i}_1\end{pmatrix} \quad (5.25)$$

denote the male and female sex determination coefficients of the genotype (5.16). Let $x(\mathbf{i}_0)$ and $y(\mathbf{i}_0)$ be the frequencies of gamete $\mathbf{i}_0$ in males and females, respectively, at the present generation and $x'(\mathbf{i}_0)$ and $y'(\mathbf{i}_0)$ the corresponding frequencies at the next generation. The transformation equations connecting the gamete frequencies over successive generations are described as follows (cf. Karlin and Liberman, 1979b, c). *Under random mating with the effects of selection and recombination in force, the gametic frequency states $\mathbf{x}'$ and $\mathbf{y}'$ at the next generation in terms of the gametic frequency states $\mathbf{x}$ and $\mathbf{y}$ at the present generation are given by*

$$w_m(\mathbf{x},\mathbf{y})x'(\mathbf{i}_0) = \sum_{\mathbf{i}_1}\sum_{\varepsilon} R(\varepsilon)w_m\begin{pmatrix}\mathbf{i}_\varepsilon\\\mathbf{i}_{1-\varepsilon}\end{pmatrix}x(\mathbf{i}_\varepsilon)y(\mathbf{i}_{1-\varepsilon}),\text{ and}$$

$$(5.26)$$

$$w_f(\mathbf{x},\mathbf{y})y'(\mathbf{i}_0) = \sum_{\mathbf{i}_1}\sum_{\varepsilon} R(\varepsilon)w_f\begin{pmatrix}\mathbf{i}_\varepsilon\\\mathbf{i}_{1-\varepsilon}\end{pmatrix}x(\mathbf{i}_\varepsilon)y(\mathbf{i}_{1-\varepsilon}),$$

where $w_m(\mathbf{x},\mathbf{y})$ and $w_f(\mathbf{x},\mathbf{y})$ calculate the male and female sex ratio, respectively, at the population state $(\mathbf{x},\mathbf{y})$ given as

$$w_m(\mathbf{x},\mathbf{y}) = \sum_{\mathbf{i}_0,\mathbf{i}_1} w_m\begin{pmatrix}\mathbf{i}_0\\\mathbf{i}_1\end{pmatrix}x(\mathbf{i}_0)y(\mathbf{i}_1),\text{ and}$$

$$(5.27)$$

$$w_f(\mathbf{x},\mathbf{y}) = \sum_{\mathbf{i}_0,\mathbf{i}_1} w_f\begin{pmatrix}\mathbf{i}_0\\\mathbf{i}_1\end{pmatrix}x(\mathbf{i}_0)y(\mathbf{i}_1),$$

respectively.

Suppose $w_m(\mathbf{i}_0,\mathbf{i}_1)$ is of the form (5.23). Then the central *population state* $\mathbf{c}^*$ exhibiting equal frequencies for the gamete types is an equilibrium for any recombination-segregation distribution. Moreover, the sex ratio at $\mathbf{c}^*$ is $w^* = (1/2^n)\sum_{\boldsymbol{\delta}}\gamma(\boldsymbol{\delta})$.

The stability conditions of $\mathbf{c}^*$ are available, invoking the results of Karlin and Liberman (1979c) and Karlin and Avni (1981). To avoid technical formulas we describe the results with two alleles per locus.

THEOREM 5.1. *Let $\gamma(\boldsymbol{\delta})$ be the probability for a genotype (5.16) having heterozygous loci at the unit components of $\boldsymbol{\delta}$ (see (5.22)) to be male. The central equilibrium $\mathbf{c}^*$ is stable if and only if the inequalities*

$$\left| \frac{1}{2^n} \frac{(1 - 2w^*)}{w^*(1 - w^*)} \sum_{\boldsymbol{\delta}} \gamma(\boldsymbol{\delta}) \sum_{\boldsymbol{\varepsilon}} R(\boldsymbol{\varepsilon}) \prod_{i=1}^{n} (-1)^{\varepsilon_i \delta_i \eta_i} \right| < 1 \quad (5.28)$$

hold for all $\boldsymbol{\eta} = (\eta_1, \ldots, \eta_n)$, $\eta_i = +1$ or 0 but $\boldsymbol{\eta} \neq (0, \ldots, 0)$ where $w^* = (1/2^n) \sum_{\boldsymbol{\delta}} \gamma(\boldsymbol{\delta})$.

COROLLARY 5.1. *For free recombination and any prescription of $\gamma(\boldsymbol{\delta})$ with $0 < w^* < 1$, the central equilibrium $\mathbf{c}^*$ is stable. The realized sex ratio at $\mathbf{c}^*$ under appropriate specifications of $\gamma(\boldsymbol{\delta})$ can attain any value between 0 and 1.*

This result contrasts sharply to Theorem 3.5 for exact sex determination at one locus where only an even sex-ratio equilibrium can be stable.

5.4. POLYMORPHISM IN SEX-DIFFERENTIATED MULTILOCUS VIABILITY SELECTION MODELS

In concluding this chapter on models of multilocus sex determination, it is relevant to describe some general principles on (gametic) polymorphisms (equilibrium states with all possible gamete types represented) of general bisexual viability systems (cf. Karlin, 1979a; Karlin and Liberman, 1979a–d; Karlin and Avni, 1981). We emphasize qualitative properties of the equilibrium patterns under conditions of tight and loose linkage, and their implications for sex ratio. Starting with some considerations on the nature of polymorphic equilibrium for monoecious multilocus selection systems, we then discuss the corresponding sex-differentiated models.

*Principle 1.** For a monoecious population associated with a multilocus trait, if there exists a stable polymorphic equilibrium under conditions of absolute linkage, then there exists a stable polymorphism for any positive recombination structure.

In line with this principle, an essential dichotomy in the nature of stable polymorphisms emerges

(i) The selection interactions are predominant, establishing the polymorphism, while the recombination mechanism exerts minor effects.

(ii) Selection forces without recombination would lead to the elimination of certain gamete types, but with some recombination the full complement of gamete types are maintained.

The characteristics of the polymorphic outcomes of (i) and (ii) markedly differ. For situation (i), usually a central globally stable polymorphism (segregating all possible gametes with about equal frequencies) is predominant for each set of recombination rates. In case (ii), a multiplicity of stable polymorphisms is possible for tight linkage. The gamete arrays divide into two groups such that the frequencies in one group are miniscule, while the gamete frequencies in the other group are more discernible. Such polymorphic states entail moderate to strong linkage disequilibrium.

A further "proposition" asserts that where a central type of polymorphism is stable at *some* recombination level, then it persists stably when "more recombination" is in force (cf. Karlin, 1979a). This property does not apply for a near-boundary equilibrium configuration, in which such polymorphisms are generally not maintained under conditions of moderate to free recombination rates.

Principle 1 can be refined as follows.

Principle 2. Consider the class of one-sex multilocus random-mating-selection viability regimes that induce a polymorphism.

* In our terminology, a theorem embodies a rigorous conclusion precluding any counterexamples. A principle (as in physics) is approximately valid and enjoys a wide scope of applications.

By superposition on such viability regimes one or several of the following effects (a) more recombination, (b) sex-differentiated viability, (c) multidemic interactions (e.g., migration and population subdivision), the dynamic system obtained usually evolves to a globally stable polymorphism.

The second principle suggests that it is easier for multilocus systems to obtain a stable polymorphism for a given recombination level with selection distributed between the two sexes, compared to a one-sex model of about the equivalent genotypic viability differences.

In applications to sex determination we would expect that the polymorphisms of type (ii) tend to show 1 : 1 sex ratios, while the central equilibrium of type (i) would present a biased sex ratio. This is confirmed in the class of models described in Theorems 5.2 and 5.3.

The contrast in the sexes is often reflected in recombination rates in which "more recombination" (diminished linkage) tends to occur more frequently for the homogametic sex than for the heterogametic sex. There are documented cases in which recombination in females is unrestricted but male recombination entails absolute linkage or limited numbers of crossover events. A classic extreme example pertains to Drosophila species in which recombination in males is rare. Genetic and cytological studies over past decades have documented significant numbers of animals with achiasmatic meiosis in one sex, usually the heterogametic sex. These include cases of protozoa, mites, copepods, mantids, grasshoppers, an assortment of Lepidoptera including *Bombyx*, and various other insects (see White, 1973). The same happens in many plant species, including hermaphroditic species in which meiosis occurs at different times with respect to male and female gamete production. In human populations, linkage mapping via segregation and pedigree analysis also confirm lower intrachromosomal recombination frequencies in males compared to females (McKusich and Ruddle, 1977).

The majority of plants and animals do not display any clear-cut sex-related linkage differentials. A recent review of contrasts in rates of recombination between the sexes is Callan and Perry (1977). Apparently, if there is a difference in crossing over between the sexes, it is lower in the heterogametic sex in those species with morphologically differentiated X and Y chromosomes (or an XO mechanism).

What is the evolutionary advantage of cases of low recombination in the heterogametic sex with higher rates in the homogametic sex? The problem is enigmatic. Two aspects of increased chiasmata are widely recognized. The first is the need for chiasmata to ensure proper disjunction of the two homologous chromosomes at meiosis. The second is the evolutionary flexibility attendant to recombination events. It is of interest to consider evolutionary advantages in sex-dependent recombination rates in terms of the nature of polymorphism for dioecious- as against monoecious-selection recombination regimes.

Multilocus theory suggests that with more recombination there is an increased tendency for the existence of polymorphism, entailing a gamete frequency configuration in near-linkage equilibrium of all orders. With tight linkage, in contrast, the nature of polymorphism (when extant) involves a partial set of gametes manifesting some strong measures of genic associations. It appears that distinct sex differences in recombination frequencies can more easily accommodate the coexistence of both these types of polymorphisms, and which is established depends on initial conditions and environmental factors. Thus, the evolutionary flexibility is certainly expanded with sexuality. Sex-dependent selection and recombination mechanisms apparently engender compensatory balances with respect to the various ranges of parameters.

It has been earlier indicated, on the basis of theoretical considerations, that sex-specific selection expression even with one-locus systems, admits greater multiplicity and variation in the nature of the stable equilibrium arrays. It is further demonstrated in our theoretical studies that sexual diversity in recombination rates engenders more scope in polymorphic expressions. This finding is consistent with the fact that the rate of recombination may be affected and/or modified by agents such as temperature, age, radiation, internal physiological conditions, general vigor, chemicals and enzymes, among others. These act differentially between the sexes to increase overall fitness and reproductive success. In this way, populations can exploit more advantageously intrinsic environmental potentialities through modifier effects and fitness plasticity.

CHAPTER SIX

Effects of Inbreeding, Population Structure, and Meiotic Drive on Sex Ratio Outcomes

Random mating and Mendelian segregation seem to be a priori conditions for 1:1 population sex ratio outcomes. A numerical parity between the sexes is then believed to be optimal, males and females having the same individual "reproductive value" (Fisher, 1930) on the basis that each sex provides half the (autosomal) genetic material to the future gene pool. On the other hand, sex-linked meiotic drive effects and population structures that impose more reproductive constraints on sibs of one sex or the other are likely to bias the sex ratio outcome.

Sex-linked modifiers of progeny sex ratio (as occur in many Drosophila species) should tend to favor their own representation over the succeeding generations leading to all male or all female populations unless other factors, e.g., viability and/or fertility selection differences, come into play (Edwards, 1961; Thomson and Feldman, 1975). Maffi and Jayakar (1981) studied a two-locus model for sex-linked meiotic drive applicable to *Aedes aegypti* populations. They suggested that recombination can be responsible for the maintenance of polymorphic equilibria.

Hamilton (1967, 1979) provided some insights pertaining to the effects of local mating constraints on the "unbeatable" sex ratio. When all females are fertilized but males have to compete locally to mate (LMC), a sex ratio bias in favor of females is predicted in order to diminish mate competition between male relatives. Colwell (1981) and Wilson and Colwell (1981) argued that group selection could explain the optimal reproductive success of the females adopting the ESS (unbeatable) sex ratio in populations subdivided into colonies with periodic dispersal characteristic of parasitic species. Sib-mating practices are also recognized to favor a female-biased sex ratio if sib-mating (partial or complete) is obligatory and all females are fertilized (cf. Maynard Smith, 1978). On the other hand, when females have to compete for local resources (LRC) and males can move freely to mate, a male-biased sex ratio is expected to occur (Clark, 1978; Bulmer and Taylor, 1980b). Charnov (1982, chap. 5) reviews various ideas and examples as to how structured populations may favor biased sex ratios.

In this chapter, we present results of global convergence for multiallele haploid, X-linked and Y-linked sex ratio distortion models incorporating partial sib-mating. (See Gregorius, 1982; and Lessard, 1984 for general two-sex haploid viability models relevant to sex ratio evolution.) A two-locus model of sex-linked meiotic drive is also studied to emphasize that an even sex-ratio equilibrium state can become unstable under the action of sex-linked modifiers.

From local stability analyses at fixation states, the optimal (ESS) sex ratio is deduced for haploid populations subdivided into mating colonies of random size (Section 6.6) and corresponding subdivided diploid populations with a constant fraction of outbreeding (Section 6.7). In this context, population (density) regulation within the local breeding groups, before dispersal but after mating, is also considered.

In this chapter, we shall make appeal to a criterion for stability at fixation states when linear approximations are not decisive, which is presented in Appendix A. In Appendices B and C we exhibit the exact recurrence equations for partial sib-mating population genetic models in haplodiploid and diplodiploid populations and indicate some consequences about classes of equilibria. The global dynamics in the case of complete selfing in diploid populations is presented in Appendix D.

6.1. RANDOM MATING HAPLOID
SEX-ALLOCATION MODELS

The beginnings of gametic sexual differentiation can be seen among Protista and Algae where the male-type and the female-type gametes are not distinguished by any morphological differences but only by differences in behavior, the male-type being more mobile.

In some organisms, including algae and fungi, two haplotypes that will be denoted α and β are produced. Random associations between these two types occur to form the next generation. In order to study the dynamics of such populations, let us assume that an A_i-gamete $(i = 1, \ldots, r)$ develops into an α-type (β-type) with probability $f_i(1 - f_i,$ respectively). Without loss of generality at the phenotypic level, we may assume $f_i \neq f_j$ for $i \neq j$. Let x_i be the frequency of A_i in the haploid phase and $2x_{ij}$ the proportion of A_iA_j in the diploid phase when $i \neq j$, and x_{ii} when $i = j$. Over two successive generations, we have the relations

$$x'_{ij} = \frac{x_ix_j[f_i(1-f_j) + (1-f_i)f_j]}{2F(1-F)}, \quad i,j = 1, \ldots, r,$$

where $F = \sum_{i=1}^{r} f_i x_i$. Then

$$x'_i = \sum_{j=1}^{r} x'_{ij} = x_i \left[\frac{f_i(1-2F) + F}{2F(1-F)} \right], \quad i = 1, \ldots, r. \quad (6.1)$$

An equilibrium of this transformation is either a fixation state (i.e., $x_i = 1$ for some i) or belongs to the equilibrium class characterized by a numerical parity between the two phenotypes segregating in the population (i.e., $F = 1/2$). We can demonstrate that this class is globally attracting in the following sense.

THEOREM 6.1. *Let* $\mathbf{x}' = (x'_1, \ldots, x'_r)$ *be the next value of the frequency vector* $\mathbf{x} = (x_1, \ldots, x_r)$ *given by the transformation (6.1) where*

183

$F = F(\mathbf{x}) = \sum_{i=1}^{r} f_i x_i$. Then $\left| F(\mathbf{x}') - 1/2 \right| \le \left| F(\mathbf{x}) - 1/2 \right|$ with equality if and only if $\mathbf{x}' = \mathbf{x}$. (For a proof, see Theorem 6.2.)

This means that $\left| F(\mathbf{x}) - 1/2 \right|$ is a strictly decreasing Lyapounov function for the transformation (6.1). Since $F(\mathbf{x})$ is a linear function and $\mathbf{x}$ is an equilibrium if and only if either $\mathbf{x}$ is a vertex of Δ, the simplex of all r-dimensional frequency vectors, or $F(\mathbf{x}) = 1/2$, the iterates of $\mathbf{x}$ will converge either to the linear manifold $F = 1/2$ or to a fixation state.

The following precise result can be proved.

COROLLARY 6.1. *For the transformation (6.1), the linear manifold defined by* $F(\mathbf{x}) = \sum_{i=1}^{r} f_i x_i = 1/2$ *is globally stable when it intersects the frequency simplex* Δ. *Otherwise, the corner* $x_{i_0} = 1$ *with* f_{i_0} *nearest to* $1/2$ *is globally stable.*

PROOF of Corollary 6.1. Let i_0 be such that f_{i_0} is nearest to $1/2$. The linear approximation of x'_{i_0} near A_i-fixation for $i \ne i_0$ reveals that

$$x'_{i_0} \cong x_{i_0} \left[\frac{f_{i_0}(1 - 2f_i) + f_i}{f_i(1 - 2f_i) + f_i} \right] > x_{i_0}$$

and convergence to A_i-fixation is therefore precluded. A similar argument shows that convergence to A_{i_0}-fixation itself is precluded if $f_{i_0} \ne 1/2$ and $F(\mathbf{x}) = 1/2$ somewhere in Δ (entailing the existence of some $f_i \ge 1/2$ and some $f_j \le 1/2$), leading to an initial increase of every coordinate corresponding to a parameter f_i on the opposite side of $1/2$ with respect to f_{i_0} at A_{i_0}-fixation.

6.2. A HAPLOID SEX-ALLOCATION MODEL WITH PARTIAL SELFING

Many plants are haploid through most of their life cycle, although union of gametes and recombination can occur. Consider the following model allowing hermaphroditism and selfing. Suppose that an A_i-gamete $(i = 1, \ldots, r)$ produces an average of $f_i n$ seeds and $(1 - f_i)\mathcal{N}$ pollen grains where n and $\mathcal{N}$ are reference constants $(\mathcal{N} \gg n)$. The quantities f_i and $(1 - f_i)$ estimate the resources allocated to female and male functions, respectively. Assume $f_i \neq f_j$ for $i \neq j$. Moreover, suppose that a constant proportion α of all seeds are subject to selfing (and/or an equivalent amount of resources are allocated to vegetative reproduction with adjusted viabilities). The recurrence equations for the frequencies x_i of A_i $(i = 1, \ldots, r)$ are

$$
\begin{aligned}
x_i' &= \alpha \left[\frac{f_i x_i}{F} \right] + (1 - \alpha) \left[\frac{f_i(1 - 2F) + F}{2F(1 - F)} \right] x_i \\
&= x_i \left[\frac{f_i(1 + \alpha - 2F) + F(1 - \alpha)}{2F(1 - F)} \right], i = 1, \ldots, r,
\end{aligned} \tag{6.2}
$$

where $F = \sum_{i=1}^{r} f_i x_i$. Apart from the fixation states, a necessary and sufficient condition for equilibrium is $F = (1 + \alpha)/2$. The following result can be readily established.

THEOREM 6.2. *Consider the transformation (6.2) of the frequency vector* $\mathbf{x} = (x_1, \ldots, x_r)$ *into* $\mathbf{x}' = (x_1', \ldots, x_r')$. *Let* $F = F(\mathbf{x}) = \sum_{i=1}^{r} f_i x_i$. *Then either* $F(\mathbf{x}) \leq F(\mathbf{x}') \leq (1 + \alpha)/2$ *or* $F(\mathbf{x}) \geq F(\mathbf{x}') \geq (1 + \alpha)/2$ *with equality if and only if* $\mathbf{x}' = \mathbf{x}$.

By analogy with Corollary 6.1, the phenotypic equilibrium $F = (1 + \alpha)/2$ will be reached or at least the system will tend toward it as far as possible. In any case, it will be ultimately reached through a series of favorable mutations (cf. Theorem 3.7). The quantity $(1 + \alpha)/2$ corresponds to an ESS for a fixation state (see, e.g., Maynard Smith, 1978; Charlesworth and Charlesworth, 1981).

PROOF of Theorem 6.2. (Take $\alpha = 0$ for Theorem 6.1.) Suppose $F = F(\mathbf{x}) < (1 + \alpha)/2$ and $\mathbf{x}$ not a vertex. Consider $F' = F(\mathbf{x}') = \sum_{i=1}^{r} f_i x_i'$ so that by (6.2)

$$
F' = \frac{(1 + \alpha - 2F) \sum_{i=1}^{r} f_i^2 x_i + (1 - \alpha)F \sum_{i=1}^{r} f_i x_i}{(1 + \alpha - 2F)F + (1 - \alpha)F}.
$$

Since we have $F^2 = \left(\sum\limits_{i=1}^{r} f_i x_i \right)^2 < \sum\limits_{i=1}^{r} f_i^2 x_i < \sum\limits_{i=1}^{r} f_i x_i = F$ (by Schwarz's inequality on the left side and the constraint $0 \leq f_i \leq 1$ on the right side), we find

$$F = \frac{(1 + \alpha - 2F)F^2 + (1 - \alpha)F^2}{(1 + \alpha - 2F)F + (1 - \alpha)F} < F'$$

$$< \frac{(1 + \alpha - 2F)F + (1 - \alpha)F^2}{(1 + \alpha - 2F)F + (1 - \alpha)F} = \frac{1 + \alpha}{2}.$$

These inequalities are reversed if $F > (1 + \alpha)/2$. The proof is complete.

6.3. Y-DRIVE OF THE SEX-RATIO: PARTIAL SIB-MATING

It has been suggested that Y-drive mechanisms of sex ratio distortion may be operative in some insect populations with an excess of males (e.g., in the mosquito *Aedes aegypti*; see Hamilton, 1967). Such models lend themselves to complete analysis under random mating and partial sib-mating.

Suppose that males are heterogametic (XY) and females homogametic (XX). Suppose that the sex ratio in the progeny of males is a Y-linked trait such that there is a proportion m_i of males in the progeny of an XY_i male ($i = 1, \ldots, r$). Consider for definiteness $m_i \neq m_j$ for $i \neq j$. Assume that all females are fertilized and a constant proportion α of offspring mate with sibs. Under these assumptions, the recurrence equations for the frequencies p_i of the mating types $XX \times XY_i$ are

$$p_i' = \alpha \left[\frac{(1 - m_i)p_i}{1 - M} \right] + (1 - \alpha) \left[\frac{m_i p_i}{M} \right]$$

$$= p_i \left[\frac{m_i(1 - \alpha - M) + \alpha M}{M(1 - M)} \right], \quad i = 1, \ldots, r \qquad (6.3)$$

where $M = M(\mathbf{p}) = \sum\limits_{i=1}^{r} m_i p_i$, i.e., M is the total proportion of

186

males produced. Parallel with Theorem 6.2, the following result is obtained:

THEOREM 6.3. *Under the transformation (6.3), the function $M =$*
$M(\mathbf{p}) = \sum_{i=1}^{r} m_i p_i$ *is monotone over successive generations with the prop-*
erty that

$$\left| M(\mathbf{p}') - (1 - \alpha) \right| \leq \left| M(\mathbf{p}) - (1 - \alpha) \right|$$

having equality when and only when $\mathbf{p}' = \mathbf{p}$. Convergence of the iterates
$\mathbf{p}^{(n)}$ *occurs to the surface $\mathscr{S}$ where $M(\mathbf{p}) = (1 - \alpha)$ if $\mathscr{S}$ intersects the*
simplex of the frequency vectors Δ and otherwise convergence occurs to
the closest point in Δ to $\mathscr{S}$ (necessarily achieved at a vertex of Δ by
the linearity of $M(\mathbf{p})$).

PROOF of pointwise convergence for transformations (6.2)
and (6.3). By analogy, we may focus on transformation (6.3)
and consider the following two cases:

Case(i): $m_1 < \cdots < m_r < 1 - \alpha$. Assume $\mathbf{p}^{(0)} \in \Delta^0$ (i.e., an
initial frequency vector with all positive components) and
let $\mathbf{p}^{(n)}$ be the nth iterate of (6.3) starting from $\mathbf{p}^{(0)}$. Suppose
$M(\mathbf{p}^{(n)}) \uparrow m_{i_0}$ as $n \uparrow \infty$ for some $i_0 \leq r - 1$. Since Δ is compact,
there exists at least one convergent subsequence of $\mathbf{p}^{(n)}$, say
$\mathbf{p}^{(n_k)} \to \mathbf{x}$ in Δ as $n_k \uparrow \infty$. By continuity of the transformation
(6.3), $\mathbf{p}^{(n_k+1)} \to \mathbf{x}'$ as $n_k \uparrow \infty$ and then $M(\mathbf{x}') = M(\mathbf{x}) = m_{i_0}$
since $M(\mathbf{x}) \leq M(\mathbf{x}')$. This implies that $\mathbf{x}$ is the equilibrium corresponding to A_{i_0}-fixation, i.e., $x_{i_0} = 1$ and $x_i = 0$ for $i \neq i_0$.
Since this holds for every convergent subsequence, we must
have $\mathbf{p}^{(n)} \to \mathbf{x}$ as $n \to \infty$, and in particular $p_r^{(n)} \to 0$. But near
A_{i_0}-fixation we have

$$p_r' \cong p_r \left\{ \frac{m_r(1 - \alpha - m_{i_0}) + \alpha m_{i_0}}{m_{i_0}(1 - \alpha - m_{i_0}) + \alpha m_{i_0}} \right\} > p_r,$$

which precludes convergence to zero. It follows that
$M(\mathbf{p}^{(n)}) \to m_r$ and $\mathbf{p}^{(n)}$ converges to A_r-fixation from any interior starting point.

Case(ii): $m_1 < \cdots < m_k \leq 1 - \alpha < m_{k+1} < \cdots < m_r$. The
arguments above ensure $M(\mathbf{p}^{(n)}) \to 1 - \alpha$ as $n \to \infty$ from any
starting point $\mathbf{p}^{(0)}$ in Δ^0. Suppose $M(\mathbf{p}^{(0)}) < 1 - \alpha$ so that $M_n =$

$M(\mathbf{p}^{(n)}) \uparrow 1 - \alpha$. (The other possibility can be treated analogously.) The recurrence relation

$$p_i^{(n+1)} = p_i^{(n)} \left\{ \frac{m_i(1 - \alpha - M_n) + \alpha M_n}{M_n(1 - \alpha - M_n) + \alpha M_n} \right\}$$

entails that $p_i^{(n)}$ will decrease for n large if $m_i < 1 - \alpha$, and increase if $m_i > 1 - \alpha$. It follows that $\mathbf{p}^{(n)}$ will converge to a point $\mathbf{x}$ in Δ such that $M(\mathbf{x}) = 1 - \alpha$.

6.4. X-DRIVE IN THE HETEROGAMETIC SEX: RANDOM MATING

X-drive at meiosis is known to occur in many species of Drosophila (see, e.g., Sturtevant and Dobzhansky, 1936). In this section, in support of Edwards (1961) and Hamilton (1967), we establish global convergence to the fixation state exhibiting the highest drive for X chromosomes when random mating takes place and no other selective forces come into play.

Let us assume that a male carrying A_i on the X-chromosome produces males with probability m_i and females with probability $(1 - m_i)$, $i = 1, \ldots, r$ and $m_i \neq m_j$ for $i \neq j$. Let q_i and p_i be the frequencies of A_i in males and females, respectively. If $2p_{ij}$ is the fraction of $A_i A_j$ females (p_{ii} when $i = j$), then $p_i = \sum_{j=1}^{r} p_{ij}$. Over two successive generations, with random mating, these variables are related as follows:

$$p_{ij}' = \frac{(1 - m_i)p_i q_j + (1 - m_j)p_j q_i}{- M)}, \quad i, j = 1, \ldots, r,$$

with $M = \sum_{i=1}^{r} m_i q_i$, so that

$$q_i' = p_i,$$

$$p_i' = \frac{(1 - m_i)q_i}{2(1 - M)} + \frac{p_i}{2}, \quad i = 1, \ldots, r.$$

(6.4)

The only equilibria are the fixation states. Moreover, the following theorem can be established.

THEOREM 6.4. *For the transformation (6.4), A_{i_0}-fixation with $m_{i_0} = \min\limits_{1 \le i \le r} m_i$ is globally stable.*

PROOF of Theorem 6.4. We will prove first that the function $M(\mathbf{x}) = \sum\limits_{i=1}^{r} m_i x_i$ satisfies

$$M\left(\mathbf{p}' + \frac{\mathbf{q}'}{2}\right) \le M\left(\mathbf{p} + \frac{\mathbf{q}}{2}\right) \tag{6.5}$$

with equality only at the fixation states of (6.4). Since $\mathbf{q}' = \mathbf{p}$ and $M(\mathbf{x})$ is a linear function, it will be sufficient to show that

$$M(\mathbf{q}'') \le \frac{M(\mathbf{q}') + M(\mathbf{q})}{2}.$$

($\mathbf{q}''$ denotes the second iterate of $\mathbf{q}$.) We find

$$M(\mathbf{q}'') = \sum_{i=1}^{r} m_i p_i'$$

$$= \frac{\sum\limits_{i=1}^{r} m_i (1 - m_i) q_i}{2(1 - M(\mathbf{q}))} + \frac{\sum\limits_{i=1}^{r} m_i q_i'}{2}$$

$$\le \frac{M(\mathbf{q}) - M(\mathbf{q})^2}{2(1 - M(\mathbf{q}))} + \frac{M(\mathbf{q}')}{2}. \tag{6.6}$$

The last inequality comes from $\sum\limits_{i=1}^{r} m_i^2 q_i \ge \left(\sum\limits_{i=1}^{r} m_i q_i\right)^2$ with equality only if $q_i = 1$ for some i. Therefore (6.5) is established and the iterates of the transformation (6.4) must converge to some fixation state. Near A_i-fixation for $i \ne i_0$ where i_0 is such that $m_{i_0} = \min\limits_{1 \le i \le r} m_i$, a linear approximation gives

$$q_{i_0}'' \cong \frac{(1 - m_{i_0})q_{i_0}}{2(1 - m_i)} + \frac{p_{i_0}}{2} > \frac{q_{i_0} + q_{i_0}'}{2}, \tag{6.7}$$

189

which precludes convergence of $q_{i_0}^{(n)}$ to zero in the vicinity of A_i-fixation. Therefore only A_{i_0}-fixation may be a limiting point from the interior of Δ. The proof is complete.

6.5. A TWO-LOCUS SEGREGATION DISTORTION MODIFICATION MODEL

The following model of sex-linked meiotic drive genes has been proposed by Maffi and Jayakar (1981) in an attempt to explain sex ratio distortions observed in *Aedes aegypti* populations in favor of males (see Wood and Newton, 1976) but without any assumption of viability (or fertility) differences as considered in Thomson and Feldman (1975) or Edwards (1961).

A primary locus with possible alleles M and m is assumed to be responsible for sex determination: males are Mm, and females mm, while a secondary locus of the male parent with possible alleles A and a governs the sex ratio in the progeny. Both loci are autosomal and a recombination event between these two loci can occur prior to the meiotic drive effects. Let the probability of recombination be r. Thus, for the two loci under consideration, the spectrum of male genotypes $\{MA/mA, MA/ma, Ma/mA, Ma/ma\}$ is composed from the gametes MA, Ma, mA, and ma, but always heterogametic at the $\{M,m\}$ locus. The female genotype array consists of $\{mA/mA, mA/ma, ma/ma\}$. Consider the case where the sex ratio expression has allele A dominant such that a male parent of genotype aa independent of his female mate genotype produces male to female offspring in a ratio $M_1:1 - M_1$, while an $\bar{A} = \{Aa, AA\}$ male parent has offspring in a sex ratio $M_2:1 - M_2$.

The recursion relations expressing the changes in male and female genotype frequencies over successive generations can be expressed with only three variables, namely,

$$u = [\text{freq}(Ma/ma) + \text{freq}(Ma/mA)]/[\text{freq}(Mm)],$$
$$v = [\text{freq}(Ma/ma) + \text{freq}(MA/ma)]/[\text{freq}(Mm)],$$
$$w = [\text{freq}(ma/ma) + \tfrac{1}{2}\text{freq}(ma/mA)]/[\text{freq}(mm)].$$

190

Under random mating, Maffi and Jayakar (1981) established the following recursion relations

$$v' = w$$

$$u' = \frac{(M_1 - M_2)uv + uM_2(1 - r) + vM_2r}{M_2 + (M_1 - M_2)uv}$$

$$2w' = w + \frac{(M_2 - M_1)uv + u(1 - M_2)r + v(1 - M_2)(1 - r)}{(1 - M_2) + (M_2 - M_1)uv}.$$

$$(6.8)$$

Maffi and Jayakar asserted that there exist only two possible equilibria, $(0,0,0)$ and $(1,1,1)$, the former being stable and the latter unstable. We now follow the analysis in Lessard and Karlin (1982). In the neighborhood of $(0,0,0)$, we have

$$(u + w + 2w)' = (u + v + 2w) + \left[\frac{(M_1 - M_2)(1 - 2M_2)}{M_2(1 - M_2)}\right]uv$$

$$+ \text{ higher order terms}, \qquad (6.9)$$

and therefore A-fixation is actually stable (unstable) if

$$(M_1 - M_2)(1 - 2M_2) < 0 \, (>0). \qquad (6.10)$$

(This is a particular application of a general criterion for stability-instability at fixation states when the leading eigenvalue of the linear approximation is 1; see Appendix A.) Moreover, in case of stability, the convergence rate is algebraic. On the other hand, writing $v = 1 - \alpha$, $u = 1 - \beta$, $w = 1 - \gamma$ when the positive quantities α, β, and γ are small, we have the linear approximation

$$\begin{bmatrix} \alpha' \\ \beta' \\ \gamma' \end{bmatrix} \cong \begin{bmatrix} 0 & 0 & 1 \\ \dfrac{rM_2}{M_1} & \dfrac{(1 - r)M_2}{M_1} & 0 \\ \dfrac{(1 - r)(1 - M_2)}{2(1 - M_1)} & \dfrac{r(1 - M_2)}{2(1 - M_1)} & \dfrac{1}{2} \end{bmatrix} \begin{bmatrix} \alpha \\ \beta \\ \gamma \end{bmatrix} \qquad (6.11)$$

Since the matrix in (6.11) is non-negative, its largest eigenvalue must be real and non-negative owing to the Perron-Frobenius theory. It can be checked that this eigenvalue will be larger than 1 if and only if $Q(1) > 0$ where $Q(\lambda)$ is the characteristic polynomial, namely

$$
Q(\lambda) = -\lambda^3 + \left[\frac{M_1 + 2(1 - r)M_2}{2M_1} \right] \lambda^2
$$

$$
+ \left[\frac{(1 - r)(M_1 - M_2)}{2M_1(1 - M_1)} \right] \lambda
$$

$$
- \frac{(1 - 2r)M_2(1 - M_2)}{2M_1(1 - M_1)}. \tag{6.12}
$$

This analysis reveals that the equilibrium $(1,1,1)$ corresponding to a-fixation is unstable if

$$
r(M_1 - M_2)(1 - 2M_2) < (M_1 - M_2)^2. \tag{6.13}
$$

The equilibrium is stable if the inequality is reversed. While the condition (6.10) for stability of A-fixation is independent of r (akin to the condition (3.11) for one-locus autosomal models), the condition (6.13) compels instability of an all aa-population if the recombination fraction is small enough. Note that this is also the case whenever A-fixation is stable. The analysis is completed by noting that under the condition

$$
(M_1 - M_2) < r(1 - 2M_2) < 0, \tag{6.14}
$$

there exists an internal equilibrium, namely

$$
u^* = \frac{(1 - 2M_2)}{2(1 - M_2)} + \frac{1}{2(1 - M_2)}
$$

$$
\times \sqrt{1 - \left\{ \frac{4M_2(1 - M_2)\left[(M_1 - M_2) + r(2M_2 - 1)\right]}{(M_1 - M_2)} \right\}} \tag{6.15}
$$

$$
v^* = w^* = \frac{(1 - M_2)u^* + (2M_2 - 1)}{M_2}.
$$

In this case, both fixation states are unstable. It is worth noting that when $M_2 = 1/2$, which is the case in the numerical example of Maffi and Jayakar, there is no polymorphic equilibrium, a-fixation is unstable, and even a quadratic analysis is not conclusive about the dynamics near A-fixation, which may account for the extremely slow rate of convergence reported. On the other hand, both fixation states are unstable if $M_1 = 1/2$ as long as $r < 1/2$. Therefore *an even sex-ratio equilibrium can be unstable if the sex ratio distortion is linked to the sex determination locus.*

6.6. OPTIMAL SEX RATIOS IN STRUCTURED POPULATIONS: BREEDING GROUPS OF RANDOM SIZE

Female-biased sex ratios are common in insects, mites, and other arthropods that mate in small colonies with episodic dispersal. Several authors have modeled the evolution of female-biased sex ratios for subdivided populations founded by inseminated females whose offspring mate locally among themselves and then disperse to found new groups. Some authors emphasize LMC (local mate competition) and the concomitant evolutionary response to reduce mating rivalry among genetic relatives, e.g., Hamilton (1967, 1979), Alexander and Sherman (1977), Taylor and Bulmer (1980), and Werren (1980). Others stress inbreeding (or sib-mating) as a determining factor, e.g., Maynard Smith (1978) and Stenseth (1978). Trivers and Hare (1976) argued that in eusocial species with haplodiploid sex-determination (e.g., *Hymenoptera*), queens gain by a $1:1$ sex ratio investment in reproductive offspring, while their sterile worker offspring gain by a $1:3$ investment for males to females among reproductive siblings based on genetic relatedness considerations; this leads to a worker-queen conflict. Colwell (1981) (see also Wilson and Colwell, 1981) accented group selection (i.e., differential productivity among groups) in the selective process responsible for sex ratio biases in subdivided populations. In the following extension for haploid populations, we introduce groups of random size and ascertain the optimal sex ratio in the ESS sense.

Suppose that an infinite number of breeding groups of random size N are formed by random sampling from an infinite

193

population. Random mating and reproduction with equal fertility take place within these groups. Then there is wide dispersal of offspring to restore the infinite pool.

Suppose that the population is haploid and two hermaphrodite types are present: types A and B that are male with probabilities a and b, respectively, and female otherwise. If the frequency of A in the whole population is p, then the proportion of breeding groups (randomly) colonized with xA and yB founders is

$$\pi_{x,y}(p) = \frac{(x+y)!}{x!y!} p^x(1-p)^y \gamma(x+y) \qquad (6.16)$$

where $\gamma(x+y)$ is the proportion of groups whose size is $\mathcal{N} = x+y$. In such breeding groups, the expected proportion of A offspring produced following random mating is

$$p_{x,y} = \frac{x^2a(1-a) + xy[a(1-b) + b(1-a)]/2}{[xa+yb][x(1-a) + y(1-b)]}, \qquad (6.17)$$

assuming that all females, whose expected number in such groups is

$$D_{x,y} = x(1-a) + y(1-b),$$

are fertilized and have equal fertility. Therefore the frequency of A after dispersal becomes

$$f(p) = \frac{\displaystyle\sum_{x,y} p_{x,y}\pi_{x,y}(p)D_{x,y}}{\displaystyle\sum_{x,y} \pi_{x,y}(p)D_{x,y}}. \qquad (6.18)$$

Differentiating with respect to p and evaluating at $p = 0$ yield

$$f'(0) = \frac{\displaystyle\sum_{y} p_{1,y}\pi'_{1,y}(0)D_{1,y}}{\displaystyle\sum_{y} \pi_{0,y}(0)D_{0,y}} \qquad (6.19)$$

since $\pi_{x,y}(0) = 0$ for all $x \geq 1$, $p_{0,y} = 0$ for all $y \geq 1$ and

$$\pi'_{x,y}(p) = \frac{(x+y)!}{x!y!} p^{x-1}(1-p)^{y-1}[x(1-p) - yp]\gamma(x+y)$$

vanishes at $p = 0$ for all $x \geq 2$. Introducing the notation

194

$\lambda(a,b) = f'(0)$ and $\varphi_y(a,b) = p_{1,y}D_{1,y}$, we have

$$\lambda(a,b) = \frac{\sum_y (y+1)\gamma(y+1)\varphi_y(a,b)}{(1-b)\sum_y y\gamma(y)} \qquad (6.20)$$

and

$$\varphi_y(a,b) = \frac{a(1-a) + y[a(1-b) + b(1-a)]/2}{[a+yb]}. \qquad (6.21)$$

The equilibrium state corresponding to B-fixation will be stable against any new mutant A if

$$\lambda(a,b) = f'(0) \leq 1$$

for all parameter values with equality only at $a = b$. Therefore we look for a parameter b such that

$$\frac{\partial}{\partial a} \lambda(a,b)\big|_{a=b} = 0.$$

Since

$$\frac{\partial}{\partial a} \varphi_y(a,b)\big|_{a=b} = \frac{(1-2b)(y+1)-1}{2b(y+1)},$$

we have

$$\frac{\partial}{\partial a} \lambda(a,b)\big|_{a=b} = \frac{(1-2b)\mu - \sum_{y\geq 1}\gamma(y)}{2b(1-b)\mu} \qquad (6.22)$$

where $\mu = \sum_y y\gamma(y)$. Hence (6.22) vanishes when

$$b = \frac{\mu^* - 1}{2\mu^*} \qquad (6.23)$$

with $\mu^* = \mu/\sum_{y\geq 1} \gamma(y)$, i.e., μ^* is the average group size among nonvoid breeding groups. Therefore we have proved the following result.

THEOREM 6.5. *In an infinite haploid hermaphrodite population with breeding groups of random size N, the fixation state corresponding to the sex ratio $(\mu^* - 1)/2\mu^*$, where μ^* is the expected value of N conditioned on $N \geq 1$, is always stable.*

195

6.7. OPTIMAL SEX RATIOS IN STRUCTURED POPULATIONS: BREEDING GROUPS WITH A CONSTANT FRACTION OF OUTBREEDING AND POPULATION REGULATION WITHIN GROUPS

The effect of partial sib-mating (or selfing) on sex ratio evolution was studied in Maynard Smith (1978), Charlesworth and Charlesworth (1981), Taylor (1981), Uyenoyama and Bengtsson (1982), and Charnov (1982), among others. When sib-mating takes place first for a fraction of female offspring and all females are fertilized, it is generally advantageous to produce more females, as shown in Section 6.2 for haploid populations. The same principle should also apply with breeding units of any finite size and a constant fraction of outbreeding irrespective of the timing of dispersal (before or after mating) if every female offspring has an equal opportunity to contribute to the next generation and the population is polygynous. But if there is population regulation within the breeding groups, such that each of them, owing to resource limitations, produce the same extent of fertilized females, then a male-biased sex ratio can evolve. This is an example of local resource competition (LRC) models which predict a bias in favor of the more mobile sex in a geographical environment (Clark, 1978; Bulmer and Taylor, 1980b; see also Cannings and Cruz Orive, 1975). The question of the optimal dispersal rate was dealt with in Motro (1982a, b) complementing Hamilton and May (1977).

In this section, we consider structured diploid populations allowing partial outbreeding and population regulation. The search for ESS sex ratios is facilitated in the diploid case (compared to the haplodiploid case) on account of the symmetry of gene transmission to sons and daughters. The following approach has been suggested by Taylor and Bulmer (1980).

Suppose that the male proportion in the progeny of RR females is r while females carrying a mutant gene S dominant to R produce a proportion s of male offspring. All broods are assumed to be of equal size and the population is polygynous. A mating pair (or fertilized female) is described by the pair

(i,j) where i is the number of S genes in the female genotype and j the corresponding number in her mate genotype. The frequency of mating pairs of type (i,j) is denoted by $x(i,j)$. Following fertilization the females randomly disperse into an infinite distribution of founding colonies with exactly $\mathcal{N}$ females per colony. If gene S is rare, virtually all colonies will contain at most one fertilized female (i,j) with $i + j \geq 1$. Subsequent to reproduction, a fraction β of female offspring mate within their own colonies while the others choose their mates at random from the population at large. All female offspring are fertilized and then disperse to found new colonies according to the previous scheme. The case of local population regulation is considered further.

Let $\psi(i,j)$ be the frequency of S genes transmitted to the next generation by all the mating pairs (i,j). A mating pair (i,j) is called mutant if $i \geq 1$. The expected number of S genes transmitted to an offspring of a mating pair (i,j) is $(i + j)/2$. Since the frequency of female offspring in the whole population is virtually $(1 - r)$ and the frequency of female offspring produced by all the mutant mating pairs (i,j) is $x(i,j)(1 - s)$, the quantity

$$\frac{1}{2}\left(\frac{i+j}{2}\right)\frac{x(i,j)(1-s)}{1-r}$$

is the frequency of S genes transmitted to the next generation through daughters issued from the mutant pairs (i,j). Their brothers can fertilize a fraction β of the female offspring from the same colony and a fraction $(1 - \beta)$ of the female offspring in the male population at large. The frequency of S genes transmitted to the next generation in the former event is

$$\beta\left[\frac{1}{2}\left(\frac{i+j}{2}\right)\left(\frac{x(i,j)[(1-s)+(\mathcal{N}-1)(1-r)]}{1-r}\right)\left(\frac{s}{s+(\mathcal{N}-1)r}\right)\right]$$

and in the latter case

$$(1-\beta)\left[\frac{1}{2}\left(\frac{i+j}{2}\right)\left(\frac{x(i,j)s}{r}\right)\right].$$

197

Therefore, for the mutant pairs (i,j), we have

$$\psi(i,j) = \left(\frac{i+j}{4}\right) x(i,j) g(s,r) \qquad (6.24)$$

where

$$g(s,r) = \frac{1-s}{1-r} + \beta \left[\frac{(1-s) + (\mathcal{N}-1)(1-r)}{1-r} \right]$$

$$\times \left[\frac{s}{s + (\mathcal{N}-1)r} \right] + (1-\beta)\frac{s}{r}.$$

In the same way (replacing s by r), we obtain

$$\psi(0,j) = \frac{j}{2} x(0,j). \qquad (6.25)$$

We conclude that the overall frequency of S genes when rare increases if and only if $g(s,r) > 1$. Thus, the population fixation state corresponding to the parameter r^* such that $g(s,r^*) < 1$ for every $s \neq r^*$ cannot be invaded by any mutant. To find r^*, we solve

$$\frac{\partial}{\partial s} g(s,r^*)\Big|_{s=r^*} = \frac{(\mathcal{N}-\beta) - 2\mathcal{N}r^*}{\mathcal{N}r^*(1-r^*)} = 0.$$

The corresponding ESS sex ratio

$$r^* = \frac{1}{2} - \frac{\beta}{2\mathcal{N}} \qquad (6.26)$$

is female-biased whenever there exists a positive degree of in-breeding (i.e., $\beta > 0$). Note also that the foregoing arguments actually hold if the mutant sex ratio modifier is recessive or dominant, active in either parent or both parents. In the case of complete selfing ($\mathcal{N} = 1, \beta = 1$) we prove in Appendix D that there must be evolution (global convergence) toward $r^* = 0$ in one-locus multiallele sex allocation models.

A main assumption in the above model is that there is no restriction on the offspring production of the breeding groups. An alternative model, taking into account limited local carrying capacities (or resources), would consider that each colony provides an equal amount of fertilized females contributing to the next generation. With such a density regulating process within groups, the previous quantities become

$$\tilde{g}(s,r) = \frac{\mathcal{N}(1-s)}{(1-s) + (\mathcal{N}-1)(1-r)} + \beta\left[\frac{\mathcal{N}s}{s + (\mathcal{N}-1)r}\right]$$

$$+ (1-\beta)\left[\frac{s}{r}\right]$$

and

$$\tilde{r}^* = \frac{1}{2} + \frac{(1-\beta)}{2(2\mathcal{N}-\beta-1)}. \tag{6.27}$$

In particular, with random mating ($\beta = 0$) the ESS sex ratio $\tilde{r}^* = \mathcal{N}/(2\mathcal{N}-1)$ is male-biased in agreement with Charnov (1982, p. 75). Actually we have proved that a male-biased sex ratio is expected whenever there is some mobility of males allowing occasional outbreeding. In the absence of outbreeding ($\beta = 1$), the ESS sex ratio is $1/2$ regardless of the frequency of sib-mating (or the degree of inbreeding). But observe in this case that the degree of competition between sons is equivalent to the degree of competition between daughters.

Wilson and Colwell (1981) simulated sex ratio evolution in structured haplodiploid populations with individuals spending a certain number of generations breeding exclusively within resource sites before randomly dispersing in search of new resource sites (Bulmer and Taylor, 1980a, considered the same model for diploid populations). If the sites are uniformly regulated, implicating an equal contribution of genetic material (same number of inseminated females if mating takes place before dispersal or the same number of individuals, males and females, if mating occurs after dispersal) to form the next generation, then an even sex ratio evolves. With or without population regulation, an even sex ratio is favored within the local sites as long as random mating takes place in agreement with Fisher's principle, but the sites where more females are produced and fertilized contribute more to the next life cycle after dispersal so an uneven sex ratio evolves.

199

Taylor and Bulmer (1980) also analyzed the case of haplodiploid populations with females dispersing after mating into new breeding areas without allowing outbreeding and density regulation. With only one breeding season between dispersal epochs, they found

$$\frac{(\mathcal{N} - 1)(2\mathcal{N} - 1)}{\mathcal{N}(4\mathcal{N} - 1)} = \frac{1}{2} - \frac{5\mathcal{N} - 2}{2\mathcal{N}(4\mathcal{N} - 1)} \qquad (6.28)$$

as the ESS male proportion in the case of mother's control and

$$\frac{\mathcal{N} - 1}{\mathcal{N}(4\mathcal{N} - 1)} = \frac{1}{2} - \frac{4\mathcal{N}^2 - 3\mathcal{N} + 2}{2\mathcal{N}(4\mathcal{N} - 1)} \qquad (6.29)$$

in the case of father's control, where $\mathcal{N}$ is the number of foundress females. This is also the optimal sex ratio expected in diploid populations with X-drive in XY male progenies. For Y-drive mechanisms, Hamilton (1967) obtained the formula $(1 - 1/\mathcal{N})$ (cf. Theorem 6.3). Bulmer and Taylor (1980b) also predicted a bias toward the more mobile sex in populations geographically structured into single cells with density regulation applicable to sex allocation in plant populations. Clark (1978) had previously proposed that local resource competition (LRC) between female relatives could explain male-biased sex ratios in prosimian primates. The results of this section also confirm these views, and more generally, that the *sex ratio should be biased so as to equalize competition between sibs for reproduction irrespective of sex.*

APPENDIX A. A CRITERION FOR STABILITY-INSTABILITY AT FIXATION STATES THAT INVOLVE AN EIGENVALUE 1

Consider an infinite population with discrete generations. Let the population states be described by frequency vectors $\mathbf{x} = (x_1, \ldots, x_n)$ such that $\mathbf{0} = (0, \ldots, 0)$ corresponds to a fixation event F. Let $T\mathbf{x} = (U_1(\mathbf{x}), \ldots, U_n(\mathbf{x}))$ be the transformation relating the population states over successive generations. Assume $T\mathbf{x} = \mathbf{0}$ if and only if $\mathbf{x} = \mathbf{0}$ and that T is smooth enough in the neighborhood of **0**. (It is sufficient that the first and second derivatives of all $U_i(\mathbf{x})$ are continuous in a neighborhood of **0**.) Let L be the gradient matrix of T at **0**, namely,

$$L = \left\| \frac{\partial U_i}{\partial x_j} (\mathbf{0}) \right\|_{i,j=1}^{n}$$

Assume that L is non-negative and denote by $\rho(L)$ its spectral radius.

By definition, the fixation event F is said to be *stable* if the iterates $T^k \mathbf{x} \to \mathbf{0}$ as $k \to \infty$ for any initial population state $\mathbf{x}$ near **0**, and *unstable* if $T^k \mathbf{x} \nrightarrow \mathbf{0}$ as $k \to \infty$ for any initial population state $\mathbf{x}$ different from **0**. Formally, the first condition corresponds to local asymptotic stability in the manner that the fixation event F is locally attracting, while the latter property is stronger than a simple negation of the stability condition since it states that the population state F cannot be reached from any nonzero state and thus is strongly (totally) repelling. It is well known that the fixation event F will be stable if $\rho(L) < 1$ and usually unstable if $\rho(L) > 1$. Actually, the second part requires some mild supplementary conditions. A brief proof of both statements is given as a remark at the end of this Appendix in the most common case where there exists $\mathbf{z} = (z_1, \ldots, z_n) > \mathbf{0}$ (i.e., $z_i > 0$ for $i = 1, \ldots, n$) such that $\mathbf{z}L = \rho(L)\mathbf{z}$. When $\rho(L) < 1$, local convergence to **0** occurs at the geometric rate $\rho(L)$, i.e., it deviates from zero after k generations to the order $[\rho(L)]^k$.

Now assume $\rho(L) = 1$ allowing L of the general form

$$L = \begin{bmatrix} A & B \\ 0 & C \end{bmatrix}, \tag{A.1}$$

where A is an irreducible aperiodic (i.e., primitive; see below) non-negative matrix of order l and $\rho(C) < 1$. We use the notation $\langle \mathbf{z}, \mathbf{w} \rangle = \sum_{i=1}^{n} z_i w_i$ for the usual inner product of two vectors $\mathbf{z}$ and $\mathbf{w}$. Define the quantity

$$S = \langle \boldsymbol{\xi}, \boldsymbol{\theta}^{(1)} + B(I - C)^{-1}\boldsymbol{\theta}^{(2)} \rangle \tag{A.2}$$

(I is the identity matrix), where $\boldsymbol{\xi}(\boldsymbol{\eta})$ is the positive left (right) principal eigenvector of A corresponding to the eigenvalue 1, that is,

$$\boldsymbol{\xi}A = \boldsymbol{\xi} = (\xi_1, \ldots, \xi_l) > \mathbf{0}, \; A\boldsymbol{\eta} = \boldsymbol{\eta} = (\eta_1, \ldots, \eta_l) > \mathbf{0} \quad (A.3)$$

and

$$\boldsymbol{\theta}^{(1)} = (\theta_1, \ldots, \theta_l), \; \boldsymbol{\theta}^{(2)} = (\theta_{l+1}, \ldots, \theta_n)$$

with

$$\theta_i = \sum_{\lambda,\mu=1}^{l} \frac{\partial^2 U_i(\mathbf{0})}{\partial x_\lambda \, \partial x_\mu} \eta_\lambda \eta_\mu \text{ for } i = 1, \ldots, n. \quad (A.4)$$

We are now prepared to state the following general criterion.

THEOREM A.1. (Lessard and Karlin, 1982.) *Consider the general nonlinear transformation T with fixation state $\mathbf{0}$ (F) and associated gradient matrix L as in ($A.1$) obeying the assumptions as stated above. The fixation event F is stable if $S < 0$ and unstable if $S > 0$. When $S < 0$, local convergence to the fixation state F occurs at an algebraic rate of degree 1, i.e., the deviation of $T^k\mathbf{x}$ from $\mathbf{0}$ after k iterations is of the precise order $1/k$.*

This may be viewed as a general multidimensional version of the condition on the second derivative in the unidimensional case. A proof will be given in the common case where the non-negative matrix $B(I - C)^{-1}$ displays no null columns, which covers most applications. In our analysis we require L to be a non-negative matrix plus other mild conditions.

For completeness and ready reference we review some definitions and results in matrix theory (see, e.g., Gantmacher, 1959).

Let $A = \|a_{ij}\|_{i,j=1}^{l}$ be a non-negative matrix (i.e., $a_{ij} \geq 0$ for $i,j = 1, \ldots, l$). Denote by $\rho(A)$ its spectral radius.

DEFINITION. A is said to be *reducible* if, by simultaneously permuting corresponding rows and columns, A can be put in the form

$$A = \begin{bmatrix} A_{11} & A_{12} \\ 0 & A_{22} \end{bmatrix}.$$

where A_{11}, A_{22} are square matrices, and 0 is a matrix with all zero entries. Otherwise, A is said to be *irreducible*.

THEOREM (Perron-Frobenius). *If A is irreducible, then $\rho(A)$ is a positive simple eigenvalue and the associated left and right eigenvectors, ξ and η, display only positive components.*

A is said to be *primitive* if A is irreducible and $\rho(A)$ is the only eigenvalue with modulus equal to $\rho(A)$. Equivalently, A is irreducible and aperiodic or there exists an integer m such that A^m exhibits all positive elements.

THEOREM (Strong Ergodic Theorem). *Let A be primitive with $\rho(A) = 1$. Then A^k converges to $U = \|\eta_i \xi_j\|_{i,j=1}^l$ as $k \to \infty$, where*

$$\xi A = \xi > 0, \; A\eta = \eta > 0, \; and \; \langle \xi, \eta \rangle = \sum_{i=1}^l \eta_i \xi_i = 1.$$

COROLLARY. *Let*

$$L = \begin{bmatrix} A & B \\ 0 & C \end{bmatrix},$$

where A satisfies the conditions of the above theorem and $\rho(C) < 1$. Then

$$L^k \to M = \begin{bmatrix} U & UB(I - C)^{-1} \\ 0 & 0 \end{bmatrix} \text{ as } k \to \infty.$$

PROOF. By induction, we get

$$L^k = \begin{bmatrix} A^k & \sum_{r=0}^{k-1} A^{k-1-r} BC^r \\ 0 & C^k \end{bmatrix}$$

We know by the strong ergodic theorem that $A^k \to U$ as $k \to \infty$. On the other hand, since $\rho(C) < 1$, we have $C^k \to 0$ and $\sum_{r=0}^{k-1} C^r \to (I - C)^{-1}$ as $k \to \infty$. It follows that

$$\sum_{r=0}^{k-1} (A^{k-1-r} - U)BC^r \to 0 \text{ as } k \to \infty$$

which completes the proof of the corollary.

PROOF of Theorem A.1. Without loss of generality, we shall assume ξ and η normalized such that $\langle \xi, \eta \rangle = 1$. Define $\mathbf{z}^* = (z_1^*, \dots z_l^*, z_{l+1}^*, \dots, z_n^*)$ taking $z_i^* = \xi_i$, $i = 1, 2, \dots, l$, and $(z^*_{l+1}, \ldots, z^*_n) = \boldsymbol{\xi} B(I - C)^{-1}$. It is easy to check that $\mathbf{z}^*L = \mathbf{z}^*$. Under the assumption that $B(I - C)^{-1}$ has no null columns, we see that the vector $\mathbf{z}^* > 0$ (has all positive components). Throughout the proof, we shall use the absolute value norm $\|\cdot\|$ weighted by the components of $\mathbf{z}^*$ defined explicitly by $\|\mathbf{y}\| = \sum_{i=1}^{n} z^*_i |y_i|$ for every $\mathbf{y} = (y_1, \ldots, y_n)$.

Consider the Taylor expansion of T about $\mathbf{0}$ evaluated at a population state $\mathbf{x} = \delta\mathbf{v} > \mathbf{0}$, where $\delta = \|\mathbf{x}\|$ is small. Thus

$$T\mathbf{x} = \delta L\mathbf{v} + \delta^2 \mathbf{Q}(\mathbf{v}) + o(\delta^2),$$

where $\mathbf{Q}(\mathbf{v}) = (Q_1(\mathbf{v}), \ldots, Q_n(\mathbf{v}))$ represents the quadratic part evaluated at $\mathbf{v} = (v_1, \ldots, v_n)$, i.e.,

$$Q_i(\mathbf{v}) = \frac{1}{2} \sum_{\lambda,\mu=1}^{n} \frac{\partial^2 U_i(\mathbf{0})}{\partial x_\lambda \partial x_\mu} v_\lambda v_\mu \text{ for } i = 1, \ldots, n.$$

Therefore, the iterates of T are given by

$$T^k\mathbf{x} = \delta L^k\mathbf{v} + \delta^2 \sum_{r=0}^{k-1} L^{k-1-r} \mathbf{Q}(L^r\mathbf{v}) + o(\delta^2).$$

Taking the scalar product with $\mathbf{z}^*$ on both sides of the above equation yields

$$\|T^k\mathbf{x}\| = \delta + \delta^2 \sum_{r=0}^{k-1} \langle \mathbf{z}^*, \mathbf{Q}(L^r\mathbf{v}) \rangle + o(\delta^2)$$

(note that all the components of $T^k\mathbf{x}$ are necessarily non-negative). Then

$$\|T^{k+1}\mathbf{x}\| - \|T^k\mathbf{x}\| = \delta^2 \langle \mathbf{z}^*, \mathbf{Q}(L^k\mathbf{v}) \rangle + o(\delta^2). \quad \text{(A.5)}$$

It should be noted that the function $o(\delta^2)$ may depend on k. Since $L^k \to M$ as $k \to \infty$ and $\mathbf{Q}$ is continuous, we have $\mathbf{Q}(L^k\mathbf{v}) \to \mathbf{Q}(M\mathbf{v})$ as $k \to \infty$ uniformly over the compact section $\mathcal{K} = \{\mathbf{v} : \|\mathbf{v}\| = 1\}$. The evaluation of $M\mathbf{v}$ is a multiple of $\boldsymbol{\eta}$ in the first l coordinates and vanishes identically in the remaining $n - l$ coordinates. Explicitly,

$$M\mathbf{v} = (\langle \boldsymbol{\xi}, \mathbf{v}^{(1)} + B(I - C)^{-1}\mathbf{v}^{(2)} \rangle \boldsymbol{\eta}; \mathbf{0}) \quad \text{(A.6)}$$

where $\mathbf{v}^{(1)} = (v_1, \ldots, v_l)$, $\mathbf{v}^{(2)} = (v_{l+1}, \ldots, v_n)$.

It follows from (A.6) that

$$\langle \mathbf{z}^*, \mathbf{Q}(M\mathbf{v})\rangle = \langle \mathbf{z}^*, \mathbf{Q}(\boldsymbol{\eta})\rangle [\langle \boldsymbol{\xi}, \mathbf{v}^{(1)} + B(I - C)^{-1}\mathbf{v}^{(2)}\rangle]^2$$
$$= \langle \mathbf{z}^*, \mathbf{Q}(\boldsymbol{\eta})\rangle [\langle \mathbf{z}^*, \mathbf{v}\rangle]^2$$
$$= \langle \mathbf{z}^*, \mathbf{Q}(\boldsymbol{\eta})\rangle \|\mathbf{v}\|^2$$
$$= \langle \mathbf{z}^*, \mathbf{Q}(\boldsymbol{\eta})\rangle.$$

Refer from (A.4) the notation

$$2\mathbf{Q}(\boldsymbol{\eta}) = (\theta_1, \dots, \theta_l; \theta_{l+1}, \dots, \theta_n) = (\boldsymbol{\theta}^{(1)}; \boldsymbol{\theta}^{(2)})$$

so that by the determination of $\mathbf{z}^*$, we have

$$2\langle \mathbf{z}^*, \mathbf{Q}(\boldsymbol{\eta})\rangle = \langle \boldsymbol{\xi}, \boldsymbol{\theta}^{(1)}\rangle + \langle \boldsymbol{\xi}B(1 - C)^{-1}, \boldsymbol{\theta}^{(2)}\rangle$$
$$= \langle \boldsymbol{\xi}, \boldsymbol{\theta}^{(1)} + B(I - C)^{-1}\boldsymbol{\theta}^{(2)}\rangle = S,$$

and the last equation is the definition of S.

The foregoing analysis produces in (A.5) for k large and $\mathbf{x} = \delta\mathbf{v}$ the equation

$$\|T^{k+1}\mathbf{x}\| - \|T^k\mathbf{x}\| = \frac{S}{2}\delta^2 + o(\delta^2), \tag{A.7}$$

with the small order term $o(\delta^2)$ generally depending on k.

Case (i). $S < 0$. We can find an integer k_0 and a positive quantity ε such that

$$\|T^{k_0+1}\mathbf{x}\| - \|T^{k_0}\mathbf{x}\| < \frac{S}{4}\|\mathbf{x}\|^2 < 0 \tag{A.8}$$

as soon as $\|\mathbf{x}\| < \varepsilon$. By continuity of T at $\mathbf{0}$, there exists $0 < \varepsilon' < \varepsilon$ such that $\|T^r\mathbf{x}\| < \varepsilon$ for $r = 1, \dots, k_0$ if $\|\mathbf{x}\| < \varepsilon'$. For such $\mathbf{x}$, it follows recursively on the basis of (A.8) that the sequence $\|T^k\mathbf{x}\|$, $k = 0,1,2,\dots$, is bounded by ε and decreasing for $k \geq k_0$. But then the inequalities

$$\|T^{k_0+r+1}\mathbf{x}\| - \|T^{k_0+r}\mathbf{x}\| < \frac{S}{4}\|T^r\mathbf{x}\|^2 < 0$$

for all $r \geq 0$ are only compatible with the convergence of $\|T^k\mathbf{x}\|$ to zero, that is, $T^k\mathbf{x} \to \mathbf{0}$ as $k \to \infty$.

Case (ii). $S > 0$. One can find an integer k_0 such that

$$\|T^{k_0+1}\mathbf{x}\| - \|T^{k_0}\mathbf{x}\| > 0$$

as soon as $\|\mathbf{x}\|$ is small enough. Assume that $T^k\mathbf{y} \to \mathbf{0}$ as $k \to \infty$ for an initial population state $\mathbf{y} \neq \mathbf{0}$. Then we should have

$$\|T^{k_0+r+1}\mathbf{y}\| > \|T^{k_0+r}\mathbf{y}\| > 0$$

for r large enough (the last inequality emanating from the fact that $T\mathbf{x} \neq \mathbf{0}$ if $\mathbf{x} \neq \mathbf{0}$). But this precludes convergence to $\mathbf{0}$.

REMARK. Assume $\rho(L) \neq 1$ and $\mathbf{z}^* > \mathbf{0}$ such that $\mathbf{z}^*L = \rho(L)\mathbf{z}^*$. Define the norm $\|\cdot\|$ as in the previous proof with respect to this $\mathbf{z}^*$. Expanding $T\mathbf{x}$ about $\mathbf{0}$ for a population state $\mathbf{x}$ and taking the scalar product with $\mathbf{z}^*$ yields

$$\|T\mathbf{x}\| = \rho(L)\|\mathbf{x}\| + o(\|\mathbf{x}\|).$$

Thus, T is locally contracting or strongly repelling according to $\rho(L) < 1$ or $\rho(L) > 1$, respectively.

APPENDIX B. RECURRENCE EQUATIONS FOR PARTIAL SIB-MATING SEX-ALLOCATION MODELS IN HAPLODIPLOID POPULATIONS

Frequency	Mating Type	Sex Allocation	
	$\female \times \male$		
$2P_{ij:k}$ $(i \neq j)$	$A_iA_j \times A_k$	$f_{ij:k}$	$\female$
$P_{ii:k}$ $(i = j)$		$1 - f_{ij:k}$	$\male$

α = proportion of sib-mating (prior to random mating).

The recurrence equations are

$$P'_{ij:k} = \alpha \left\{ \frac{P_{ik:j}f_{ik:j} + P_{jk:i}f_{jk:i} + \delta_{ik}z_{i:j} + \delta_{jk}z_{j:i}}{4F} \right\}$$

$$+ (1 - \alpha) \left\{ \frac{(z_{i:j} + z_{j:i})(p_k - z_{k\cdot})}{2F(1 - F)} \right\} \tag{B.1}$$

where

$$z_{i:j} = \sum_l P_{il:j}f_{il:j}$$

$$z_{i\cdot} = \sum_j z_{i:j}, \qquad z_{\cdot j} = \sum_i z_{i:j}$$

$$F = \sum_i z_{i\cdot} = \sum_j z_{\cdot j} = \text{female sex ratio among offspring}$$

$$P_{ij} = \sum_l P_{ij:l}$$

$$p_i = \sum_j P_{ij} = \text{frequency of } A_i \text{ in mothers}$$

$$q_k = \sum_{i,j} P_{ij:k} = \text{frequency of } A_k \text{ in fathers}$$

$$\delta_{ij} = \begin{cases} 1 & \text{if } i = j \\ 0 & \text{if } i \neq j \end{cases}$$

We deduce the following relations:

$$P'_{ij} = \alpha \left\{ \frac{z_{i:j} + z_{j:i} + z_{i:j} + z_{j:i}}{4F} \right\} + (1-\alpha) \left\{ \frac{z_{i:j} + z_{j:i}}{2F} \right\}$$

$$= \frac{z_{i:j} + z_{j:i}}{2F}$$

$$p'_i = \frac{z_{i\cdot} + z_{\cdot i}}{2F}$$

$$q'_k = \alpha \left\{ \frac{z_{k\cdot}}{F} \right\} + (1-\alpha) \left\{ \frac{p_k - z_{k\cdot}}{1 - F} \right\} = \frac{(\alpha - F)z_{k\cdot} + (1-\alpha)Fp_k}{F(1 - F)}$$

$$\text{(B.2)}$$

Special Case: $f_{ij:k} = f_k$ for all i,j.
We have

$$z_{\cdot j} = f_j q_j$$

so that at equilibrium

$$z_{i\cdot} = 2Fp_i - f_i q_i$$

and then

$$F(1 - F)q_k = (\alpha - F)\{2Fp_k - f_k q_k\} + (1-\alpha)Fp_k$$

i.e.,

$$q_k \{F(1 - F) + (\alpha - F)f_k\} = p_k \{2F(\alpha - F) + (1-\alpha)F\}$$
$$= p_k \{F(1 - F) + (\alpha - F)F\} \quad \text{(B.3)}$$

207

APPENDIX C. RECURRENCE EQUATIONS FOR PARTIAL SIB-MATING SEX-ALLOCATION MODELS IN DIPLODIPLOID POPULATIONS

Frequency	Mating Type	Sex Allocation	

$4P_{ij:kl}\ (i \neq j, k \neq l)$

$2P_{ii:kl}\ (k \neq l)$

$2P_{ij:kk}\ (i \neq j)$ $\qquad A_iA_j \times A_kA_l \qquad \left\{ \begin{array}{l} f_{ij:kl} \qquad ♀ \\ 1 - f_{ij:kl} \qquad ♂ \end{array} \right.$

$P_{ii:kk}$

Pure sib-mating:

$$P'_{ij:kl} = \frac{P_{ik:jl}f_{ik:jl} + P_{il:jk}f_{il:jk} + P_{jk:il}f_{jk:il} + P_{jl:ik}f_{jl:ik}}{16F}$$

$$+ \delta_{ik}\left\{\frac{z_{i:jl} + z_{jl:i}}{16F}\right\} + \delta_{il}\left\{\frac{z_{i:jk} + z_{jk:i}}{16F}\right\}$$

$$+ \delta_{jk}\left\{\frac{z_{j:il} + z_{il:j}}{16F}\right\} + \delta_{jl}\left\{\frac{z_{j:ik} + z_{ik:j}}{16F}\right\}$$

$$+ \delta_{ik}\delta_{jl}\left\{\frac{z_{i:j} + z_{j:i}}{16F}\right\} + \delta_{il}\delta_{jk}\left\{\frac{z_{i:j} + z_{j:i}}{16F}\right\} \qquad (C.1)$$

Random mating:

$$P'_{ij:kl} = \left\{\frac{z_{i:j} + z_{j:i}}{2F}\right\} \times \left\{\frac{P_{k:l} - z_{k:l} + P_{l:k} - z_{l:k}}{2(1 - F)}\right\} \qquad (C.2)$$

Partial sib-mating (α = proportion of sib-mating):

$$P'_{ij:kl} = \alpha\{\text{sib-mating}\} + (1 - \alpha)\{\text{random mating}\} \qquad (C.3)$$

In particular, we have

$$p'_i = \frac{z_{i\cdot} + z_{\cdot i}}{2F}$$

$$q'_i = \alpha\left\{\frac{z_{i\cdot} + z_{\cdot i}}{2F}\right\} + (1 - \alpha)\left\{\frac{p_i - z_{i\cdot} + q_i - z_{\cdot i}}{2(1 - F)}\right\}$$

$$= \frac{(1 - \alpha)(p_i + q_i) + 2(\alpha - F)p'_i}{2(1 - F)}$$

so that at equilibrium

$$q_i = \frac{(1 + \alpha - 2F)p_i + (1 - \alpha)q_i}{(1 + \alpha - 2F) + (1 - \alpha)} \tag{C.4}$$

Either $F = (1 + \alpha)/2$ or $p_i = q_i$ for all i.

Notation:

$$z_{i:jl} = \sum_v P_{iv:jl} f_{iv:jl}$$

$$z_{ij:k} = \sum_\mu P_{ij:k\mu} f_{ij:k\mu}$$

$$z_{i:j} = \sum_{v,\mu} P_{iv:j\mu} f_{iv:j\mu}$$

$$z_{i\cdot} = \sum_j z_{i:j}, \, z_{\cdot j} = \sum_i z_{i:j}$$

$$F = \sum_{i,j} z_{i:j} = \text{female sex ratio in offspring}$$

$$P_{k:l} = \sum_{v,\mu} P_{kv:l\mu}$$

$$p_k = \sum_l P_{k:l} = \text{frequency of } A_k \text{ in females}$$

$$q_l = \sum_k P_{k:l} = \text{frequency of } A_l \text{ in males}$$

APPENDIX D. PURE SELFING MODELS IN DIPLOID POPULATIONS

Consider a population of diploid organisms reproducing only by selfing. Suppose that fertility is controlled at an autosomal locus with possible alleles $A_1, A_2, \ldots, A_r$. Let f_{ij} measure the relative progeny size of an $A_i A_j$ genotype whose frequency in the whole population is $2p_{ij}$ when $i \neq j$ but p_{ii} when $i = j$. Assume that the fertility parameters satisfy $0 < f_{ij} < 1$ for $i, j = 1, \ldots, r$ and are all different from each other. Over two successive generations, we have the recurrence relations

$$Fp'_{ij} = \frac{1}{2} f_{ij} p_{ij} \text{ for all } i \neq j,$$

$$Fp'_{ii} = f_{ii} p_{ii} + \frac{1}{2} \sum_{k \neq i} f_{ik} p_{ik} \text{ for } i = 1, \ldots, r$$

where $F = \sum\limits_{i,j=1}^{r} f_{ij}p_{ij}$, or, in a more compact notation

$$p'_{ij} = \frac{f_{ij}p_{ij} + \delta_{ij}z_i}{2F} \text{ for } i,j = 1, \ldots, r, \qquad (D.1)$$

where $z_i = \sum\limits_{k=1}^{r} f_{ik}p_{ik}$ and $\delta_{ij} = 0$ if $i \neq j$ and 1 if $i = j$. Using the following ordering

$$\underline{11 \ 12 \cdots 1r}_{} \ \underline{22 \ 21 \ 23 \cdots 2r}_{} \cdots \underline{rr \ r1 \cdots r(r-1)}_{},$$

the recurrence system (D.1) in vector notation can be written in the form

$$\mathbf{p}' = \frac{L\mathbf{p}}{2F} \qquad (D.2)$$

where L is a block matrix whose klth matrix is null if $k \neq l$ and

$$(L)_{kk} = L_k = \begin{bmatrix} 2f_{kk} & f_{k1} & \cdots & f_{k(k-1)} & f_{k(k+1)} & \cdots & f_{kr} \\ & f_{k1} & \ddots & & & & \\ & & & f_{k(k-1)} & & & \\ & & & & f_{k(k+1)} & & \\ & & & & & \ddots & \\ & & & & & & f_{kr} \end{bmatrix}$$

$$\text{for } k = 1, \ldots, r. \qquad (D.3)$$

Successive iterations of (D.3) give

$$\mathbf{p}^{(N)} = \frac{L^N \mathbf{p}^{(0)}}{W^{(N)}} \qquad (D.4)$$

where $W^{(N)}$ is a normalizing constant to make $\mathbf{p}^{(N)}$ a frequency vector. The matrix L^N is a diagonal block matrix (i.e., $(L^N)_{kl} = 0$ for all $k \neq l$) with

$$(L^N)_{kk} = L_k^N = \begin{bmatrix} (2f_{kk})^N & a_{k1} & \cdots & a_{k(k-1)} & a_{k(k+1)} & \cdots & a_{kr} \\ & f_{k1}^N & \ddots & & & & \\ & & & f_{k(k-1)}^N & & & \\ & & & & f_{k(k+1)}^N & & \\ & & & & & \ddots & \\ & & & & & & f_{kr}^N \end{bmatrix}$$

$$\text{for } k = 1, \ldots, r, \qquad (D.5)$$

where $a_{kl} = \sum\limits_{j=1}^{N} (2f_{kk})^{N-j} f_{kl}^{j}$. Two cases (modulo relabeling of alleles) have to be considered:

Case (i): $2f_{11} = \max\limits_{k,l} (2f_{kk}, f_{kl})$.

A direct inspection of (D.5) reveals that

$$
\frac{L_1^N}{(2f_{11})^N} \xrightarrow[N \to \infty]{}
\begin{bmatrix}
1 & \dfrac{f_{12}}{2f_{11} - f_{12}} & \cdots & \dfrac{f_{1r}}{2f_{11} - f_{1r}} \\
0 & 0 & & 0 \\
\vdots & \vdots & \cdots & \vdots \\
0 & 0 & & 0
\end{bmatrix}
\tag{D.6}
$$

while for every $k \neq 1$

$$
\frac{L_k^N}{(2f_{11})^N} \xrightarrow[N \to \infty]{} 0 \text{ (a matrix with all zero entries)}.
$$

In the limit, A_1-fixation occurs, i.e., $p_{11}^{(\infty)} = 1$ and $F^{(\infty)} = f_{11}$.

Case (ii): $f_{12} = \max\limits_{k,l} (2f_{kk}, f_{kl})$.

The following limits are easily deduced:

$$
\frac{L_k^N}{f_{12}^N} \xrightarrow[N \to \infty]{}
\begin{bmatrix}
0 & \dfrac{f_{12}}{f_{12} - 2f_{kk}} & 0 & \cdots & 0 \\
0 & 1 & 0 & \cdots & 0 \\
0 & 0 & & & \\
\vdots & \vdots & \vdots & & \vdots
\end{bmatrix}
\tag{D.7}
$$

for $k = 1$ and 2. The limit matrix is null otherwise. Then the limiting frequency distribution of genotypes is given by the equilibrium values

$$
Dp_{11}^{(\infty)} = \frac{f_{12}}{f_{12} - 2f_{11}}
$$

$$
Dp_{22}^{(\infty)} = \frac{f_{12}}{f_{12} - 2f_{22}}
\tag{D.8}
$$

$$
Dp_{12}^{(\infty)} = Dp_{21}^{(\infty)} = 1
$$

where

$$D = \frac{f_{12}}{f_{12} - 2f_{11}} + \frac{f_{12}}{f_{12} - 2f_{22}} + 2.$$

In this case it is readily verified that $F^{(\infty)} = \sum_{i,j=1}^{2} f_{ij}p_{ij}^{(\infty)} = f_{12}/2$. Note that $F^{(\infty)}$ is the mean fertility in the population at equilibrium.

Therefore, we have proved the following result.

THEOREM D.1. *With pure selfing in diploid populations, there is global convergence to the equilibrium that maximizes the mean fertility.*

Note that the fertility parameters $\{f_{ij}\}$ can incorporate viability differences. They can also be viewed as proportions of fixed reproductive resources allocated to the female function in hermaphrodite species or plant populations in which male gametes are in excess. In this context, Theorem D.1 asserts that such populations should tend to *all female* progenies as predicted by Maynard Smith (1978) and others.

CHAPTER SEVEN

General Two-Sex Two-Allele Viability Models

In Chapter 2, we proposed that sex determination models should be viewed as special cases of sex-differentiated viability models. In the one-locus multiallele random mating case, they correspond to the situation where the viability matrices M and F, for males and females, respectively, obey the relationship $F = U - M$ where U is the matrix of all unit entries. However, such a relationship usually does not hold when environmental factors or other effects are involved in sex determination. For example, sex-specific viability differences may be associated with sex-determining genes, as occurs in the platyfish with sex-linked color genes (Kallman, 1970). In the case of hermaphrodites, Charlesworth and Charlesworth (1981) and Charnov (1982, chap. 14) proposed measures of sex allocation in the form $m(r) = r^{\alpha}$ for the resources allocated to male function and $f(r) = (1 - r)^{\beta}$ for the female function, where r may be genotypically determined. In general, $f(r) \neq 1 - m(r)$ unless the constants α and β are 1. Moreover, when gender (male versus female) fitness varies spatially with sex determination (see, e.g., Bull, 1981a; Charnov, 1979a), a general two-sex selection model may be operating in the form (2.19) or (2.20). (These models are discussed in detail in Section 7.5.) This is also the case but in a different form when differential fertilities as well as progeny sex ratios are under maternal (or paternal) genotypic control, as in Model II' of Chapter 2. See Section 7.3 for elaborations on this theme.

What are the principles underlying the dynamics of these more complex models? Is it possible to characterize the equilibria that are stable as it is for pure sex determination models?

This chapter suggests some answers by presenting a complete analysis of general two-sex two-allele viability models. (See, e.g., Bodmer, 1965; Karlin, 1972, 1978; Uyenoyama and Bengtsson, 1979, for some previous analyses.) The basic equilibrium configurations are described in Sections 7.1 and 7.2 with applications and further developments including multiallele models in Sections 7.3 and 7.5 and more technical proofs in Appendices A–C. In Section 7.4, we discuss a maximization characterization of an ESS in general two-sex models introduced in MacArthur (1965) and developed in Charnov (1982) that is based on the product of the mean fitnesses in males and females. Finally, the effects of age-specific variations on sex expression are considered in Section 7.6.

7.1. CONDITIONS FOR THE EXISTENCE OF POLYMORPHIC EQUILIBRIA

The two-sex viability model (2.5) in the case of two alleles A_1 and A_2 generalizing the sex determination model (4.2) can be written in the form

$$p' = \frac{(f_1 - 1)pq + (p + q)/2}{(f_1 + f_2 - 2)pq + (1 - f_2)(p + q) + f_2}, \quad (7.1a)$$

$$q' = \frac{(m_1 - 1)pq + (p + q)/2}{(m_1 + m_2 - 2)pq + (1 - m_2)(p + q) + m_2}, \quad (7.1b)$$

where p and q are the frequencies of A_1 in the female and male populations, respectively, in which the fitnesses of the genotypes A_1A_1, A_1A_2, and A_2A_2 have been assumed to be f_1, 1, f_2 and m_1, 1, m_2, respectively. Defining the non-negative variables $x = p/(1 - p)$ and $y = q/(1 - q)$, the recurrence equations (7.1) are equivalent to

$$x' = \frac{f_1xy + (x + y)/2}{f_2 + (x + y)/2} = g(x,y), \quad (7.2a)$$

$$y' = \frac{m_1xy + (x + y)/2}{m_2 + (x + y)/2} = h(x,y). \quad (7.2b)$$

Therefore a polymorphic equilibrium of the two-sex two-allele viability model (7.1) corresponds to a finite positive solution of $x = g(x,y)$ and $y = h(x,y)$.

It will prove convenient to adopt another equivalent formulation to express the conditions for the existence of polymorphic equilibria. We define the variable

$$t = \frac{y}{x}$$

in which case the equations (7.2) at equilibrium entail

$$x = \frac{t - \alpha_2}{1 - \alpha_1 t}, y = \frac{\beta_2 t - 1}{\beta_1 - t}, \tag{7.3}$$

where the parameters α_1, α_2, β_1, β_2 are defined by the relations

$$\alpha_1 = 2f_1 - 1, \alpha_2 = 2f_2 - 1, \beta_1 = 2m_1 - 1, \beta_2 = 2m_2 - 1. \tag{7.4}$$

Moreover, the variable t at equilibrium must be a root of the cubic equation

$$f(t) = t^3 - 3\alpha t^2 + 3\beta t - 1, \tag{7.5}$$

where

$$\alpha = \tfrac{1}{3}(\alpha_1 \beta_2 + \alpha_2 + \beta_1), \beta = \tfrac{1}{3}(\alpha_2 \beta_1 + \alpha_1 + \beta_2). \tag{7.6}$$

For a positive solution of (7.5) to be admissible as a polymorphism of (7.2), it is necessary and sufficient that the relations

$$\frac{t - \alpha_2}{1 - \alpha_1 t} > 0 \text{ and } \frac{\beta_2 t - 1}{\beta_1 - t} > 0 \tag{7.7}$$

be obeyed. All admissibility regions are given in Table 7.1 with respect to the parameters of the model. Note that in all circumstances, at most three polymorphisms can exist, since the third-degree polynomial $f(t)$ can have at most three roots.

215

TABLE 7.1. Requirments for a Positive Root of $f(t)$ of (7.5) to Determine a Polymorphism of (7.2)

	Sign of α_1, β_2	Relatives values of $\alpha_2, \dfrac{1}{\alpha_1}$ and $\dfrac{1}{\beta_2}, \beta_1$	Range for $t > 0$ to be admissible
(1)	$\alpha_1 < 0, \beta_2 < 0$		$t > \alpha_2, t > \beta_1$
(2)	$\alpha_1 > 0, \beta_2 < 0$	$\alpha_2 < \dfrac{1}{\alpha_1}$	$\alpha_2 < t < \dfrac{1}{\alpha_1}, t > \beta_1$
(3)	$\alpha_1 > 0, \beta_2 < 0$	$\dfrac{1}{\alpha_1} < \alpha_2$	$\dfrac{1}{\alpha_1} < t < \alpha_2, t > \beta_1$
(4)	$\alpha_1 < 0, \beta_2 > 0$	$\dfrac{1}{\beta_2} < \beta_1$	$\dfrac{1}{\beta_2} < t < \beta_1, t > \alpha_2$
(5)	$\alpha_1 < 0, \beta_2 > 0$	$\beta_1 < \dfrac{1}{\beta_2}$	$\beta_1 < t < \dfrac{1}{\beta_2}, t > \alpha_2$
(6)	$\alpha_1 > 0, \beta_2 > 0$	$\alpha_2 < \dfrac{1}{\alpha_1}, \dfrac{1}{\beta_2} < \beta_1$	$\alpha_2 < t < \dfrac{1}{\alpha_1}, \dfrac{1}{\beta_1} < t < \beta_1$
(7)	$\alpha_1 > 0, \beta_2 > 0$	$\alpha_2 < \dfrac{1}{\alpha_1}, \beta_1 < \dfrac{1}{\beta_2}$	$\alpha_2 < t < \dfrac{1}{\alpha_1}, \beta_1 < t < \dfrac{1}{\beta_2}$
(8)	$\alpha_1 > 0, \beta_2 > 0$	$\dfrac{1}{\alpha_1} < \alpha_2, \dfrac{1}{\beta_2} < \beta_1$	$\dfrac{1}{\alpha_1} < t < \alpha_2, \dfrac{1}{\beta_2} < t < \beta_1$
(9)	$\alpha_1 > 0, \beta_2 > 0$	$\dfrac{1}{\alpha_1} < \alpha_2, \beta_1 < \dfrac{1}{\beta_2}$	$\dfrac{1}{\alpha_1} < t < \alpha_2, \beta_1 < t < \dfrac{1}{\beta_2}$

7.2. A CLASSIFICATION OF EQUILIBRIUM CONFIGURATIONS

For simplicity, write T for the mapping defined in (7.2). We wish to ascertain the character of all equilibria of T and their domains of attraction. The analysis of T and its iterates is made easier than it might be by the fact that T is monotone. That is, where $\mathbf{z} = (x,y) < \tilde{\mathbf{z}} = (\tilde{x},\tilde{y})$ (the ordering signifies the inequality for each coordinate), we have that

$$T\mathbf{z} < T\tilde{\mathbf{z}} \text{ with strict inequality in each coordinate.} \quad (7.8)$$

This guarantees that the iterates of T converge from any starting point. (See Appendix A for more details on the properties of monotone transformations.) Condition (7.8) can be checked for the case at hand by direct inspection of T. Note also that all partial derivatives of first order of the transformation equations (7.2) are positive, e.g.,

$$\frac{\partial g}{\partial x}(x,y) = \frac{2f_1 y^2 + 4f_1 f_2 y + 2f_2}{(2f_2 + x + y)^2} > 0. \qquad (7.9)$$

The stability properties of an equilibrium of T are usually determined by analyzing the local linear approximation of the nonlinear mapping T in the neighborhood of the fixed point. More specifically, we examine the matrix transformation given by the gradient matrix

$$L(\mathbf{z}^*) = \begin{bmatrix} \dfrac{\partial g}{\partial x}(\mathbf{z}^*) & \dfrac{\partial g}{\partial y}(\mathbf{z}^*) \\[2ex] \dfrac{\partial h}{\partial x}(\mathbf{z}^*) & \dfrac{\partial h}{\partial y}(\mathbf{z}^*) \end{bmatrix} \qquad (7.10)$$

evaluated at the fixed point $\mathbf{z}^* = T\mathbf{z}^*$. If both eigenvalues of the gradient matrix at $\mathbf{z}^*$ are in magnitude less than 1, then $\mathbf{z}^*$ is locally stable. If at least one eigenvalue exceeds 1 in magnitude, then $\mathbf{z}^*$ is unstable.

The conditions for local stability of the pure equilibria $\mathbf{0} = (0,0)$ and $\boldsymbol{\infty} = (\infty,\infty)$ corresponding to the fixation states of alleles A_2 and A_1, respectively, are determined by invoking the local linear analysis just described. We find that $\mathbf{0}$ (fixation of allele A_2) is stable if and only if

$$\frac{1}{2f_2} + \frac{1}{2m_2} < 1, \qquad (7.11a)$$

and $\boldsymbol{\infty}$ (pure $A_1 A_1$ population) is stable if and only if

$$\frac{1}{2f_1} + \frac{1}{2m_1} < 1. \qquad (7.11b)$$

Using the monotonicity property of T and the fact that at most three polymorphisms can exist, it is possible to determine all a

217

priori possible equilibrium configurations based on the stability properties of the fixation states and the number of existing polymorphic equilibria. The results are presented in Figure 7.1 and complemented by the following general rules about the equilibrium structures (as proved in Appendix A).

THEOREM 7.1. i) *The equilibria of the bisexual two-allele selection model (7.2) are ordered in the sense of (7.8), i.e., coordinate by*

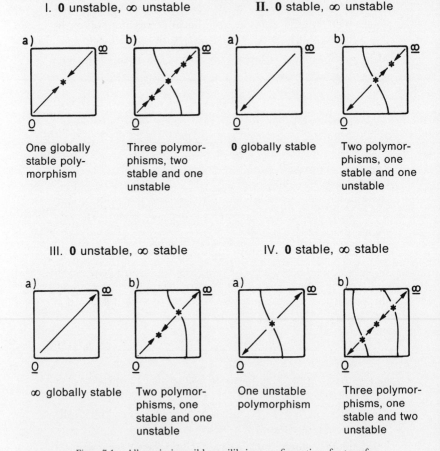

I. **0** unstable, ∞ unstable

a) One globally stable polymorphism

b) Three polymorphisms, two stable and one unstable

II. **0** stable, ∞ unstable

a) **0** globally stable

b) Two polymorphisms, one stable and one unstable

III. **0** unstable, ∞ stable

a) ∞ globally stable

b) Two polymorphisms, one stable and one unstable

IV. **0** stable, ∞ stable

a) One unstable polymorphism

b) Three polymorphisms, one stable and two unstable

Figure 7.1. All a priori possible equilibrium configurations for transformation (7.2)

coordinate; ii) *stable and unstable equilibria alternate with respect to the order (7.8);* iii) *convergence to a stable equilibrium occurs from any initial point except on curves separating the domains of attraction and containing the unstable equilibria.*

We will now consider specific viability regimes in each sex. In the female population (and analogously in the male population), there are essentially four possibilities. These are:

   i) overdominance, i.e., $1 > f_1, f_2$
   ii) underdominance, i.e., $1 < f_1, f_2$
   iii) directed selection in favor of A_1, i.e., $f_1 > 1 > f_2$
   iv) directed selection in favor of A_2, i.e., $f_1 < 1 < f_2$

We can prove the following assertions about the possible equilibrium configurations with different combinations of viability regimes (see Appendix B for proofs):

1. With overdominance in both sexes, there is a unique (globally) stable polymorphism (Case Ia).

2. With underdominance in both sexes, there is a unique unstable polymorphism (Case IVa).

3. With directed selection in favor of the same A_i in both sexes, there is global convergence to an all A_iA_i population for $i = 1, 2$ (Cases IIa and IIIa).

4. In the case of directed selection in opposite directions (e.g., in favor of A_1 in one sex and in favor of A_2 in the other), there cannot coexist three polymorphisms (Cases Ib and IVb precluded).

5. In the case of underdominance in one sex and directed selection in the other, the fixation state favored by directed selection is stable (Cases I and II or III precluded).

6. With overdominance in one sex and directed selection in the other, the fixation state disfavored by directed selection is unstable (Cases IV and II or III precluded).

7. In all circumstances, three polymorphisms are precluded when both fixation states are stable (Case IVb never possible).

All possibilities are summarized in Table 7.2. Note that three polymorphisms (Case Ib) is possible only with overdominance in one (and only one) sex.

Two cases of the bisexual selection model (7.1) that have some biological interest lend themselves to complete analysis.

Case of reciprocal symmetry of homozygote fitnesses between males and females. Consider the following scheme of fitnesses in the two-sex two-allele model:

	Females			Males		
Genotypes	A_1A_1	A_1A_2	A_2A_2	A_1A_1	A_1A_2	A_2A_2
Fitnesses	f_1	1	f_2	f_2	1	f_1

$$(7.12)$$

The male homozygote A_1A_1 has the same fitness as the female homozygote A_2A_2, and vice versa. From a biological point of view, the alleles A_1 and A_2 are exchanging their homozygotic roles with respect to the two sexes in a completely symmetrical manner.

In the circumstances (7.12), we can have overdominance in both sexes, or underdominance in both sexes, or directed selection in opposite directions. In every case, three polymorphisms are precluded. Since the fixation states must be both stable or both unstable, we are left with cases Ia or IVa of Figure 7.1, i.e., one and only one polymorphism. This unique polymorphism is $p = q = 1/2$ in the notation used in (7.1) and is stable if and only if the fixation states are unstable, i.e.,

$$\frac{1}{2f_1} + \frac{1}{2f_2} > 1. \tag{7.13}$$

The condition (7.13) is also the exact condition for the product of the mean fitnesses in males and females at the polymorphic equilibrium to exceed the corresponding product at the fixation states, namely

$$\left[\frac{(f_1 + f_2 - 2)}{4} + 1 \right]^2 > f_1 f_2.$$

220

TABLE 7.2. All Possible Equilibrium Structures for the Two-Sex Two-Allele Viability Model (7.2) According to the Viability Regime in Each Sex with Reference to Figure 7.1.

♀ \ ♂	Overdominance	Underdominance	Directed selection in favor of A_1	Directed selection in favor of A_2
Overdominance	Ia	Ia, b IIa, b IIIa, b IVa, b	Ia, b IIIa, b	Ia, b IIa, b
Underdominance	Ia, b IIa, b IIIa, b IVa	IVa	IIIa, b IVa	IIa, b IVa
Directed selection in favor of A_1	Ia, b IIIa, b	IIIa, b IVa	IIIa	Ia IIa, b IIIa, b IVa
Directed selection in favor of A_2	Ia, b IIa, b	IIa, b IVa	Ia IIa, b IIIa, b IVa	IIa

Case of sex-independent degrees of homozygote fitness. Consider a fitness scheme of the following form:

	Females		
Genotypes	A_1A_1	A_1A_2	A_2A_2
Fitnesses	f_1	1	$1 - \gamma + \gamma f_1$

	Males		
Genotypes	A_1A_1	A_1A_2	A_2A_2
Fitnesses	m_1	1	$1 - \gamma + \gamma m_1$

$$(7.14)$$

Intrinsic to the model is the fact that γ, which can be interpreted as the ratio of the selection parameters for the homozygotes, is the same for each sex. Indeed,

$$\gamma = \frac{1 - f_2}{1 - f_1} = \frac{1 - m_2}{1 - m_1} \tag{7.15}$$

The parameter γ can be regarded as a measure of the asymmetry with respect to the homozygote fitnesses. The model (7.14) therefore exhibits a sex-independent degree of homozygote fitness asymmetry. Such a situation occurs whenever 1 is a root of the polynomial $f(t)$ in (7.5).

In matrix notation, the representation (7.14) is equivalent to

$$F = aU + bM \tag{7.16}$$

with viability matrices

$$F = \begin{bmatrix} f_1 & 1 \\ 1 & 1 - \gamma + \gamma f_1 \end{bmatrix}, M = \begin{bmatrix} m_1 & 1 \\ 1 & 1 - \gamma + \gamma m_1 \end{bmatrix}$$

and coefficients

$$a = \frac{f_1 - m_1}{1 - m_1}, b = \frac{1 - f_1}{1 - m_1}. \tag{7.17}$$

(Recall that U has all unit entries.) Since $a + b = 1$, the coefficients a and b cannot both be negative. We consider all possible cases allowing any number of alleles with the condition (7.16) satisfied.

Case: $a > 0$, $b > 0$. Rescaling the viability matrices in the manner $\tilde{F} = (1/a)F$ *and* $\tilde{M} = (b/a)M$, we have

$$\tilde{F} = U + \tilde{M} \tag{7.18}$$

It is routine (by adapting the arguments used for the sex determination model (2.4) as in Karlin, 1978) to check that the equilibria of the two-sex viability model (2.6) with viability matrices $\tilde{M}$ and $\tilde{F}$ in males and females, respectively, correspond to the equilibria for the one-sex viability model (2.4) with viability matrix $\tilde{M}$ with exactly the same stability conditions (see Chapter 2 for an account of these conditions). Therefore, when a previous stable equilibrium is rendered unstable by the introduction of a mutant allele, the only stable equilibrium in the augmented system is such that the quantity

$$\langle \mathbf{p}, \tilde{M}\mathbf{p} \rangle = \frac{b}{a} \langle \mathbf{p}, M\mathbf{p} \rangle \tag{7.19}$$

is larger than it was (see Appendix D of Chapter 3), i.e., the number of males (and concomitantly of females) increases. In the case of two alleles, global convergence can be proved using standard monotonicity arguments.

Case: $a < 0$, $b > 0$. Writing

$$M = -\frac{a}{b} U + \frac{1}{b} F, \tag{7.20}$$

we are in the previous case but with the roles of the sexes interchanged.

Case: $a > 0$, $b < 0$. Redefining the viability matrices in females and males, respectively, as follows:

$$\tilde{\tilde{F}} = \frac{1}{a} F \text{ and } \tilde{\tilde{M}} = -\frac{b}{a} M,$$

we have

$$\tilde{\tilde{F}} = U - \tilde{\tilde{M}} \tag{7.21}$$

and we are in the case of the sex-determination model (2.4). Hence, from one equilibrium to the next one attainable in the sense of Theorem 3.7, the quantity

$$\langle \mathbf{p}, \tilde{\tilde{M}} \mathbf{q} \rangle = -\frac{b}{a} \langle \mathbf{p}, M \mathbf{q} \rangle \tag{7.22}$$

comes closer to $1/2$, i.e., the optimal number of males is $-a/2b$. The global dynamics according to this scheme is confirmed in Appendix A of Chapter 4 in the case of two alleles, for which convergence over successive generations can be ascertained because of monotonicity properties. The above general arguments are valid with any number of alleles as long as the relationship (7.16) holds.

THEOREM 7.2. *Consider the r-allele two-sex viability model (2.6) with $F = aU + bM$. The quantity*

$$(a + bw)w \tag{7.23}$$

where $w = \langle \mathbf{p}, M \mathbf{q} \rangle$, which represents the product of the mean fitnesses in females and males, is maximized over successive stable equilibria: It increases from one equilibrium to the next in the sense of Theorem 3.7 when new alleles are introduced one at a time.

7.3. GENERAL MATERNAL FERTILITY SCHEMES WITH SEX DIFFERENCES

Suppose that mothers of genotype $A_i A_j$ conceive male and female offspring with equal frequencies but their female offspring have fitness f_{ij} while their male offspring have fitness m_{ij} $(i,j = 1, \ldots, r)$. Note that the fitnesses are entirely determined by the mother's genotype and may be interpreted as sex-differentiated fertility parameters. Let q_i be the frequency of allele A_i paternally transmitted and x_i, y_i the frequencies of A_i in females and males, respectively, maternally transmitted, i.e.,

$$x_i = \sum_j f_{ij} p_{ij} \Big/ \sum_{k,l} f_{kl} p_{kl}, \tag{7.24a}$$

$$y_i = \sum_j m_{ij} p_{ij} \Big/ \sum_{k,l} m_{kl} p_{kl}, \tag{7.24b}$$

where $2p_{ij}$ $(p_{ii}$ when $i = j)$ is the frequency of A_iA_j mothers. Defining the maternal inheritance viability matrices $F = \|f_{ij}\|$ and $M = \|m_{ij}\|$, the transformation for the frequency vectors $\mathbf{q} = (q_1, \ldots, q_r)$, $\mathbf{x} = (x_1, \ldots, x_r)$, and $\mathbf{y} = (y_1, \ldots, y_r)$ over two successive generations formed by random mating is

$$\mathbf{q}' = \frac{\mathbf{q}}{2} + \frac{\mathbf{y}}{2} \tag{7.25a}$$

$$\mathbf{x}' = \frac{\mathbf{x} \circ F\mathbf{q} + \mathbf{q} \circ F\mathbf{x}}{2\langle \mathbf{x}, F\mathbf{q} \rangle} \tag{7.25b}$$

$$\mathbf{y}' = \frac{\mathbf{x} \circ M\mathbf{q} + \mathbf{q} \circ M\mathbf{x}}{2\langle \mathbf{x}, M\mathbf{q} \rangle}. \tag{7.25c}$$

(Cf. (2.16).) At equilibrium $\mathbf{q}^* = \mathbf{y}^*$, and the couple $\{\mathbf{x}^*, \mathbf{y}^*\}$ is an equilibrium for the general two-sex viability model (2.6) with fitness values genotypically determined by the offspring. Moreover, using the methods developed in Chapter 3, it can be shown that the stability properties correspond for both models, at least in the case $F = aU + bM$.

In the case of two alleles, introducing the variables

$$x = x_1/x_2, y = y_1/y_2, q = q_1/q_2, \tag{7.26}$$

and the parameters

$$f_{11} = f_1, \quad f_{12} = f_{21} = 1, \quad f_{22} = f_2,$$
$$m_{11} = m_1, m_{12} = m_{21} = 1, m_{22} = m_2, \tag{7.27}$$

the recurrence equations (7.25) can be put into the form

$$q' = \frac{yq + (y + q)/2}{1 + (y + q)/2} = v(y,q) \tag{7.28a}$$

$$x' = \frac{f_1xq + (x + q)/2}{f_2 + (x + q)/2} = g(x,q) \tag{7.28b}$$

$$y' = \frac{m_1xq + (x + q)/2}{m_2 + (x + q)/2} = h(x,q). \tag{7.28c}$$

It is routine to check that the transformation at hand is monotone in the sense of (7.8). Moreover, an equilibrium is

225

stable if and only if the corresponding equilibrium in (7.2) is stable (see Appendix C).

The following conclusion ensues:

THEOREM 7.3. *The equilibrium structures for the general two-sex two-allele maternal inheritance model (7.28) coincide with the equilibrium structures for the general two-sex two-allele viability model (7.2). The same is true for the multiallele versions (7.25) and (2.6) at least in the case where the viability matrices F and M satisfy the relation* $F = aU + bM$.

We surmise that Theorem 7.3 is of general validity with any number of alleles. Theorem 7.3 includes the case studied in Uyenoyama and Bengtsson (1979) with fertility parameters σ_{ij} and progeny sex ratios α_{ij} maternally determined. It suffices to take $m_{ij} = \sigma_{ij}\alpha_{ij}$ and $f_{ij} = \sigma_{ij}(1 - \alpha_{ij})$ as fitness values for the male and female offspring of an A_iA_j mother. Moreover, it can be checked that, for haplodiploid populations, the same fitness parameters lead to the same basic dynamic properties. Consequently, the optimality principle highlighted in Theorem 7.2 is applicable to both cases if a linear relationship in the form $f_{ij} = a + bm_{ij}$ between the fitness parameters is satisfied. Such an optimality principle may not hold in general since the product of the mean fitnesses in males and females does not necessarily increase from generation to generation, as confirmed by numerical simulations in Speith (1974). This product may not even be maximized at a nonsymmetric equilibrium as suggested by some analytical results in Uyenoyama and Bengtsson (1979).

7.4. THE PRODUCT CHARACTERIZATION OF AN ESS IN GENERAL TWO-SEX MODELS

MacArthur (1965) introduced an optimality criterion (in the ESS sense) for general models of sex-differentiated fertility selection. The criterion is called the product theorem in Charnov (1982). Given a fitness set S defined as the set of all feasible brood types (m,f) where m is the expected number of mature males and f the expected number of mature females in the

226

progeny of a mating type, the basic problem is to find a population state that is monomorphic for a certain brood type (m^*, f^*) and not invadable by any other brood type in S, at least when rare. The solution proposed is to maximize the product $m \times f$ in S.

In the setting of Section 7.3 (the same arguments would apply in the context of Section 7.1 for sex-differentiated allocation of resources in hermaphrodites), let $m_{ij} > 0$ and $f_{ij} > 0$ be the expected numbers of mature males and females, respectively, in the progeny of a mother of genotype $A_i A_j$ ($i,j = 1,2$). Suppose first a functional relationship between the fertility parameters in the form $f_{ij} = f(m_{ij})$ where $f(m)$ continuously decreases to zero from $m = 0$ to $m = m_{\max}$ and is twice-differentiable for $0 < m < m_{\max}$. The interval $(0, m_{\max})$ with the functional relationship $f(m)$ defines the feasible types. According to the condition (7.11) and owing to Theorem 7.3, the fixation state of allele A_1 is stable against the introduction of allele A_2 in a randomly mating population if

$$\phi(m_{11}, m_{12}) = \frac{m_{12}}{m_{11}} + \frac{f(m_{12})}{f(m_{11})} < 2 \qquad (7.29)$$

and unstable if the inequality above is reversed. The function ϕ is a particular form of the Shaw-Mohler equation (Shaw and Mohler, 1953; see Charnov, 1982). We look for a parameter m_{11}^* in $(0, m_{\max})$ such that $\phi(m_{11}^*, m_{12}) \leq 2$ with equality only at $m_{12} = m_{11}^*$. Such an m_{11}^* must satisfy

$$\frac{\partial}{\partial m_{12}} \phi(m_{11}^*, m_{11}^*) = \frac{1}{m_{11}^*} + \frac{f'(m_{11}^*)}{f(m_{11}^*)} = 0 \qquad (7.30a)$$

and

$$\frac{\partial^2}{\partial m_{12}^2} \phi(m_{11}^*, m_{11}^*) = \frac{f''(m_{11}^*)}{f(m_{11}^*)} < 0, \qquad (7.30b)$$

where f' and f'' indicate the first and second derivatives of f. The conditions (7.30) imply that m_{11}^* is a maximum point of

227

the product $m \times f(m)$. However, the converse is not true since a maximum point of the product $m \times f(m)$ may not satisfy (7.30b). A fixation state corresponding to a global maximum of $m \times f(m)$ may even be unstable against the introduction of any new feasible type (take, for example, $f(m) = 1 - m/2 + 1/2m$ for $0 < m < (1 + \sqrt{2})/2$). Actually a necessary and sufficient condition for m_{11}^* to satisfy (7.30) is to be a point of local maximum for $m \times f(m)$ such that $f''(m_{11}^*) < 0$.

When $f(m)$ is a linear function, i.e., in the form $f(m) = a + bm$, its second derivative vanishes everywhere and a parameter m_{11}^* satisfying (7.30a), i.e., $m_{11}^* = -a/2b$, is such that

$$\frac{m_{12}}{m_{11}^*} + \frac{f(m_{12})}{f(m_{11}^*)} = 2 \qquad (7.31)$$

for every m_{12}. On this basis, it is impossible to decide about the stability of the fixation state corresponding to m_{11}^*. We must resort to a local quadratic analysis as described in Appendix A of Chapter 6 or an indirect approach by looking at the stability properties of adjacent equilibria and using monotonicity properties as in Appendix A of this chapter. In any circumstances, the rate of approach to or departure from the fixation state can be at best algebraic.

Consider next a bounded convex fitness set S and let (m^*, f^*) be a point of global maximum for $m \times f$ on S. It is possible to find a function $f(m)$ such that $f(m^*) = f^*$ and in general $f(m) = cf$ for some multiplier $c \geq 1$ for every (m, f) in S. The function $f(m)$ can be chosen decreasing with $f'' \leq 0$ everywhere and such that $m \times f(m)$ is globally maximized at $m = m^*$. Then, for every (m, f) in S, we have

$$\frac{m}{m^*} + \frac{f}{f^*} \leq \frac{cm}{m^*} + \frac{cf}{f^*} = \frac{cm}{m^*} + \frac{f(cm)}{f(m^*)} \leq 2. \qquad (7.32)$$

The pair (m^*, f^*) is called a *pure* ESS with respect to S.

For a general bounded fitness set S, we consider the r-allele fertility model (7.25) with all pairs of fertility parameters (m_{ij}, f_{ij}) in S. Introducing a new allele A_{r+1} in small frequency into this system at a symmetric equilibrium $(\hat{\mathbf{q}},\hat{\mathbf{x}},\hat{\mathbf{y}})$, i.e., $\hat{\mathbf{y}}=\hat{\mathbf{x}}=\hat{\mathbf{q}}=(\hat{q}_1,\ldots,\hat{q}_r)$, a local linear analysis reveals that the allele A_{r+1} cannot go extinct near the equilibrium if

$$\frac{a}{\hat{f}}+\frac{b}{\hat{m}}>2 \tag{7.33}$$

where

$$\hat{f}=\sum_{i,j=1}^{r}f_{ij}\hat{q}_i\hat{q}_j, \quad a=\sum_{j=1}^{r}f_{r+1,j}\hat{q}_j,$$

$$\hat{m}=\sum_{i,j=1}^{r}m_{ij}\hat{q}_i\hat{q}_j, \quad b=\sum_{j=1}^{r}m_{r+1,j}\hat{q}_j.$$

If $\hat{m}\times\hat{f}$ is a global maximum for the product of the components of any convex combination of points in S, the inequality (7.33) cannot hold and $(\hat{m},\ \hat{f})$ is said to be a *mixed* ESS for S.

Although a maximization procedure based on (7.33) may lead to explicit results, we have some reservations on their meaning. Symmetric equilibria apart from fixation states are unlikely to occur in general two-sex models since they do not represent generic cases under small perturbations of the parameters. Moreover equality in (7.33) is quite possible, in which case the fate of A_{r+1} is undetermined.

7.5. SEX RATIO PATTERNS WHEN MALE VERSUS FEMALE FITNESS VARIES SPATIALLY

Charnov (1979a, 1982) and Bull (1981a) developed theoretical models for sex ratio patterns found in parasitic species (e.g., some nematodes and wasps) associated with hosts (insects, plants, or crevices) varying in size (or nutrient). With the view that a larger body size related to more host nutrient may be more beneficial to females than males in these species, it may be adaptive to produce sons in small hosts and daughters in large hosts—as tends to occur in natural populations.

This section bears on the general consequences that spatially heterogeneous environments may have on sex ratio evolution.

(Temporal variations are dealt with in the next section.) We focus on the case where sex is determined by the offspring genotype, but a similar analysis would be possible with control by the mother's genotype.

Consider N habitats and let $c_k > 0$ be the proportion of offspring that settle in habitat $\mathscr{D}_k$ where the fitness of a male is σ_k and that of a female is τ_k $(k = 1, \ldots, N)$. Suppose that an offspring of genotype $A_i A_j$ $(i,j = 1,2)$ is a male with probability $m_{ij}^{(k)}$ and a female with probability $1 - m_{ij}^{(k)}$ in habitat $\mathscr{D}_k$ $(k = 1, \ldots, N)$. Assuming non-overlapping generations and random mating in a common area, the recurrence equations for the frequencies p and q of allele A_1 in females and males, respectively, are those in (7.1) with

$$m_i = \frac{\sum_{k=1}^{N} c_k \sigma_k m_{ii}^{(k)}}{\sum_{k=1}^{N} c_k \sigma_k m_{12}^{(k)}}, \; i = 1,2,$$

$$f_i = \frac{\sum_{k=1}^{N} c_k \tau_k (1 - m_{ii}^{(k)})}{\sum_{k=1}^{N} c_k \tau_k (1 - m_{12}^{(k)})}, \; i = 1,2. \tag{7.34}$$

(See (2.19) for the general case with r alleles.) Referring to (7.11), the fixation state of allele A_1 is stable against the introduction of allele A_2 if and only if

$$\frac{\sum_{k=1}^{N} c_k \sigma_k m_{12}^{(k)}}{\sum_{k=1}^{N} c_k \sigma_k m_{11}^{(k)}} + \frac{\sum_{k=1}^{N} c_k \tau_k (1 - m_{12}^{(k)})}{\sum_{k=1}^{N} c_k \tau_k (1 - m_{11}^{(k)})} < 2. \tag{7.35}$$

The sign of the derivative of the expression on the left side of (7.35) with respect to $m_{12}^{(k)}$ $(k = 1, \ldots, N)$ is the sign of the expression

$$\sigma_k \sum_{j=1}^{N} c_j \tau_j (1 - m_{11}^{(j)}) - \tau_k \sum_{j=1}^{N} c_j \sigma_j m_{11}^{(j)}. \tag{7.36}$$

In particular, only one partial derivative can vanish for any given set of parameters $\{m_{11}^{(k)}\}_{k=1}^{N}$ if the ratios σ_k/τ_k $(k = 1, \ldots, N)$ are all different. Suppose, without loss of generality, that

$$\frac{\sigma_1}{\tau_1} < \frac{\sigma_2}{\tau_2} < \cdots < \frac{\sigma_N}{\tau_N}. \qquad (7.37)$$

Then the inequality (7.35) (with equality permitted) is satisfied for every set of parameters $\{m_{12}^{(k)}\}_{k=1}^{N}$ if $m_{11}^{(1)} = \cdots = m_{11}^{(k-1)} = 0$ and $m_{11}^{(k+1)} = \cdots = m_{11}^{(N)} = 1$ for some k with

$$m_{11}^{(k)} = \begin{cases} 0 & \text{if } m_k^* \le 0 \\ m_k^* & \text{if } 0 < m_k^* < 1 \\ 1 & \text{if } m_k^* \ge 1 \end{cases} \qquad (7.38)$$

where

$$m_k^* = \frac{\sigma_k \sum_{j=1}^{k} c_j \tau_j - \tau_k \sum_{j=k+1}^{N} c_j \sigma_j}{2 c_k \sigma_k \tau_k}.$$

This is the explicit form of a result mentioned in Bull (1981a) and Charnov (1982, chap. 4). It gives the exact optimal (in the ESS sense) sex determination pattern in a spatially heterogeneous environment. Small deviations from this pattern allowing a gradual transition from 0 to 1 for the frequency of males in the different habitats ordered according to (7.37) are generally attributed to random fluctuations in gender fitnesses and/or host size distributions. Note that the population sex ratio at birth (before the effects of local viability selection) corresponding to (7.38) is

$$\sum_{j=1}^{N} c_j m_{11}^{(j)} = c_k m_{11}^{(k)} + c_{k+1} + \cdots + c_n \qquad (7.39)$$

which is in general different from $1/2$.

With the assumption that c_k is the proportion of adults that come from habitat $\mathcal{D}_k$ $(k = 1, \ldots, N)$, the recurrence equations are those in (2.20) with $M^{(k)} = \|m_{ij}^{(k)}\|_{i,j=1}^{2}$, $\mathbf{p} = (p_1, p_2)$,

and $\mathbf{q} = (q_1, q_2)$ in the case of two alleles. A linear approximation for these equations when the frequencies p_2 and q_2 are
small yields

$$p_2' + q_2' \approx \left[\frac{\sum\limits_{k=1}^{N} c_k s_k \sigma_k m_{12}^{(k)}}{2 \sum\limits_{k=1}^{N} c_k s_k \sigma_k m_{11}^{(k)}} + \frac{\sum\limits_{k=1}^{N} c_k s_k \tau_k (1 - m_{12}^{(k)})}{2 \sum\limits_{k=1}^{N} c_k s_k \tau_k (1 - m_{11}^{(k)})} \right] (p_2 + q_2)$$

where

$$s_k = \frac{1}{\sigma_k m_{11}^{(k)} + \tau_k (1 - m_{11}^{(k)})}, \quad k = 1, \ldots, N. \qquad (7.40)$$

By analogy with (7.35), the quantity in brackets in (7.40) will
never exceed one for any set of parameters $\{m_{12}^{(k)}\}_{k=1}^{N}$ if the set
of parameters $\{m_{11}^{(k)}\}_{k=1}^{N}$ satisfies (7.38) with c_j replaced by $c_j s_j$
for $j = 1, \ldots, N$. Therefore the conclusions are similar for both
models, which may be surprising if we compare with the results
of Section 6.7, which qualitatively differ when there is local
regulation of population size.

7.6. AGE-SPECIFIC VARIATIONS, OVERLAPPING GENERATIONS, AND SEX REVERSAL

In age-structured populations, the environmental and physiological conditions may vary over the whole life cycle of an individual and may be sex-dependent. If a genetic mechanism exists
to control sex expression over the lifetime, what patterns of sex
allocation are expected to evolve? Many authors discussed the
adaptiveness of offspring sex determination according to the
mother's condition in mammals (e.g., Trivers and Willard,
1973), the distribution of the sex ratio under demographic
effects (e.g., Charlesworth, 1977), the optimal time of sex reversal for sequential hermaphrodites (e.g. Leigh et al., 1976;
Charnov, 1979a), the best timing for eclosion according to sex
in insects (e.g. Bulmer, 1983; Iwasa et al., 1983), the condi-
tions for protandry ($\male \rightarrow \female$) or protogyny ($\female \rightarrow \male$) to evolve.
In this section, we shall illustrate typical problems and solutions

of sex allocation with age-specific variations by considering a population of hermaphrodites with two breeding seasons. (See sources cited above for other hypotheses.)

Consider a large population with two age classes $\mathscr{A}_1$ and $\mathscr{A}_2$ corresponding to two overlapping generations. Let N_1 and N_2 be the total numbers of individuals in $\mathscr{A}_1$ and $\mathscr{A}_2$, respectively, before the effects of any selective pressures. The population is monomorphic for some sex-determining allele A such that every individual in $\mathscr{A}_i$ is a male with probability $m^{(i)}$ and a female with probability $1 - m^{(i)}$ for $i = 1, 2$. In the age class $\mathscr{A}_i$ ($i = 1,2$), a male has viability α_i while a female has viability β_i and fertility f_i. Following random mating and reproduction, the numbers of individuals in $\mathscr{A}_1$ and $\mathscr{A}_2$ at the beginning of the next season will be

$$\begin{bmatrix} N'_1 \\ N'_2 \end{bmatrix} = \begin{bmatrix} a & b \\ c & 0 \end{bmatrix} \begin{bmatrix} N_1 \\ N_2 \end{bmatrix} \tag{7.41}$$

where

$$a = f_1\beta_1(1 - m^{(1)}), \, b = f_2\beta_2(1 - m^{(2)}),$$
$$c = \alpha_1 m^{(1)} + \beta_1(1 - m^{(1)}).$$

If we assume that several generations have elapsed, we have the approximations

$$\frac{N'_1}{N_1} \cong \lambda, \frac{N_1}{N_2} \cong \frac{\lambda}{c} \tag{7.42}$$

where λ is the largest eigenvalue of the matrix in (7.41), namely,

$$\lambda = \frac{a + \sqrt{a^2 + 4bc}}{2}. \tag{7.43}$$

We suppose $bc \geq 1 - a$ in order to guarantee $\lambda \geq 1$. The eigenvalue λ is the asymptotic growth rate of the population and the eigenvector (λ,c) gives the stable age distribution.

Now introduce into this population a mutant allele B such that the individuals of genotype AB are male with probability $\tilde{m}^{(1)}$ in $\mathscr{A}_1$ and $\tilde{m}^{(2)}$ in $\mathscr{A}_2$. Let $\tilde{N}_1$ and $\tilde{N}_2$ be the numbers of AB individuals in $\mathscr{A}_1$ and $\mathscr{A}_2$, respectively. When these numbers are small compared to $\mathcal{N}_1$ and $\mathcal{N}_2$, a first-order approximation neglecting the effects of the BB individuals gives the recurrence equations

$$\begin{bmatrix} \tilde{\mathcal{N}}_1' \\ \tilde{\mathcal{N}}_2' \end{bmatrix} \simeq \begin{bmatrix} \tilde{a} & \tilde{b} \\ \tilde{c} & 0 \end{bmatrix} \begin{bmatrix} \tilde{\mathcal{N}}_1 \\ \tilde{\mathcal{N}}_2 \end{bmatrix} \tag{7.44}$$

where

$$\tilde{c} = \alpha_1 \tilde{m}^{(1)} + \beta_1 (1 - \tilde{m}^{(1)}),$$
$$\tilde{b} = \tfrac{1}{2}[f_2 \beta_2 (1 - \tilde{m}^{(2)}) + d\alpha_2 \tilde{m}^{(2)}],$$
$$\tilde{a} = \tfrac{1}{2}[f_1 \beta_1 (1 - \tilde{m}^{(1)}) + d\alpha_1 \tilde{m}^{(1)}],$$

with

$$d = \frac{f_1 \beta_1 (1 - m^{(1)})\lambda + f_2 \beta_2 (1 - m^{(2)})c}{\alpha_1 m^{(1)} \lambda + \alpha_2 m^{(2)} c}.$$

The quantity d can be interpreted as the mean fertility of a male at the stable age distribution (7.42). After several generations, we will have the recurrence relationship $\tilde{\mathcal{N}}_1' \cong \tilde{\lambda} \tilde{\mathcal{N}}_1$, where

$$\tilde{\lambda} = \frac{\tilde{a} + \sqrt{\tilde{a}^2 + 4\tilde{b}\tilde{c}}}{2}. \tag{7.45}$$

The allele B will spread into the population if the growth rate $\tilde{\lambda}$ exceeds λ. This will never occur if $\tilde{\lambda}$ is maximized at $\tilde{m}^{(1)} = m^{(1)}$ and $\tilde{m}^{(2)} = m^{(2)}$. The signs of the derivatives of $\tilde{\lambda}$ with respect to $\tilde{m}^{(1)}$ and $\tilde{m}^{(2)}$ evaluated at these values are the signs of the expressions

$$[\lambda d + f_2 \beta_2 (1 - m^{(2)}) + d\alpha_2 m^{(2)}]\alpha_1$$
$$- [\lambda f_1 + f_2 \beta_2 (1 - m^{(2)}) + d\alpha_2 m^{(2)}]\beta_1 \tag{7.46a}$$

and

$$d\alpha_2 - f_2 \beta_2, \tag{7.46b}$$

respectively. Note that the expressions (7.46) represent differences in expected numbers of offspring produced by males and females in $\mathscr{A}_1$ and $\mathscr{A}_2$, respectively.

234

We make the assumptions

$$\beta_1 < \alpha_1 \text{ and } \frac{f_1\beta_1}{\alpha_1} < \frac{f_2\beta_2}{\alpha_2}. \tag{7.47}$$

Three cases can occur in maximizing $\tilde{\lambda}$.

i. *(7.46a) is positive and (7.46b) vanishes.* Then $m^{(1)} = 1$ and $m^{(2)}$ is the solution between 0 and 1/2 of the equation

$$\alpha_1 f_2 \beta_2 (1 - m^{(2)}) = \alpha_2^2 (1 - 2m^{(2)})^2 \tag{7.48}$$

with the condition $\alpha_1 f_2 \beta_2 < \alpha_2^2$.

ii. *(7.46a) vanishes and (7.46b) is negative.* Then $m^{(2)} = 0$ and $m^{(1)}$ is the solution between 1/2 and 1 to the equation

$$f_1 \beta_1 (2m^{(1)} - 1)\lambda = f_2 \beta_2 [2m^{(1)}(\beta_1 - \alpha_1) - \beta_1] \tag{7.49}$$

with the condition $\alpha_1 f_1^2 \beta_1^2 > f_2 \beta_2 (2\alpha_1 - \beta_1)^2$.

iii. *(7.46a) is positive and (7.46b) is negative.* Then $m^{(1)} = 1$ and $m^{(2)} = 0$.

Therefore the conditions (7.47) should favor the evolution of protandry i.e., male to female sex reversal

APPENDIX A. MONOTONICITY PROPERTIES OF THE BISEXUAL TWO-ALLELE VIABILITY MODEL

We start by establishing the notation and terminology that will be used in this section. If $\mathbf{x} = (x_1, \ldots, x_n)$ and $\mathbf{y} = (y_1, \ldots, y_n)$ are n-tuples, we write

$$\mathbf{x} < \mathbf{y} \tag{A.1}$$

if and only if the inequality holds for each component, i.e., $x_i < y_i$ for every $i = 1, \ldots, n$.

A transformation T in n-tuples is said to be *monotone* if

$$T\mathbf{x} < T\mathbf{y} \text{ where } \mathbf{x} < \mathbf{y}. \tag{A.2}$$

It is a routine exercise to check that the transformation (7.2) for the general bisexual two-allele viability model is monotone with respect to two-dimensional positive vectors.

A simple property of monotone transformations is that their Jacobian matrices (the matrices of the first-order partial derivatives) are non-negative, usually positive. We will consider throughout only the case of irreducible Jacobian matrices, i.e., those with at least some power positive. In such circumstances, the local stability analysis of an equilibrium reduces to determining whether the Frobenius eigenvalue (equal to the spectral radius) exceeds 1.

The monotonicity property has important implications for the equilibrium and dynamic properties of the bisexual two-allele selection model (7.2). It is worth emphasizing that monotone mappings occur in many genetic and ecological models.

We recall that the local stability properties of an equilibrium can be determined from the absolute values of the eigenvalues of the relevant Jacobian matrix. In the general situation where some of the eigenvalues have absolute value exceeding 1 and some have absolute value less than 1, there will exist lower dimensional manifolds on which convergence to and divergence from the equilibrium will take place. It follows then that even if an equilibrium is not stable, it may still be approached asymptotically in some direction.

An interesting property of the fixation states $(0,0)$ and (∞,∞) in the two-sex two-allele model (7.2) is that it will be *strongly repelling* (meaning that it will not be approached from any direction) if the spectral radius exceeds 1. In this event, the allele absent from the unstable fixation state is said to be *protected* since it cannot go extinct. This property is a direct consequence of monotonicity. (A general proof is given in Appendix A of Chapter 6.)

Suppose for definiteness that $\mathbf{0} = (0,0)$ is unstable. Let $\lambda > 1$ be the leading eigenvalue of the (positive) gradient matrix L evaluated at $\mathbf{0}$. An associated positive right eigenvector $\boldsymbol{\xi}$ can be chosen of norm sufficiently small so that the linear approximation

$$T\boldsymbol{\xi} \cong L\boldsymbol{\xi} = \lambda\boldsymbol{\xi} \qquad (A.3)$$

236

guarantees the inequality

$$T\xi > \xi. \tag{A.4}$$

Successive applications of the monotonicity property of T to (A.4) give

$$T^n\xi > T^{n-1}\xi. \tag{A.5}$$

Since the sequence $T^n\xi$ is increasing in each coordinate, it must converge to a fixed point $\mathbf{x}^*$. Moreover, for every $\mathbf{x}$ satisfying

$$\xi < \mathbf{x} < \mathbf{x}^*, \tag{A.6}$$

we have

$$T^n\xi < T^n\mathbf{x} < T^n\mathbf{x}^* = \mathbf{x}^*, \tag{A.7}$$

and consequently $T^n\mathbf{x}$ must converge to $\mathbf{x}^*$. Hence, the equilibrium $\mathbf{x}^*$ is stable (otherwise the same arguments applied to $\mathbf{x}^*$ but in the opposite direction would lead to a contradiction).

In the case where $\mathbf{0}$ is stable, define

$$D = \{\mathbf{x} > \mathbf{0}: T^n\mathbf{x} \to \mathbf{0} \text{ as } n \to \infty\}. \tag{A.8}$$

The domain of attraction D of $\mathbf{0}$ is star-shaped, i.e., if $\mathbf{x}$ belongs to D, then $t\mathbf{x}$ belongs to D for every $0 < t < 1$. This is a direct consequence of the monotonicity property. Let ∂D be the frontier of D with its extremities included (see Figure A.1). Using the continuity of T, it can be shown that the curve ∂D is T-invariant, i.e., $T\mathbf{x}$ belongs to ∂D if $\mathbf{x}$ belongs to ∂D. By the Brouwer's fixed point theorem, there must exist a fixed point $\mathbf{x}^*$ lying on the curve ∂D. Actually, this fixed point cannot be at either extremity of ∂D different from $\infty = (\infty, \infty)$ by direct inspection of the equations (7.2). In any case, the equilibrium $\mathbf{x}^* > \mathbf{0}$ is unstable and unique on ∂D. (The only alternative would be three fixed points on ∂D as shown in Figure A.1 with the corner states $\mathbf{0}$ and ∞ stable, which is precluded by the analysis in Appendix B.)

Similar arguments hold for any equilibrium interior or on the boundary. This leads to the general rules on the equilibrium structures stated in Theorem 7.1.

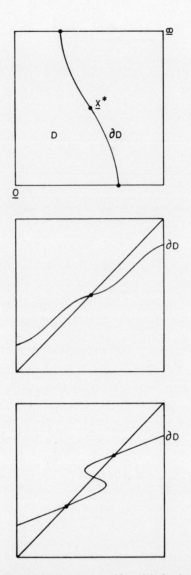

Figure A.1. The domain of attraction of $\mathbf{0} = (0,0)$ for the bisexual two-allele viability model (7.2)

a) The star-shaped domain of attraction D and its frontier ∂D
b) Possible graphs of the transformation (7.2) on the invariant curve ∂D: (i) A single equilibrium on ∂D, (ii) Two equilibria on ∂D.

NOTE: More than one equilibrium on ∂D is precluded as shown in Appendix B

238

APPENDIX B. THE EQUILIBRIUM STRUCTURES OF THE BISEXUAL MODEL UNDER SPECIFIC VIABILITY REGIMES

Overdominance in Both Sexes

The algebraic conditions for such a viability regime are given by the inequalities

$$1 > f_1, f_2, m_1, m_2 \tag{B.1a}$$

or equivalently

$$1 > \alpha_1, \alpha_2, \beta_1, \beta_2 > -1, \tag{B.1b}$$

using the notation (7.4).

By monotonicity arguments given in Appendix A, there exists at least one internal equilibrium due to the instability of the fixation states **0** and ∞. To prove uniqueness and stability, it is sufficient to show that multiple *positive* roots of $f(t) = t^3 - 3\alpha t^2 + 3\beta t - 1$ introduced in (7.5) cannot hold. Since the roots of the derivative $f'(t) = 3t^2 - 6\alpha t + 3\beta$ are $\alpha \pm \sqrt{\alpha^2 - \beta}$, it will be sufficient to show that $f(\alpha - \sqrt{\alpha^2 - \beta}) < 0$ in the case $\alpha > 0$, $\beta > 0$, and $\alpha^2 - \beta > 0$. Moreover, with the conditions (B.1) in force, we have

$$|\alpha| = \tfrac{1}{3}|\alpha_1 \beta_2 + \alpha_2 + \beta_1| < 1. \tag{B.2}$$

In these circumstances, we find

$$
\begin{aligned}
f(\alpha - \sqrt{\alpha^2 - \beta}) &= 2(\alpha^2 - \beta)^{3/2} - 2\alpha^3 + 3\alpha\beta - 1 \\
&< (\alpha^2 - \beta)[2\sqrt{\alpha^2 - \beta} - 3\alpha] \\
&< -\alpha(\alpha^2 - \beta) \\
&< 0
\end{aligned}
\tag{B.3}
$$

and the proof is complete.

Underdominance in Both Sexes

The conditions

$$1 < f_1, f_2, m_1, m_2, \tag{B.4a}$$

239

i.e.,

$$1 < \alpha_1, \alpha_2, \beta_1, \beta_2, \tag{B.4b}$$

ensure stability to the fixation states **0** and **∞**, in which case we may deduce by virtue of the monotonicity property (7.8) the existence of at least one internal unstable equilibrium. To show that this internal equilibrium is the only one, it suffices to eliminate the only alternative case of three polymorphisms.

With reference to Table 7.1, every polymorphism must correspond to a positive root of $f(t)$ defined in (7.5) satisfying $1/\alpha_1 < t < \alpha_2$ and $1/\beta_2 < t < \beta_1$. Since $f(0) = -1 < 0$, while the conditions (B.4) imply

$$f\left(\frac{1}{\beta_2}\right) = \frac{(1 - \alpha_2\beta_2)(1 - \beta_1\beta_2)}{\beta_2^3} > 0 \tag{B.5}$$

and

$$f\left(\frac{1}{\alpha_1}\right) = \frac{(1 - \alpha_1\beta_1)(1 - \alpha_1\alpha_2)}{\alpha_1^3} > 0, \tag{B.6}$$

there exists at least one inadmissible root and three polymorphisms are precluded.

Directed Selection in Favor of
A_2 (or A_1) in Both Sexes

Under the conditions

$$f_1 < 1 < f_2 \text{ and } m_1 < 1 < m_2, \tag{B.7}$$

adding the recurrence equations in (7.2) and using obvious inequalities we get

$$x' + y' < 2\left[\frac{xy + (x + y)/2}{1 + (x + y)/2}\right]. \tag{B.8}$$

Since $4xy \leq (x + y)^2$, we see that $x' + y' < x + y$. It follows that the sum of the iterates $x^{(n)} + y^{(n)}$ decreases in n and the limit is necessarily zero, indicating that $(0,0)$ is globally stable. With directed selection in favor of A_1 in both sexes, we would get global convergence to A_1-fixation.

*Directed Selection in Favor of A_1 in One Sex and
in Favor of A_2 in the Other*

Without loss of generality, we consider the case

$$f_1 > 1 > f_2 \text{ and } m_1 < 1 < m_2, \qquad \text{(B.9a)}$$

which translates into

$$\alpha_1 > 1 > \alpha_2 > -1 \text{ and } \beta_2 > 1 > \beta_1 > -1. \qquad \text{(B.9b)}$$

Since $\alpha_1 > 0$ and $\beta_2 > 0$, we need only consider cases 6, 7, 8, 9 in Table 7.1. By the fundamental theorem of algebra, the product of the roots of the polynomial $f(t) = t^3 - 3\alpha t^2 + 3\beta t - 1$ must be equal to 1.

For the cases at hand, the upper bounds α_2, $1/\alpha_1$, $1/\beta_2$, β_1 for the admissible positive roots are all less than 1 precluding three admissible roots. Therefore the possibility of three polymorphisms is precluded.

Underdominance in One Sex and Directed Selection in the Other

Consider the case

$$1 < f_1, f_2 \text{ and } m_1 < 1 < m_2. \qquad \text{(B.10)}$$

Since

$$\frac{1}{2f_2} + \frac{1}{2m_2} < 1 \qquad \text{(B.11)}$$

the fixation state of allele A_2 is stable. As a general rule, the fixation state favored by directed selection in one sex when there is underdominance in the other is always stable.

Overdominance in One Sex and Directed Selection in the Other

Take, for example

$$1 > f_1, f_2 \text{ and } m_1 > 1 > m_2. \qquad \text{(B.12)}$$

Then we have

$$\frac{1}{2f_2} + \frac{1}{2m_2} > 1 \qquad \text{(B.13)}$$

and A_2-fixation is unstable. In general, the fixation state disfavored by directed selection in one sex when there is overdominance in the other is unstable.

Three Polymorphisms is Never Possible with Two Stable Fixation States

Given the above results, it remains to consider the possibility of underdominance in one sex with overdominance or directed selection in the other. Suppose

$$1 < f_1, f_2, \text{ i.e., } 1 < \alpha_1, \alpha_2. \tag{B.14}$$

Assume that both fixation states are stable, i.e., $\alpha_1 \beta_1 > 1$ and $\alpha_2 \beta_2 > 1$. (These are the conditions (7.11) in the notation (7.4).) Then all the parameters α_1, α_2, β_1, β_2 must be positive and since $\alpha_1 \alpha_2 > 1$, we have only to consider cases 8 and 9 in Table 7.1. In these cases, a positive root of $f(t)$ must at least exceed $1/\alpha_1$ to be admissible as a polymorphism. Since $f(0) = -1 < 0$ and

$$f\left(\frac{1}{\alpha_1}\right) = \frac{(1 - \alpha_1 \beta_1)(1 - \alpha_1 \alpha_2)}{\alpha_1^3} > 0, \tag{B.15}$$

there exist at least one inadmissible root out of three, i.e., at most two polymorphisms. But at this point the stability of the fixation state and the monotonicity property of the two-sex two-allele viability model compels only one polymorphism, which is unstable (see Figure 7.1).

APPENDIX C. ANALYSIS OF THE GENERAL TWO-SEX TWO-ALLELE MATERNAL INHERITANCE MODEL

The matrix of the partial derivatives of transformation (7.28) evaluated at an equilibrium $\{q^*, x^*, y^*\}$ with $q^* = y^*$ is

$$D = \begin{bmatrix} \frac{1}{2} & 0 & \frac{1}{2} \\ g_q(x^*, y^*) & g_x(x^*, y^*) & 0 \\ h_q(x^*, y^*) & h_y(x^*, y^*) & 0 \end{bmatrix} \tag{C.1}$$

where the partial derivatives g_q, h_q, and g_x, h_x with respect to q and x, respectively, are all positive.

Since D is non-negative and D^2 is positive, the leading eigenvalue λ of D (governing the stability properties of the equilibrium at hand) is positive admitting an eigenvector (a,b,c) with all positive components. Actually, we have

$$\lambda = \max\{\mu > 0: D\mathbf{x} \geq \mu\mathbf{x} \text{ for some vector } \mathbf{x} \text{ with all} \quad \text{(C.2)}$$
positive components}

(see, e.g., Gantmacher, 1959). Let $\tilde{\lambda}$ and $(\tilde{b}, \tilde{a})$ be the corresponding quantities for the positive matrix

$$\tilde{D} = \begin{bmatrix} g_x(x^*,y^*) & g_q(x^*,y^*) \\ h_x(x^*,y^*) & h_q(x^*,y^*) \end{bmatrix} \quad \text{(C.3)}$$

that governs the dynamical behavior of the two-sex viability model (7.2) near the equilibrium $\{x^*,y^*\}$. We show that $\lambda > 1$ if and only if $\tilde{\lambda} > 1$.

Indeed, we have

$$D \begin{bmatrix} \tilde{a} \\ \tilde{b} \\ \left(\dfrac{1+\tilde{\lambda}}{2}\right)\tilde{a} \end{bmatrix} = \begin{bmatrix} \left(\dfrac{3+\tilde{\lambda}}{4}\right)\tilde{a} \\ \tilde{\lambda}\tilde{b} \\ \tilde{\lambda}\tilde{a} \end{bmatrix} > \begin{bmatrix} \tilde{a} \\ \tilde{b} \\ \left(\dfrac{1+\tilde{\lambda}}{2}\right)\tilde{a} \end{bmatrix} \quad \text{(C.4)}$$

if $\tilde{\lambda} > 1$, compelling $\lambda > 1$.

Conversely,

$$D \begin{bmatrix} a \\ b \\ c \end{bmatrix} = \lambda \begin{bmatrix} a \\ b \\ c \end{bmatrix} \quad \text{(C.5)}$$

implies $c = (2\lambda - 1)$ and

$$\tilde{D} \begin{bmatrix} b \\ a \end{bmatrix} = \lambda \begin{bmatrix} b \\ (2 - 1)a \end{bmatrix} > \begin{bmatrix} b \\ a \end{bmatrix} \quad \text{(C.6)}$$

if $\lambda > 1$ entailing $\tilde{\lambda} > 1$.

CHAPTER EIGHT

Some Incompatibility Models

Incompatibility systems can be described by specifying which among the types of mating considered possible are allowed to occur. In particular, the dichotomous sex-determination systems introduced in Chapters 2–5 are of this kind. There are many other subtle incompatibility features in nature, especially in plant populations in which several mechanisms exist to promote outcrossing. Heteromorphic incompatibility systems in which differences in flower morphology determine the inter-compatible types generally involve sporophytic mechanisms under maternal control. Homomorphic incompatibility systems with no differences in the flower are usually dependent on the pollen grain's own gene and therefore are gametophytically determined. It is also necessary to distinguish between pollen elimination and zygote elimination. In the former case, the plant producing the ovum is drawn at random and then the pollen is supplied by a plant selected at random among those able to fertilize the first one. In the latter case, all matings take place at random but the prohibited ones are infertile. (See Lewis, 1979, for more biological details.)

Whatever the incompatibility mechanism is, the same question can be asked: What polymorphic equilibria can be expected to occur? How can we characterize these equilibria and what are the conditions for their existence?

Theoretical incompatibility models and negative assortative mating systems were studied in Finney (1952), Fisher (1941), Workman (1964), Karlin and Feldman (1968a, b), Speith (1971), Karlin (1968a, 1972), Lloyd (1974), and Heuch (1979), among others. We present here a collection of results with emphasis on polymorphic equilibria exhibiting equal frequencies of types (or alleles). The occurrence or stability of such equilibria is confirmed in cases of heteromorphic (Section 8.1) and homomorphic (Section 8.2) self-incompatibility systems, a case where mating between certain like genotypes cannot happen (Section 8.3), and a case of incompatibility between haplotypes that are identical at two specific loci (Section 8.4).

8.1. SELF-INCOMPATIBILITY CLASSES AND ISOPLETHIC EQUILIBRIA

Examples of self-incompatibility classes are provided by heterostylous plants. Distylous plants for which fertilization is possible only between two different types (the long-styled plants, called pins, and the short-styled plants or thrums) include *Primula vulgaris*, where pins are *ss* and thrums *Ss*. Tristylous plants are divided into long-, mid-, and short-styled plants. In *Oxalis valdiviensis* two loci are involved: longs are *ss/mm*, mids *ss/Mm* or *ss/MM*, and shorts *Ss/mm* or *Ss/Mm* or *Ss/MM*. In *Lythrum salicaria*, a tetraploid species, longs are *ssss/mmmm* and mids are *ssss/Mmmm* or *ssss/MMmm* or *ssss/MMMm*, the other possibilities being all short. For *Theobroma cacao*, a model with r alleles, $A_1, \ldots, A_r$, and self-incompatibility classes $S_i = \{A_iA_j$ for all $j \geq i\}$ for $i = 1, \ldots, r$ (which reflects a serial dominance) has been postulated.

For a theoretical study, assume r self-incompatibility classes denoted by $S_1, \ldots, S_r$ with frequencies $u_1, \ldots, u_r$, respectively, in an infinite population. These incompatibility classes correspond to disjoint collections of genotypes determined at one or several autosomal loci assuming any underlying recombination scheme. Mating is possible or fertile only between members of different classes. In this framework, a *polymorphic equilibrium* is an equilibrium with all incompatibility classes represented, i.e., $u_1 > 0, \ldots, u_r > 0$. Of prime interest are the conditions that would ensure that the self-incompatibility classes should be present in equal frequency at a polymorphic equilibrium, i.e., $u_1 = u_2 = \cdots = u_r = 1/r$. Such populations are called *isoplethic*.

245

With no fitness differences and Mendelian segregation, the following general result can be proved.

THEOREM 8.1. (Heuch, 1979) *If a particular allele A_v can only be found in the self-incompatibility class S_v and perhaps also in $S_{v-1}, \ldots, S_1$ for $v = 1, \ldots, r$ at a polymorphic equilibrium, then this equilibrium is isoplethic.*

This result is valid with the usual mating systems corresponding to pollen or zygote elimination. The genetic assumption made about the underlying self-incompatibility determination systems is fulfilled in practically all relevant cases of heterostylous plants as exemplified above. However, Theorem 8.1 does not guarantee the *existence* of isoplethic equilibria with all incompatibility classes equally represented since the existence of polymorphic equilibria has to be demonstrated first. We know even less about the stability of such equilibria. From a biological point of view, it would be surprising if there were no stable internal equilibria but an exact population genetic analysis is required in each case. We proceed with the proof of Theorem 8.1.

Consider an equilibrium with $u_1 > 0, \ldots, u_r > 0$. Choose one gene at random (i.e., one plant from the population at equilibrium and then one gene from that plant). The probability for a particular allele A to be chosen is

$$P(A) = \sum_{v=1}^{r} P(A|S_v)u_v \tag{8.1}$$

where $P(A|S_v)$ is the probability for A to be chosen if the plant chosen belongs to S_v and u_v is the frequency of the self-incompatibility class S_v at equilibrium. An alternative procedure is to choose one gene at random in the parental generation. Therefore we also have

$$P(A) = \sum_{v=1}^{r} P(A|S_v)v_v \tag{8.2}$$

where v_v is the probability that one parent chosen at random belongs to S_v. Consequently

$$\sum_{v=1}^{r} P(A \,|\, S_v)\,(u_v - v_v) = 0. \qquad (8.3)$$

If an allele A_1 present in S_1 and not in $S_2, \ldots, S_r$ exists in the population at equilibrium, then we must have $u_1 = v_1$. Recursively if $u_i = v_i$ for $i = 1, \ldots, v - 1$ and an allele A_v present in S_v and not in $S_{v+1}, \ldots, S_r$ exists at equilibrium, (8.3) implies $u_v = v_v$. Therefore, the assumptions of Theorem 8.1 entail $u_v = v_v$ for all v.

In all cases, we have

$$v_v = \frac{1}{2} \sum_{\mu \neq v} P(S_v \times S_\mu) \quad v = 1, \ldots, r \qquad (8.4)$$

where $P(S_v \times S_\mu)$ is the proportion of parental types $S_v \times S_\mu$. Two different incompatibility mechanisms will now be considered.

i) *Zygote elimination*. In this case, matings between plants of the same class occur but are infertile. Therefore

$$P(S_v \times S_\mu) = \frac{2u_v u_\mu}{\sum\limits_{k \neq l} u_k u_l} \qquad \text{for } v \neq \mu \qquad (8.5)$$

Hence

$$v_v = \frac{u_v \sum\limits_{\mu \neq v} u_\mu}{\sum\limits_{k \neq l} u_k u_l} = \frac{u_v(1 - u_v)}{1 - \sum\limits_k u_k^2}. \qquad (8.6)$$

Since $u_v = v_v$, we conclude that $u_v = \sum\limits_k u_k^2$ for all v entailing an isoplethic equilibrium.

ii) *Pollen elimination*. In this model, pollination is possible only between different classes. An ovum produced by a plant of type S_v (whose frequency is u_v) will be fertilized by a pollen grain from a plant of type S_μ ($\mu \neq v$) with probability $u_\mu/(1 - u_v)$. Therefore

$$P(S_v \times S_\mu) = u_v\left(\frac{u_\mu}{1 - u_v}\right) + u_\mu\left(\frac{u_v}{1 - u_\mu}\right) \text{ for } v \neq \mu. \quad (8.7)$$

247

Hence

$$v_v = \frac{1}{2}\left[\frac{u_v}{1-u_v}\sum_{\mu\neq v}u_\mu + u_v\sum_{\mu\neq v}\frac{u_\mu}{1-u_\mu}\right]$$

$$= \frac{u_v}{2}\left[1 + \sum_\mu \frac{u_\mu}{1-u_\mu} - \frac{u_v}{1-u_v}\right]. \qquad (8.8)$$

Since $u_v = v_v$, we have $u_v/(1-u_v) = \sum_\mu u_\mu/(1-u_\mu) - 1$ for all v and then u_v must take the same value for all v. The proof of Theorem 8.1 is complete.

8.2. A CASE OF SELF-STERILITY ALLELES

In the genus *Nicotiana* (the tobacco plant), there is a series of alleles at a single locus that determine an incompatibility system of the pollen-elimination type. The mature plant is diploid with no differences in flower morphology and an ovum accepts only pollen that contains an allele foreign to the style (East and Mangelsdorf, 1925). In the following treatment of self-sterility alleles it is assumed that all ova are fertilized. In practice, the number of alleles is large. In fact, as many as thirty- five alleles have been identified in a sample of 500 plants of *Oenothera organensis*, an evening primrose from New Mexico having this kind of self-incompatibility (Darlington and Mather, 1949). The common reason ascribed to the existence of many alleles in this context is that a new mutant has a selective advantage when its frequency is small. We show next that a central equilibrium where all alleles are equally represented should be stable.

Assume r alleles denoted by $A_1, \ldots, A_r$. The model postulates that an ovum produced by an A_1A_2 plant may only be fertilized by $A_3, \ldots, A_r$ pollen, etc. Possible genotypes are then A_iA_j for all $i \neq j$ (for homozygotes have only transient existence and cannot last through a complete reproductive cycle). Let $2s_{ij}$ be the frequency of A_iA_j. Therefore $q_j = \sum_{i\neq j} s_{ij}$ is the frequency of A_j and an ovum produced by an A_iA_k plant $(i, k \neq j)$

will be fertilized by A_j pollen with probability $q_j/(1 - q_i - q_k)$. Then the following recurrence equations are easily deduced:

$$N s'_{ij} = q_j \sum_{k \neq i,j} \frac{s_{ik}}{1 - q_i - q_k} + q_i \sum_{k \neq i,j} \frac{s_{jk}}{1 - q_j - q_k} \text{ for all } i \neq j$$

$$(8.9)$$

where N is a normalizing factor. There is no general solution for the fixed points of this system of equations. However, we can directly verify that the central point

$$s_{ij} = \frac{1}{r(r - 1)} \text{ for all } i \neq j \qquad (8.10)$$

is a fixed point. The corresponding equilibrium allelic frequencies are $q_i = 1/r$ for all i.

Using a method due to Wright, we may trace the approximate change in allele frequency from generation to generation. The results obtained should be reasonably accurate in the neighborhood of an equilibrium state. Consider once again alleles $A_1, A_2, \ldots, A_r$. We now assume that the genotype frequencies are very nearly identical, and that the pollen production is comparable for each genotype (or allele). We further assume that all permissible fertilizations are equally likely. Given an ovum produced by a plant of genotype $A_k A_l$ ($k \neq l$), there are $r - 2$ pollen types compatible with it. If we were interested in pollination by a certain allele, say $A_i (i \neq k,l)$ we should observe that there exist $r - 3$ alleles competing directly with pollen A_i for fertilization. Also, to each allele in the pollen, there are $r - 1$ remaining alleles (in gamete form). Thus, a pollen grain of a particular type in an attempt to fertilize a specific ovum has $r - 3$ competitions out of $r - 1$ alternative alleles. The effectiveness of these competitions is in direct proportion to their frequency. Therefore the relative fertilization (contact) frequency of pollen A_i is given by

$$\frac{q_i}{q_i + \dfrac{r - 3}{r - 1}(1 - q_i)} = \frac{(r - 1)q_i}{(r - 3) + 2q_i}. \qquad (8.11)$$

Moreover, the frequency of all genotypes $A_k A_l$ for $k,l \neq i$ is given by $1 - 2q_i$. This is the frequency of ova that can be fertilized by pollen A_i.

The next frequency q_i' of allele A_i is given by one-half the fertilization frequency of pollen A_i with a compatible ovum plus one-half the frequency of all ova that contain an A_i allele, i.e.,

$$q_i' = \left[\frac{(r-1)q_i}{(r-3) + 2q_i} \right] \times \left[\frac{1 - 2q_i}{2} \right] + \frac{q_i}{2} = \frac{(r-2)q_i(1 - q_i)}{r - 3 + 2q_i}.$$

$$(8.12)$$

The change in frequency over one generation is

$$q_i' - q_i = \frac{q_i(1 - rq_i)}{r - 3 + 2q_i}.$$

$$(8.13)$$

Therefore the frequency of A_i increases or decreases accordingly as $q_i < 1/r$ or $q_i > 1/r$. Hence, for a population at a state near the central fixed point, the perturbed gene frequencies tend to return to the fixed point. Thus we have a stable equilibrium when $q_i = 1/r$ for all i. This approximate result also shows that any gene that arises by mutation has initially an effective selective advantage which is numerically large compared with the usual sort of value of most selective differences.

8.3. A MODEL OF NEGATIVE ASSORTATIVE MATING

Workman (1964) introduced a class of pollen elimination models involving what is commonly called *negative assortative mating*. That is, matings between certain like genotypes are prohibited. Such models were analyzed in detail by Karlin and Feldman (1968a, b). One case is particularly interesting because it is not very intuitive. Assume that the matings $AA \times AA$ and $Aa \times Aa$ are prohibited in a two-allele one-locus system. (Generally pollen elimination entails several alleles. The analysis of the present model may still be suggestive of more general mechanisms). Letting u, v, and w be the current frequencies of

250

genotypes AA, Aa, and aa, respectively, the recursion relations are

$$u' = \frac{uv(1 + w)}{2(w + uv)},$$

$$v' = \frac{1 - w^2}{2} + \frac{uw}{2}\left(\frac{2 - u}{1 - u}\right),$$

$$w' = w\left[\frac{v}{2(1 - v)} + \frac{v}{2} + w\right]. \qquad (8.14)$$

There are two equilibria obtained by direct inspection, namely, $u = 0$, $v = 0$, $w = 1$ and $u = 1/2$, $v = 1/2$, $w = 0$. The latter equilibrium is locally stable since a local linear approximation of the transformation (8.14) with $u = 1/2 - \varepsilon$, $v = 1/2 - \eta$, $w = \varepsilon + \eta$, where ε and η are small perturbations, reduces to the matrix transformation

$$\begin{bmatrix} \varepsilon' \\ \eta' \end{bmatrix} \approx \begin{bmatrix} \frac{3}{2} & \frac{3}{2} \\ -\frac{3}{4} & -\frac{3}{4} \end{bmatrix} \begin{bmatrix} \varepsilon \\ \eta \end{bmatrix} \qquad (8.15)$$

whose eigenvalues are less than 1 (precisely, 0 and 3/4). Therefore convergence occurs geometrically fast near that equilibrium. On the other hand, if u and v are small, we have the approximations

$$u' = uv + \text{higher-order terms},$$

$$v' = 2u + v - \frac{2u^2 + 4uv + v^2}{2} + \text{higher-order terms}.$$

Therefore after one generation

$$(2u + v)' = (2u + v) - \frac{v^2}{2} + \text{higher-order terms}. \qquad (8.16)$$

(See Appendix A of Chapter 6 for a general procedure for ascertaining local stability or instability of a fixation state when a local linear analysis is not decisive and contributions from quadratic terms must be taken into account.) We conclude that fixation of allele a is also stable, but the rate of convergence

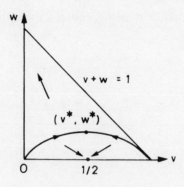

Figure 8.1. Equilibrium configuration for the negative assortative mating model (8.14)

is algebraic. Since the domains of convergence to the respective fixed points must be open sets, they must be separated by a curve as in Figure 8.1. On this curve, there is an unstable interior equilibrium that can be obtained from (8.14) as the solution v^* of the cubic $v^3 + 8v^2 - 12v + 4 = 0$ (v^* is approximately 0.55). The corresponding equilibrium value of w is $w^* = 1 - v^*(2 - v^*)/2(1 - v^*)$. Above the curve the population converges to $w = 1$ corresponding to a-fixation, while below it there is convergence to $v = 1/2$, $w = 0$ where the genotypes AA and Aa are equally represented. From any point on the curve there would be convergence to (v^*, w^*). See Karlin and Feldman (1968a, b) or Karlin (1968a) for elaborations on other one-locus two-allele models of negative assortative mating.

8.4. INCOMPATIBILITY GENETIC SYSTEM DETERMINED AT TWO LOCI

We will discuss two incompatibility models exhibiting a unique stable polymorphic equilibrium involving two loci. (Prime examples of incompatibility mechanisms determined at two loci are found in grasses and in the fungus *Schizophyllum commune*.) Consider the case of a haploid population with the usual four gametic types *AB, Ab, aB, ab*. Incompatibility is usually considered to occur in two ways, by the so-called zygote-elimination or pollen-elimination alternative.

Zygote-Elimination Model

In this case unions (or matings) occur at random but the unions $AB \times AB$, $Ab \times Ab$, $aB \times aB$, and $ab \times ab$ are affected and infertile, while all other matings produce equal relative numbers of offspring. Specifically, only the unions $AB \times Ab$, $AB \times ab$, and $Ab \times aB$ can contribute to the AB haplotype of the next generation, etc. We assume in this treatment a recombination rate r between the two loci but no differential viability among haplotypes.

Let $\mathbf{x} = (x_1, x_2, x_3, x_4)$ be the frequency array for the AB, Ab, aB, and ab haplotypes in this order. Then the change to $\mathbf{x}' = (x_1', x_2', x_3', x_4')$ for the next generation is determined by the recursion relations

$$x_i' = \frac{x_i(1 - x_i) - \varepsilon_i r D}{1 - \sum_{j=1}^{4} x_j^2}, \quad i = 1,2,3,4 \qquad (8.17)$$

where $\varepsilon_1 = \varepsilon_4 = -\varepsilon_2 = -\varepsilon_3 = 1$ and $D = x_1 x_4 - x_2 x_3$.

We prove the following convergence theorem:

THEOREM 8.2. *For any initial frequency vector* $\mathbf{x}^{(0)} = (x_1^{(0)}, x_2^{(0)}, x_3^{(0)}, x_4^{(0)})$, $x_i^{(0)} > 0$, *the iterates* $\mathbf{x}^{(n)}$ *of the transformation (8.17) with* $0 \leq r \leq 1/2$ *converge to* $\mathbf{x}^* = (1/4, 1/4, 1/4, 1/4)$.

REMARK. A local stability analysis of the central equilibrium is given in Feldman (1971). The rate of convergence is easily computed to be $2/3$ (the largest eigenvalue of the gradient matrix of the transformation (8.17) evaluated at $\mathbf{x}^*$).

PROOF. Without loss of generality (relabel the types, if necessary), we may assume $x_1 > x_4$ and $x_2 > x_3$. Direct evaluation produces

$$x_1' - x_4' = \frac{(x_1 - x_4)(x_2 + x_3)}{1 - \sum_{i=1}^{4} x_i^2} \qquad (8.18)$$

showing that $x_1^{(n)} > x_4^{(n)}$ for all $n \geq 1$. Similarly, we deduce $x_2^{(n)} > x_3^{(n)}$ for all $n \geq 1$. Observe next that

$$x_1' + x_4' = \frac{(x_1 + x_4) - (x_1 + x_4)^2 + 2(1 - r)x_1 x_4 + 2rx_2 x_3}{1 - \sum x_i^2}$$

$$> \frac{(x_1 + x_4)(x_2 + x_3)}{1 - \sum x_i^2}. \qquad (8.19)$$

Combining (8.18) and (8.19) yields

$$\frac{x_1' - x_4'}{x_1' + x_4'} < \frac{x_1 - x_4}{x_1 + x_4} \text{ with equality iff } x_1 = x_4. \qquad (8.20)$$

Thus $(x_1^{(n)} - x_4^{(n)})/(x_1^{(n)} + x_4^{(n)})$ decreases to the limit 0. A direct check shows that after the initial generation $x_i^{(n)} < 1/2$, $i = 1,2,3,4$. In fact, that $x_1' < 1/2$ holds if $2(x_1 - x_1^2 + x_2 x_3) < 1 - x_1^2 - x_2^2 - x_3^2 - x_4^2$ is correct and the latter inequality is equivalent to $(1 - x_1)^2 > x_2^2 + x_3^2 + x_4^2 + 2x_2 x_3$, which is obvious. It follows that $x_1^{(n)} - x_4^{(n)} \to 0$ and by analogous means we infer $x_2^{(n)} - x_3^{(n)} \to 0$. Since $x_1^{(n)} + x_2^{(n)} + x_3^{(n)} + x_4^{(n)} = 1$, we also obtain $x_1^{(n)} - (1/2 - x_i^{(n)}) \to 0$, $i = 2,3$, for $n \to \infty$.

Inserting the limit values into the first relation of (8.17) produces the recursion formula

$$x_1^{(n+1)} = f(x_1^{(n)}) + \varepsilon_n$$

where

$$f(\xi) = \frac{\xi(1 - \xi) + r(\tfrac{1}{4} - \xi)}{\tfrac{1}{2} + 2\xi - 4\xi^2} \text{ and } \varepsilon_n \to 0 \qquad (8.21)$$

The functional equation $f(\xi) = \xi$ has a unique fixed point $\xi = 1/4$ in the range $0 < \xi < 1/2$ provided $0 \leq r \leq 1/2$. For $\xi < 1/2$, the iteration $f_{(n)}(\xi) = f_{(n-1)}(f(\xi))$ converges to $1/4$. We conclude from the representation (8.21) where $\varepsilon_n \to 0$ that $x_1^{(n)} \to 1/4$ and the proof of the theorem is complete in this case.

Pollen-Elimination Model

We assume that no ova can be fertilized by pollen of the same haplotype but that all ova are fertilized. This is a common

254

phenomenon in many plant and insect populations. The sex allocation is stipulated as $1:1$. We keep track of the frequencies in the female population. In this setup the transformation equations of the haplotype frequencies in successive generations become

$$x_1' = \frac{x_1}{2}\left(1 + \frac{x_2}{1 - x_2} + \frac{x_3}{1 - x_3} + \frac{x_4}{1 - x_4}\right) - rD$$

$$x_2' = \frac{x_2}{2}\left(1 + \frac{x_1}{1 - x_1} + \frac{x_3}{1 - x_3} + \frac{x_4}{1 - x_4}\right) + rD$$

$$x_3' = \frac{x_3}{2}\left(1 + \frac{x_1}{1 - x_1} + \frac{x_2}{1 - x_2} + \frac{x_4}{1 - x_4}\right) + rD \qquad (8.22)$$

$$x_4' = \frac{x_4}{2}\left(1 + \frac{x_1}{1 - x_1} + \frac{x_2}{1 - x_2} + \frac{x_3}{1 - x_3}\right) - rD$$

where

$$D = \frac{x_1 x_4}{2}\left(\frac{1}{1 - x_4} + \frac{1}{1 - x_1}\right) - \frac{x_2 x_3}{2}\left(\frac{1}{1 - x_3} + \frac{1}{1 - x_2}\right).$$

THEOREM 8.3. *For $0 \le r \le 1/2$ and any initial frequency vector $\mathbf{x}^{(0)}$ with positive components, we have $\mathbf{x}^{(n)} = T^n \mathbf{x}^{(0)} \to \mathbf{x}^* = (1/4, 1/4, 1/4, 1/4)$ where T is the transformation (8.22).*

The proof will be done in a series of steps. Without loss of generality we may assume $x_1^{(0)} > x_4^{(0)}$, $x_2^{(0)} > x_3^{(0)}$.

Claim (i): $x_1^{(n)} > x_4^{(n)}$, $x_2^{(n)} > x_3^{(n)}$. We compute

$$x_1' - x_4' = \frac{(x_1 - x_4)}{2}\left[1 + \frac{x_2}{1 - x_2} + \frac{x_3}{1 - x_3} - \frac{x_1 x_4}{(1 - x_1)(1 - x_4)}\right]$$

and note that the expression in brackets is positive.

A Lyapounov function that will serve to establish global convergence is

$$L(\mathbf{x}) = L(x_1, x_2, x_3, x_4) = \max_{i,j} |x_i - x_j|. \qquad (8.23)$$

We will validate $L(T\mathbf{x}) < L(\mathbf{x})$ for all positive frequency vectors $\mathbf{x}$ except if $\mathbf{x} = \mathbf{x}^* = T\mathbf{x}^*$.

Claim (ii): For n large enough, $x_1^{(n)} + x_2^{(n)} \leq 2/3$ and $x_i^{(n)} \leq 1/2$, $i = 1, 2, 3, 4$. Consider

$$x_3' + x_4' = \frac{(x_3 + x_4)}{2}\left[1 + \frac{x_1}{1 - x_1} + \frac{x_2}{1 - x_2}\right]$$

$$+ \frac{x_3 x_4}{2}\left[\frac{1}{1 - x_3} + \frac{1}{1 - x_4}\right]$$

$$> \frac{b}{2}\left[1 + \frac{x_1}{1 - x_1} + \frac{x_2}{1 - x_2}\right] \text{ where } b = x_3 + x_4.$$

Plainly,

$$\min_{x_1 + x_2 = 1 - b}\left(1 + \frac{x_1}{1 - x_1} + \frac{x_2}{1 - x_2}\right) = 1 + \frac{2(1 - b)}{(1 + b)}$$

and

$$\frac{b}{2}\left(1 + \frac{2(1 - b)}{1 + b}\right) > b$$

provided $b < 1/3$. Thus if $x_3 + x_4 < 1/3$, then $x_3' + x_4' > x_3 + x_4$. Moreover, where $x_3 + x_4 > 1/3$ the same argument shows that $x_3' + x_4' > 1/3$. With this information the first conclusion of (ii) is readily established. In order to show $x_1' \leq 1/2$ we consider two cases:

(α) $D \geq 0$. Then $x_1' \leq \frac{x_1}{2}\left(1 + \sum_{2}^{4} \frac{x_i}{1 - x_i}\right)$. The expression $\sum_{i=2}^{4} \frac{x_i}{1 - x_i}$ is convex on the linear section $x_2 + x_3 + x_4 = 1 - x_1$ with $x_2, x_3, x_4 \geq 0$. Thus its maximum is achieved at a corner and consequently $x_1' \leq \frac{x_1}{2}\left(1 + \frac{1 - x_1}{x_1}\right) = \frac{1}{2}$.

(β) For $D < 0$, a little manipulation using $r \leq \frac{1}{2}$ leads to

$$x_1' + x_4' \leq \frac{1}{2}\left[\frac{x_2(x_1 + x_3 + x_4)}{1 - x_2} + \frac{x_3(x_1 + x_4 + x_2)}{1 - x_3}\right] \leq \frac{1}{2}, \text{ and,}$$

a fortiori, $x_1' \leq \frac{1}{2}$.

To prove Theorem 8.3 we treat two cases:

Case (a): $x_1 > x_2 > x_3 > x_4$; and *Case (b):* $x_1 > x_2, x_4 > x_3$. There are two other cases that reduce to those above by re-

256

labelings. Proceeding with *case* (*a*), we will show that

$$\max_{i,j} |x_i' - x_j'| < x_1 - x_4, \qquad (8.24)$$

unless there is equilibrium. This is done by verifying all the specific inequalities of (8.24). We illustrate with a number of these:

(a1). $x_1' - x_4' < x_1 - x_4$. We obtain

$$x_1' - x_4' < \frac{(x_1 - x_4)}{2} \left[1 + \frac{x_2}{1 - x_2} + \frac{x_3}{1 - x_3} \right].$$

But with $x_3 \le x_2 \le 1/3$ since $x_1 + x_2 \le 2/3$ by (ii) and $x_1 \ge x_2$ we find that

$$\frac{x_2}{1 - x_2} + \frac{x_3}{1 - x_3} \le 1$$

and (a1) is proved.

(b1). $x_2' - x_3' < x_1 - x_4$. Consider

$$x_2' - x_3' - (x_1 - x_4)$$
$$< \frac{(x_2 - x_3)}{2} \left[1 + \frac{x_1}{1 - x_1} + \frac{x_4}{1 - x_4} \right] - (x_1 - x_4)$$
$$= f(x_1, x_2, x_3, x_4). \qquad (8.25)$$

By differentiation and the fact that $x_2 < x_1 \le 1/2$, we find that f is decreasing as a function of x_1 for $x_1 \le 1/2$. This easily implies $x_2' - x_3' - (x_1 - x_4) < 0$, by evaluating (8.25) at $x_1 = x_2$, and (b1) is proved.

(c1). $|x_2' - x_4'| < x_1 - x_4$. We examine for $D < 0$,

$$x_2' - x_4' - (x_1 - x_4)$$
$$< \frac{(x_2 - x_4)}{2} \left(1 + \frac{x_1}{1 - x_1} + \frac{x_3}{1 - x_3} \right) - (x_1 - x_4)$$

and the right-hand expression decreases in x_1 for $x_1 \le 1/2$. Substituting x_2 for x_1 gives on the right

$$\frac{(x_2 - x_4)}{2} \left(1 + \frac{x_2}{1 - x_2} + \frac{x_3}{1 - x_3} \right) - (x_2 - x_4) \le 0.$$

Thus (c1) is proved when $D < 0$.

257

For $D > 0$ and $r \leq 1/2$ we have

$$x_2' - x_4' - (x_1 - x_4)$$

$$< \frac{(x_2 - x_4)}{2}\left[1 + \frac{x_1}{1 - x_1} + \frac{x_3}{1 - x_3}\right] + D - (x_1 - x_4)$$

$$= g(x_1, x_2, x_3, x_4). \tag{8.26}$$

A straight calculation shows that

$$\frac{\partial g}{\partial x_1} = \frac{x_2}{2}\left[\frac{1}{(1 - x_1)^2}\right] - 1 + \frac{x_4}{2(1 - x_4)}.$$

Since $x_1 > x_2 > x_3 > x_4$ and the sum adds to 1, we infer $x_2 + x_4 < 1/2$. Hence, using the fact that $x_1 \leq 1/2$, we find that

$$\frac{x_2}{2(1 - x_1)^2} + \frac{x_4}{2(1 - x_4)} \leq \frac{1}{2}\left[\frac{x_2 + (1 - x_1)x_4}{(1 - x_1)^2}\right] < 1$$

and therefore g is decreasing with respect to x_1. Put x_1 equal to x_2 in (8.26) to get

$$x_2' - x_4' - (x_1 - x_4)$$

$$< \frac{(x_2 - x_4)}{2}\left[1 + \frac{x_2}{1 - x_2} + \frac{x_3}{1 - x_3}\right] - (x_2 - x_4) \leq 0.$$

If $x_2' - x_4' < 0$, then $|x_2' - x_4'| \leq D < x_1 - x_4$ for $r \leq 1/2$. Thus in all circumstances (c1) is proved.

By analogous methods, we check that all the separate differences $|x_i' - x_j'|$ are smaller than $x_1 - x_4$.

Case (b): $x_1 > x_2$, $x_4 > x_3$. The arguments parallel case (a) to prove

$$\max_{i,j} |x_i' - x_j'| \leq x_1 - x_3.$$

with equality only at equilibrium.

CHAPTER NINE

Multifactorial Sex Determination (MSD) Models

9.1. BACKGROUND

Multifactorial sex determination (MSD) denotes a mechanism of sex expression involving many factors of small effects. These include a hierarchy of multiple gene determinants often interacting with environmental factors. In this perspective, multifactorial sex determination also encompasses the prospects of environmental sex determination (ESD). ESD generally signifies sex determination after conception, where the individual genotype has less influence on the sex phenotype, although genes are certainly involved in sexual development and in other life stages. Sex conversion (sequential hermaphroditism) also relates to a multiplicity of environmental and genetic factors and can be considered in the context of MSD.

As contrasted with sex determination under chromosomal or major gene control, Winge (1934) proposed a mechanism of small additive effects contributed from many loci leading to a male (or female) sex phenotype when the cumulative genic effects exceed a threshold. In an earlier study, Winge (1932) considered a heterogametic XX/XY chromosomal mechanism allowing for manifold autosomal modifiers that can override the major factor, producing, for example, XX males. Such possibilities exist (cf. Bull, 1983, chap. 8; and Section 9.7 below) but generally heterogametic determinants are not much influenced by exogeneous factors.

Bull (1983, chap. 8) discusses the practical difficulties in demonstrating the existence of multifactorial sex determination. Comparisons of between-family sex ratio variation and the

259

observation of paternal as against maternal effects on brood composition are among the criteria used in testing MSD.

Examples of MSD with mixed genetic factors include poeciliid fish. Swordtails presumably involve at least two or three gene factors (Kallman, 1983). However, it is generally difficult to distinguish two- or three-factor sex determination structures from multifactorial forms.

Environmental Sex Determination (ESD)

Some authors (e.g., Ohno, 1967; Jones, 1983) postulate that the basic (primitive) sex type was environmentally determined (e.g., temperature-sensitive) that later specialized to genetic (probably multiple-factored) controls that in turn coalesced mediated by reduced recombination, inversions, insertions and transpositions to chromosomal determination of male or female heterogamety.

Bull (1983, chap. 9) reviews various actual and putative cases of ESD. Examples are widespread, including sex expression in response to a temperature state at some developmental stage, sex phenotype ascertained in relation to prenatal or neonatal conditions, sex function dependent on birth size or influenced by physiological health and resource availabilities of host, and sex changes linked to conspecific interactions (cf. Charnov, 1982, and references therein). ESD can also be affected by "water potential," meaning the degree of dry versus wet ambient environment. Other influences (especially in fish populations) include pH, photo period, and social milieu.

Bacci (1965) describes some fish and worm examples subscribing to the rule that the first larva to settle on a fish becomes female while any subsequent become male. For example, the marine worms *Echiurid* and *Boxellia virelis* are sessile as adults. Moreover, their larvae in settling on an adult female develop as males, but when they settle in isolation they tend to emerge as females. In other situations, if raised in isolation, they usually function as females, but when raised in association with adult females they develop as males (see also Leutert, 1975).

Sex ratio in some parasitic cases varies as a function of host size. Daughter versus son fitness comparisons can also change with host size. Nematodes (e.g., mermithids) parasitize insects as larvae, thereby receiving nourishment by host insect. The sex is determined by the amount of food available to developing worms. Increased nutritional resources cause bias toward femaleness. Concomitantly, under crowded conditions, the worm responds as a male. In plant parasitic nematodes, under crowding and stressful conditions a bias toward maleness is observed. Maternal choice of nest or host site may play a role in environmental sex determination.

Sex expression can be temperature (at time of incubation, nesting, or at a later developmental stage)-sensitive. Several authors highlight cases of ESD based on temperature conditions, suggesting temperature-sensitive controlling genes for sex expression. Established examples founded on incubation temperatures feature an assortment of reptiles and amphibians. The following systems occur: there exist one or two threshold levels resulting in a male or female sex phenotype depending on the incubation temperature conditions relative to the threshold levels. Thus, the schemes depicted in the diagram below are relevant. In case (a) encountering the environmental variable $V > v^*$ induces maleness while $V < v^*$ induces femaleness.

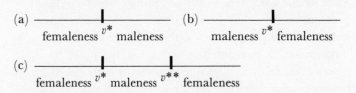

The threshold values are not absolute (cf. Bull, 1983, pp. 117–119). Sex expression of certain lizards and alligators correspond approximately to the ESD structure of (a) while a number of turtle species conform to (b). In the case of crocodiles the environmental stimulus is the nesting temperature, subscribing to the two-threshold criterion as in (c). By contrast to the models (a)–(c), sex determination for most lizards and snakes is not sensitive to the incubation temperature.

A physiological variable (e.g., egg size) can underlie ESD. Sex determination patterns following (b) show cases where large eggs tend to be female and small to be male (e.g., some Annelids) with subsidiary cytoplasmic controls.

Natural selection may act against ESD in view of physiological complications. If gender fitness varies spatially or temporally, then an advantage to ESD can result whenever the embryo can increase its self-fitness. For ESD superimposed on GSD or interactions thereof the sex determination mechanism effectively assigns probabilities to different genotypes (cf. Chapter 2).

Sex Conversion

There are many similarities in processes of sex conversion and ESD. The timing and order of sequential hermaphroditism (sex conversion) in part relate to environmental stimuli that activate specific sensory areas that subsequently affect the maturation of appropriate gonadal organs. The phenomenon of sequential hermaphroditism is widespread (see Policansky, 1982, for a recent review).

Sexual conversion has been particularly strong in fishes inhabiting the coral reef seas. For example, in *Sparus aurata*, a protandrous fish, early periods of development are characterized by individuals of male sex that within a year are generally transformed to functioning females. On the other hand, in protogynous fish (e.g., *Authias squamipinnis*) early development is female converting subsequently to a male phenotype. The sex phenotype at each stage is modulated with genetic multifactors, the stimulation of certain hormonal secretions, and the general environmental state reflecting on physical, biotic, and social factors. Social conditions can cause conversions in all life stages (e.g., wrasses). For example, the absence of females can stimulate a change of the dominant female into a functioning male (Fishelson, 1970). Similar changes are documented for Australian coral fish. In *Sparus aurata*, the presence of several males in the same locality causes a reversal of sex in the nondominant males. Accordingly, sex is not fixed in the genome in these fish populations. Thus, sex tendencies are subject to *multifactorial* controls that can change in response to environmental conditions. The appearance of male phenotypes (for platyfish) among female genotypes is sometimes explained as due to autosomal modifier genic effects (see Kallman, 1983). Other examples are discussed in Charnov (1982).

Sex phenotypes in genetically identical *Revulus* fishes can be mediated by different temperature conditions. Overall, there are multiple mechanisms, genetic and nongenetic, the latter including social and physical environmental factors that can cause sexual succession in protogynous or protandrous fishes. It is interesting that population density can apparently affect the sex ratio. In this vein, preliminary studies on the sex ratio in *Sparus aurata* grown in cages at high density showed a surprising 1:4 or 1:5 sex ratio, while these fishes grown in open pools exhibited an even sex ratio (Dvorah Patinkin, private communication). Dense growth conditions in this case seem to favor a preponderance of females.

9.2. FORMULATION AND ISSUES OF MULTIFACTORIAL SEX DETERMINATION MODELS

The study of multifactorial sex determination is relevant in at least two contexts. First, the sex phenotype may be influenced by many loci (e.g., mediated through some developmental process) in association with biochemical, physiological, ecological, and behavioral variables, as is probable with various fish, amphibian, reptile, and invertebrate species. Second, environmental sex determination probably reflects a complex situation of gene-environment interactions. In this perspective the environmental variable can be regarded as an autocorrelated multifactorial trait obeying certain familial phenotypic transmission rules.

The concept of a multifactorial (polygenic) model is intrinsically unclear. Many genes are interacting, and this may be coupled to environmental stimuli in complex ways. In models of this chapter we assume that sex expression is conditioned in response to a phenotype value (or vector value) X (e.g., size, some physiological covariate, exposure to sunlight) or is determined in response to a variable X that indexes an environmental state (e.g., temperature, humidity) such that an offspring endowed with or encountering the state $X = x$ becomes male with probability $p(x)$ and female with probability $1 - p(x)$. We do not attempt to make precise the underlying genetics. The changes of the variable X are described by a transmission process in which the offspring value receives some "blend" of the parental values modified by independent individual residual terms.

The main conclusion that emerges from our studies of multifactorial sex determination is that a non-even population sex ratio is expected when environmental pressures affect the sexes asymmetrically, or when paternal and maternal genetic or phenotypic contributions are unequal.

Let the density of index values at generation t be $f_t(x)$, i.e., the fraction of the population encountering index values in the range $(x, x + \Delta x)$ is approximately $f_t(x)\,\Delta x$. The sex ratio at conception is

$$r_t = \int p(x) f_t(x)\, dx. \tag{9.1}$$

The density of X in the male population is

$$f_t^*(x) = \frac{p(x) f_t(x)}{r_t} \tag{9.2a}$$

and that of females

$$f_t^{**}(x) = \frac{[1 - p(x)] f_t(x)}{1 - r_t}. \tag{9.2b}$$

We display the important case

$$p(x) = \begin{cases} 1, & x > v^*, \\ 0, & x \leq v^*, \end{cases} \tag{9.3}$$

of a threshold criterion.

At the adult stage the proportions of males and females can be adjusted in several ways. We describe two cases:

Case I. At mating, the proportions of males and females are equalized, their trait distributions being those of (9.2).

Case II. Sex-differentiated viability selection acts with relative factors $\gamma(x)$ and $\delta(x)$ producing the new adult male and female phenotype distributions

$$g_t^*(x) = \frac{\gamma(x)p(x)f_t(x)}{\int \gamma(\xi)p(\xi)f_t(\xi)\,d\xi} \tag{9.4}$$

$$g_t^{**}(x) = \frac{\delta(x)[1 - p(x)]f_t(x)}{\int \delta(\xi)[1 - p(\xi)]f_t(\xi)\,d\xi}. \tag{9.5}$$

Henceforth, unless stated otherwise, we concentrate on Case I.

We assume blending transmission with the offspring index value inheriting a mixture of the parental index values, allowing for an individual residual addend. More specifically, the index variable endowed to the offspring is taken to be

$$X_{t+1} = \lambda X_t^* + \mu X_t^{**} + \nu \varepsilon_t \tag{9.6}$$

where X_t^* (X_t^{**}) is the paternal (maternal) contribution and ε_t is an independent environmental term following a density function $e(x)$ independent of the generation epoch. The random variables X_t^*, X_t^{**}, ε_t are independent and governed by the density functions $f_t^*(x)$, $f_t^{**}(x)$, and $e(x)$, respectively.

Generally, λ and μ are non-negative scale constants. For $\lambda = \mu$ the parameter can be construed as a heritability coefficient. When $\lambda \neq \mu$ the parents transmit asymmetrically. For $e(x)$ a degenerate distribution concentrating at zero, i.e., $\varepsilon_t = 0$

and $\lambda = \mu = 1/2$, we obtain the Galtonian blending recursion:

$$X_{t+1} = \frac{X_t^* + X_t^{**}}{2} \qquad (9.7a)$$

which we write for convenience as

$$\tilde{X} = \frac{X^* + X^{**}}{2} \qquad (9.7b)$$

suppressing the t index. The symmetric transmission model of (9.7) was introduced in Bulmer and Bull (1982).

More general formulations allow $p(x)$ to vary systematically or stochastically in time and/or space. The perturbation terms ε_t may be taken as a random process involving some dependence relationships over time and may even be density-dependent.

The mean male index value ($f(x) = f_t(x)$ suppressing the t index) is

$$m^* = \frac{\int xp(x)f(x)\,dx}{\int p(x)f(x)\,dx} = m + \frac{\int (x-m)p(x)f(x)\,dx}{r}, \qquad (9.8a)$$

where $m = \int xf(x)\,dx$ and $r = \int p(x)f(x)\,dx$. Similarly, in females

$$m^{**} = \frac{\int x(1-p(x))f(x)\,dx}{\int (1-p(x))f(x)\,dx} = m - \frac{\int (x-m)p(x)f(x)\,dx}{1-r}.$$

$$(9.8b)$$

The mean value of $\tilde{X}$ defined in (9.7b) is

$$\tilde{m} = m + \frac{1-2r}{2r(1-r)} \int (x-m)p(x)f(x)\,dx. \qquad (9.9)$$

By virtue of the Tchebycheff rearrangement inequality (provided $p(x)$ is an increasing function of x) we have

$$\int (x-m)p(x)f(x)\,dx \geq \left(\int p(x)f(x)\,dx \right)\left(\int (x-m)f(x)\,dx \right) = 0,$$

$$(9.10)$$

266

and therefore,

$$\tilde{m} - m > 0 \text{ when } r < 1/2,$$
$$\tilde{m} - m < 0 \text{ when } r > 1/2.$$

(9.11)

We could expect that the sex ratio r over successive generations tends to $1/2$. This is generally *not* true even for the model (9.6) involving environmental terms.

The density of $\bar{X}$ defined in (9.7b) is

$$\tilde{f}(x) = 2 \int_{-\infty}^{\infty} f^*(\xi) f^{**}(2x - \xi) \, dx.$$

(9.12)

But following this transformation, the relation over successive generations is formidable. Nonetheless, various models are worked out in the later sections of this chapter.

Bulmer and Bull (1982) consider equation (9.9) where $f(x)$ is specified as a normal density. However, when $f(x)$ is normal, (9.12) is no longer a normal density for the progeny generation. Thus the assumption is not useful for theoretical studies nor is it realistic in terms of data on ESD; Bull (1983) calls attention to this problem.

Issues of MSD

We conclude this section with a brief discussion of several objectives in the study of multifactorial sex determination models. The hope is to get qualitative insights into some of the following questions:

(i) Under what conditions for MSD does an even sex ratio evolve? How is it dependent on environmental versus parental transmission components? What is the effect of asymmetry in the maternal versus paternal contributions to the progeny index value?

(ii) How does a mixed-major-gene plus MSD model influence sex ratio evolution? In what respects does MSD differ from a two- or three-factor sex-determining mechanism?

(iii) To what degree does an environmental sex determination (ESD) mechanism conform to MSD? What are the implications for sex conversion?

267

In the following section, we present a number of results pertaining to the temporal changes of the index value distribution that underlies multifactorial sex determination. Section 9.3 concentrates on the case of MSD having the index value transformation dominated by parental transmission effects. Section 9.4 develops further convergence and equilibrium results. The formulation of MSD reflecting mother control in diploid organisms and in cases of haploid-diploid species are set forth in Section 9.6. Section 9.7 describes some mixed models allowing for major gene control coupled to multifactorial sex determinants (e.g., modifiers). These models relate to sex conversion system.

9.3. MULTIFACTORIAL SEX-RATIO DETERMINATION MODELS WITH RANDOM RESIDUAL INFLUENCES

This section discusses some results bearing on the dynamics of sex ratio evolution for the symmetric model described by the recursion relation

$$X_{t+1} = \lambda \frac{X_t^* + X_t^{**}}{2} + (1 - \lambda)\varepsilon_t \qquad (9.13)$$

where $\lambda : (1 - \lambda)$, $0 < \lambda < 1$, scales the relative contribution to the offspring index value from parental transmission compared to an independent environmental term having ε_t distributed with density $e(x)$ of mean $1/2$ and positive variance, A more general transformation rule (cf. (9.6)) has the form

$$X_{t+1} = \lambda X_t^* + \mu X_t^{**} + (1 - \lambda - \mu)\varepsilon_t \qquad (9.14)$$

where λ and μ can be construed as non-negative sex-dependent "heritability" coefficients obeying $0 \leq \lambda + \mu \leq 1$.

We consider first the case of the index range $0 < x < 1$ with probability of femaleness determined by the function $p(x) = x$. We also assume $0 < \varepsilon_t < 1$. In the following theorem we use the notation E for expectation and Var for variance.

THEOREM 9.1. *Consider the model (9.13) where all variables are* between 0 and 1, $E(\varepsilon_t) = 1/2$, $p(x) = x$ for $0 < x < 1$. Suppose $0 < \lambda < 1$. Let $f_t(x)$, $0 < x < 1$, be the population index value density function at generation t with cumulative distribution $F_t(x)$. Then independent of the initial distribution, $F_t(x)$ converges to a limiting distribution $F_\infty(x)$ (i.e., $X_t \to X_\infty$ in law) where the mean converges to $1/2$, i.e., $m_t = E[X_t] \to 1/2$, and $\lim_{t \to \infty} \mathrm{Var}[X_t] = \sigma^*$ where σ^* is the unique positive solution of the quadratic equation

$$\sigma = \frac{\lambda^2}{2}[\sigma - 4\sigma^2] + (1 - \lambda)^2 b, \; b = \mathrm{Var}[\varepsilon_t]. \quad (9.15)$$

For $E[\varepsilon_t] = a$ and the recursion (9.13), any equilibrium sex ratio falls strictly between $\lambda/(\lambda + \mu)$ and a.

A proof of the above theorem is given in Appendix A.

Purely Environmental Index Determination

The extreme cases of (9.14), $\lambda + \mu = 0$ and $\lambda + \mu = 1$, deserve special attention. Obviously, if $\lambda + \mu = 0$ (entailing no parental transmission), the index variable X_t is distributed as the environmental variable ε_t in each generation. The sex ratio is accordingly invariant with value $r^* = \int p(x)e(x)\, dx$ which can take on any value between 0 and 1.

Purely Parental Transmission

When $\lambda + \mu = 1$, then (9.14) reduces to

$$X_{t+1} = \lambda X_t^* + (1 - \lambda)X_t^{**}. \quad (9.16)$$

For an initial population state satisfying $E[X_0] < \lambda$ we infer $E(X_t) < E(X_{t+1}) < \lambda$ (cf. Appendix A), implying $\lim_{t \to \infty} E(X_t) = m_\infty$ with $m_\infty \le \lambda$. Also, $\lim_{t \to \infty} E(X_t^2) = m_\infty^2$ obtains. The equality $E(X_\infty^2) = (EX_\infty)^2$ implies that X_∞ concentrates at the index value m_∞ with no variance.

When $E(X_0) > \lambda$, then $E(X_t)$ strictly decreases to m_∞ and again $\lim_{t \to \infty} E(X_t^2) = m_\infty^2$. Thus, in the model of purely parental transmission the limiting population index variable coalesces

269

to a single value that depends on the initial X_0-population distribution, whereas Theorem 9.1 indicates that the population has a limiting stationary stable distribution independent of the initial population index composition.

Note in the model of Theorem 9.1 that a *biased* sex ratio will evolve under either of the following conditions: (a) The environmental trait is asymmetric relative to sex determination (it suffices that the environmental mean index value differs from $1/2$); or (b) the paternal versus maternal transmission contributions to the offspring index are asymmetric (e.g., as when cytoplasmic factors are decisive).

There is a temptation to try to interpret the strength of asymmetry in the parental transmission components in accordance with the predictions on "parental investment" and the ratio of relatedness measures (e.g., see Trivers and Hare, 1976; Hamilton, 1972). It is doubtful that such concepts can be made meaningful in the context of MSD.

General Sex Determination Probability Function

Let the random variables X_t^* and X_t^{**} follow the densities $f_t^*(x) = p(x)f_t(x)/\int p(\xi)f_t(\xi)\,d\xi$ and $f_t^{**}(x) = [1 - p(x)]f_t(x)/\int [1 - p(\xi)]f_t(\xi)\,d\xi$, respectively, for a general sex determination probability function $p(x)$ subject to the transformation law

$$X_{t+1} = \lambda X_t^* + \mu X_t^{**} + \nu\varepsilon_t \qquad (9.17)$$

where λ, μ, ν are each non-negative and $\lambda^2 + \mu^2 < 1$.

For the special case $\nu = 0$ we expect $X_\infty = \lim_{t \to \infty} X_t$ to degenerate to a single value which will ordinarily depend on the initial density $f_0(x)$. However, for $\nu > 0$ and $\text{Var}[\varepsilon_t] > 0$ the recursion (9.17) tends to a stationary density $f_\infty(x)$.

For $\lambda^2 + \mu^2 > 1$, the index distribution $F_t(x)$ will diverge as $t \to \infty$. This is a general phenomenon inherent to multifactorial transmission models (cf. Karlin, 1979b) when the parental contributions dominate the offspring index value and tend to spread it out.

270

9.4. LIMIT SEX RATIOS FOR PURE BLENDING INHERITANCE WITH THRESHOLD SEX DETERMINATION CRITERION*

For the index variable transformation

$$X_{t+1} = \frac{X_t^* + X_t^{**}}{2}$$

(X_t of density $f_t(x)$ and the random variables X_t^*, X_t^{**} of densities f_t^* and f_t^{**} calculated as in (9.2)) we might expect that the sex ratio r_t converges to $1/2$ with the population index distribution reducing to a single value. Indeed, for a threshold criterion of the form (9.3) we have the following theorem:

THEOREM 9.2. *Let sex determination as a function of the index variable X be governed by the threshold criterion (9.3). When $f_0(x)$ (the population index variable density of the initial generation) is a log-concave function (see REMARK 9.2 below), then* $\mathrm{Var}(X_t) \to 0$ *at a geometric rate and the population sex ratio as $t \to \infty$ approaches $1/2$.*

Moreover, the population index variable distribution converges to the degenerate distribution with all individuals in the population showing index value v^*.

The proof of Theorem 9.2 is given in Appendix B. We *conjecture* that the conclusion holds without any restrictions on the density $f(x)$ and generally with respect to any sex determination probability function $p(x)$, where always $p(x) = 0$ or 1.

REMARK 9.1. The variance of X_t goes to zero also for the recursion

$$X_{t+1} = \alpha X_t^* + \beta X_t^{**}$$

provided $\alpha^2 + \beta^2 < 1$. The corresponding variance inequality (B.3) in Appendix B becomes

$$\mathrm{Var}(X_t) < (\alpha^2 + \beta^2)^t \mathrm{Var}(X_0).$$

---

* This and the following section deal with more technical formulations and may be omitted by the biologically oriented reader without hindrance in understanding the later sections.

271

REMARK 9.2. A density function $f(x)$ is log concave if it is of the form

$$f(x) = e^{-\varphi(x)}$$

where $\varphi(x)$ is convex. This class of densities includes most common densities (Normal; Gamma family — $G(\alpha, \lambda)$ with parameters $\alpha \geq 1$, $\lambda > 0$; all the order statistics associated with the uniform density; the t-density; and for the case of discrete valued densities: the Binomial, Poisson, Negative Binomial, among others; see Karlin, 1968b, chaps. 4, 7, and 8 for other examples).

9.5. SEX DETERMINATION PROBABILITY FUNCTION SYMMETRICAL WITH RESPECT TO A CENTRAL INDEX VALUE*

A number of generalizations of Theorem 9.1 are discussed in this section that suggest wide scope for the existence of a stable stationary index value population distribution. The expected sex ratio is generally not $1:1$ except under special symmetry conditions with respect to parental transmission and environmental effects.

An increasing probability sex determination function is said to be symmetric with respect to the point v^* if

$$p(x + v^*) = 1 - p(v^* - x) \text{ for all } x. \qquad (9.18)$$

For example, on the range $0 < x < 1$, $p(x) = x$ is symmetrical with respect to $v^* = 1/2$. A cumulative distribution function $p(x) = \int_{-\infty}^{x} u(\xi) \, d\xi$ where $u(\xi)$ is a symmetric density provides a class of examples of (9.18) with $v^* = 0$.

For ease of exposition we assume in this section that $v^* = 0$ and focus on the symmetrical transmission law

$$X_{t+1} = \lambda \left(\frac{X_t^* + X_t^{**}}{2} \right) + (1 - \lambda) \varepsilon_t \qquad (9.19)$$

subject to the usual interpretation, having X_t^* and X_t^{**} determined with respect to $p(x)$ by the formula (9.2), with $p(x)$

272

obeying (9.18) and $v^* = 0$. We further assume that the density $e(x)$ of ε_t is also symmetrical with respect to $v^* = 0$.

It is easy to verify inductively that when $f_t(x)$ is symmetric with respect to $v^* = 0$, then

$$r_t = \int p(x)f_t(x)\,dx = \frac{1}{2} \text{ and } \int xf_t(x)\,dx = 0.$$

Indeed, a change of variable gives

$$\int p(x)f_t(x)\,dx = \int p(-x)f_t(-x)\,dx$$

$$= \int [1 - p(x)]f_t(x)\,dx$$

$$= 1 - \int p(x)f_t(x)\,dx,$$

and therefore $r_t = 1/2$. The random variable X_t^{**} can be realized as an independent sample of the variable $-X_t^*$ and therefore $X_t^* + X_t^{**}$ is a symmetric random variable. Since ε_t is also symmetric (by assumption) it follows that X_{t+1} is symmetric.

These facts together imply that if $f_0(x)$ is symmetric, then $m_t = \int xf_t(x)\,dx = 0$ and $r_t = \int p(x)f_t(x)\,dx = 1/2$ in every generation. The recursion for the variance $\sigma_t^2 = \int x^2 f_t(x)\,dx$ based on (9.19) becomes

$$\sigma_{t+1}^2 = \left(\frac{\lambda}{2}\right)^2 \left[\frac{\int x^2 p(x) f_t(x)\,dx}{\frac{1}{2}} + \frac{\sigma_t^2 - \int x^2 p(x) f_t(x)\,dx}{\frac{1}{2}} \right] + (1-\lambda)^2 b$$

$$= \frac{\lambda^2}{2}\sigma_t^2 + (1-\lambda)^2 b \text{ where } b = \text{Var}(\varepsilon_t)$$

and therefore σ_{t+1}^2 converges to $(1-\lambda)^2 b/(1 - \lambda^2/2)$ at a geometric rate (with factor $\lambda^2/2$). The higher central moments converge by similar analysis (cf. Appendix A). The foregoing results entail that the limiting distribution exists and is independent of the initial distribution, provided $f_0(x)$ is symmetric.

Consider next an initial density of the form $f_0(x) = \phi(x)\psi(-x)$ where $\phi(x)/\psi(x)$ is monotone. We construct the symmetric density $\tilde{f}_0(x) = \kappa \psi(x)\psi(-x)$ (κ serves as a normalizing constant to ensure that $\tilde{f}$ is a density). Since $f_0(x)/\tilde{f}_0(x)$ is monotone, say for definiteness increasing, we have the ordering $f_0(x) \overset{\text{st}}{\gtrsim} \tilde{f}_0(x)$, also written symbolically $X_0 \overset{\text{st}}{\gtrsim} \tilde{X}_0$ in terms of the corresponding random variables. The notation $\overset{\text{st}}{\gtrsim}$ stands for stochastically larger, signifying that for any increasing function $u(x)$

$$\int u(x)f_0(x)\,dx \geq \int u(x)\tilde{f}_0(x)\,dx.$$

The ratio of the associated densities of X_0^* and $\tilde{X}_0^*$ are monotone in the same direction as X_0 relative to $\tilde{X}_0$, implying that $X_0^* \overset{\text{st}}{\gtrsim} \tilde{X}_0^*$. Similarly, we have $X_0^{**} \overset{\text{st}}{\gtrsim} \tilde{X}_0^{**}$. The stochastic relationship is preserved for independent summands and therefore $(X_0^* + X_0^{**})/2 \overset{\text{st}}{\gtrsim} (\tilde{X}_0^* + \tilde{X}_0^{**})/2$ and finally $X_1 \overset{\text{st}}{\gtrsim} \tilde{X}_1$. Iterating this procedure entails the relationship $X_t \overset{\text{st}}{\gtrsim} \tilde{X}_t$ for all t. Because $p(x)$ is increasing by assumption, we have

$$r_t = \int p(x)f_t(x)\,dx \geq \int p(x)\tilde{f}_t(x)\,dx = \frac{1}{2} \text{ for all } t.$$

We further deduce

$$m_t = \int x f_t(x)\,dx \geq \int x\tilde{f}_t(x)\,dx = 0 \text{ for all } t.$$

The fact of $r_t \geq 1/2$ and the inequality $\gamma_t \geq m_t r_t$ where $\gamma_t = \int x p(x) f_t(x)\,dx$ (cf. (9.10)) gives

$$0 < m_{t+1} = \frac{\lambda}{2}\left[\frac{\gamma_t}{r_t} + \frac{m_t - \gamma_t}{1 - r_t}\right] = \frac{\lambda}{2}\left[\frac{m_t}{1 - r_t} + \frac{(1 - 2r_t)\gamma_t}{r_t(1 - r_t)}\right] \leq \lambda m_t.$$

$$(9.20)$$

Because $\lambda < 1$, it follows that $m_t \to 0$.

The environmental contribution does not permit $f_t(x)$ to degenerate and therefore strict inequality $\gamma_t - m_t r_t > 0$ prevails also for any limit density of any subsequence from $f_t(x)$. The convergence of m_t to zero would be violated in (9.20) unless $r_t \to 1/2$. The proof of the convergence of the higher moments ensues, paraphrasing the analysis of Theorem 9.1 elaborated in Appendix A.

If $f_0(x)$ is a translate i.e., $f_0(x - c) = g(x)$ where $g(x)$ is a symmetric log concave density, then $f_0(x)/g(x)$ is monotone. The argument implemented in the case $f_0(x) = \kappa\phi(x)\psi(x)$ with $\phi(x)/\psi(x)$ monotone works to yield the same limit distribution as for an initial symmetric density. By a more intricate analysis the convergence of $f_t(x)$ can be established for any $f_0(x)$ that is merely log concave, not necessarily a translate of a symmetric density.

We sum up the preceding analysis as a theorem:

"THEOREM" 9.3. *For the multifactorial model (9.19) based on the symmetric probability sex determination function (9.18), there exists a globally stable population equilibrium.*

We have not achieved an encompassing proof, but on account of the established cases and augmented numerical computations we may expect the validity of the theorem.

9.6. MULTIFACTORIAL SEX RATIO EVOLUTION UNDER MOTHER CONTROL AND IN THE CASE OF HAPLODIPLOID SYSTEMS

Mother Control Model

We assume that the determination of the offspring sex phenotype depends only on the mother index value. This is accomplished by assuming that $p(x)$ and $1 - p(x)$ is the probability that a mother of index value x bears a son and daughter, respectively. Let the index variable of the adult female population at generation t be described by the density function $g_t(y)$ with associated random variable Y_t. The conditional density of the mother index contribution to a son (daughter) is

$$g_t^*(y) = \frac{p(y)g_t(y)}{\int p(\eta)g_t(\eta)\,d\eta}, \quad g_t^{**}(y) = \frac{[1 - p(y)]g_t(y)}{\int [1 - p(\eta)]g_t(\eta)\,d\eta}. \quad (9.21)$$

We denote the corresponding random variables by Y_t^* and Y_t^{**}, respectively. Let the adult male index variable density in generation t be $h_t(x)$ with associated random variable X_t. Under

275

blending transmission a female offspring carries index value

$$Y_{t+1} = \lambda Y_t^{**} + \mu X_t + (1 - \lambda - \mu)\varepsilon_t' \qquad (9.22a)$$

while that of a male is

$$X_{t+1} = \alpha Y_t^* + \beta X_t + (1 - \alpha - \beta)\varepsilon_t \qquad (9.22b)$$

with X_t, Y_t^*, and Y_t^{**} independent random variables. The parameters λ, μ, $1 - \lambda - \mu$, α, β, and $1 - \alpha - \beta$ reflect the relative strengths of the different parental contributions and the residual environmental contributions.

The analysis of the system (9.22) is formidable. Numerical studies suggest convergence of $g_t(x)$ and $h_t(x)$.

Without environmental terms the distributions of Y_t and X_t of the index value for female and male individuals, respectively, tend to concentrate at a single value, the limit value depending on the initial distribution.

The corresponding result to Theorem 9.2 for the mother control sex determination model (Section 9.4) is

THEOREM 9.4. *Let* $Y_{t+1} = (Y_t^{**} + X_t)/2$, $X_{t+1} = (Y_t^* + X_t)/2$ *(see (9.22)). If* $g_0(x)$ *is a log-concave density and* $p(x)$ *is a generalized threshold sex determination function of the form (9.3), then* $\mathrm{Var}[Y_{t+1}]$ *and* $\mathrm{Var}[X_{t+1}]$ *converge to zero geometrically fast.*

Consider the symmetric mother control model subject to the transmission relations

$$Y_{t+1} = \frac{Y_t^{**} + X_t}{2}, \; X_{t+1} = \frac{Y_t^* + X_t}{2}.$$

Let m_t and μ_t be the average character values among males and females of the tth generation. The recursions of the mean values in this case become

$$m_{t+1} = \frac{1}{2}\left[\frac{m_t - \gamma_t}{1 - r_t} + \mu_t\right], \; \mu_{t+1} = \frac{1}{2}\left[\frac{\gamma_t}{r_t} + \mu_t\right]$$

where

$$\gamma_t = \int xp(x)\phi_t(x) \, dx \text{ and } r_t = \int p(x)\phi_t(x) \, dx.$$

Assuming all limits exist as $t \to \infty$, $\mu^* = \lim_{t \to \infty} \mu_t$, $m^* = \lim_{t \to \infty} m_t$, $\gamma^* = \lim_{t \to \infty} \gamma_t$, $r^* = \lim_{t \to \infty} r_t$, we get

$$\mu^* = \frac{\gamma^*}{r^*} \text{ and } m^* = \frac{1}{2}\left[\frac{m^*}{1 - r^*} + \frac{(1 - 2r^*)\gamma^*}{r^*(1 - r^*)}\right].$$

If the limiting distribution is nondegenerate and $p(x)$ is increasing, then $\gamma^* > m^*r^*$ (see (9.10)). In the case $r^* < 1/2$ ($>1/2$) then the second equation above implies $m^* < m^*$ ($> m^*$), which is absurd. Thus for $r^* \neq 1/2$ the limit distribution is necessarily degenerate and then $\mu^* = m^*$. For $r^* = 1/2$ and $\gamma^* > m^*r^*$ we get $\mu^* > m^*$ (actually always $\mu_t \geq m_t$ for all $t \geq 2$) and no other restriction holds. Under the conditions of Theorem 9.4, necessarily $\gamma^* = m^*r^*$ and then also $\mu^* = m^*$. It appears in all cases that $r^* = 1/2$.

Haplodiploid Models

Model I: Mother control. We first consider the model where sex determination is under mother control. Let the index value density of the female population at generation t be $g_t(y)$ (with corresponding random variable Y_t). Let $p(y)$ be the probability of producing a fertilized (therefore female) offspring contributed by a mother of index value y. The probability that a mother of index value y produces a viable unfertilized (male) offspring is $1 - p(y)$. In this context, the density

$$g_t^*(y) = \frac{p(y)g_t(y)}{\int p(\eta)g_t(\eta)\,d\eta} \tag{9.23}$$

(with the associated random variable denoted by Y_t^*) is that of the conditional index value contributed by a mother to a daughter, and that for a son is

$$g_t^{**}(y) = \frac{[1 - p(y)]g_t(y)}{\int [1 - p(\eta)]g_t(\eta)\,d\eta}$$

(for the random variable Y_t^{**}). Let the male index density function at maturity be $h_t(x)$ (corresponding to the random

277

variable X_t). The dynamics connecting $g_{t+1}(y)$ and $g_t(y)$ over successive generations expressed in terms of random variables (assuming "blending" transmission with $1:1$ adult sex ratio) if we allow for residual (environmental) addends is described by the relations

$$Y_{t+1} = \alpha \frac{Y_t^* + X_t}{2} + \beta A_t$$

$$X_{t+1} = \gamma Y_t^{**} + \delta B_t \tag{9.24}$$

where the residual terms A_t and B_t are independent random variables. For the haplodiploid model (9.24), $\alpha = \gamma = 1$, $\beta = \delta = 0$ with a threshold sex determination probability as in (9.2), the analog of Theorem 9.4 holds.

Model II: Offspring control. Let $f_t(y)$ be the index variable density for diploids of generation t at conception. Denote the corresponding random variable by $\mathcal{Z}_t$. Let $p(y)$ $(1 - p(y))$ be the probability of a diploid zygote being male (female). Thus the female index density is

$$g_t(y) = \frac{[1 - p(y)]f_t(y)}{\int [1 - p(\eta)]f_t(\eta)\,d\eta} \tag{9.25}$$

(associated with the random variable Y_t) and that of male diploids

$$h_t(x) = \frac{p(x)f_t(x)}{\int p(\xi)f_t(\xi)\,d\xi} \tag{9.26}$$

(corresponding to the random variable X_t).

Let $q(y)$ be the proportion of offspring fertilized. Therefore $1 - q(y)$ is the proportion of unfertilized eggs that develop into males. The index variable distribution of male haploids is therefore

$$k_t(x) = \frac{(1 - q(x))g_t(x)}{\int (1 - q(\xi))g_t(\xi)\,d\xi} \tag{9.27}$$

with corresponding random variable X_t^*.

278

The conditional index variable distribution of female adults is

$$g_t^*(y) = \frac{q(y)g_t(y)}{\int q(\eta)g_t(\eta)\,d\eta} \qquad (9.28)$$

with associated random variable Y_t^*.

We assume the relative viability of haploid to diploid males is $1:s$ independent of their index values. Now define the adult male representation resulting by random sampling as

$$U_t^* = \begin{cases} X_t & \text{with probability } \dfrac{s}{1+s} \\[2ex] X_t^* & \text{with probability } \dfrac{1}{1+s} \end{cases} \qquad (9.29)$$

Assume $1:1$ adult sex ratio. A diploid offspring of the next generation is taken to be

$$Z_{t+1} = \lambda \frac{Y_t^* + U_t^*}{2} + (1 - \lambda)\varepsilon_t. \qquad (9.30)$$

The index density h_{t+1}^* of haploids in the next generation is that of X_{t+1}^*, determined following (9.26). The dynamic behavior of the system (9.30) is unknown. In general, we expect convergence to a stationary stable distribution.

9.7. MIXED MAJOR GENE AND MULTIFACTORIAL SEX DETERMINATION MODEL

We consider a model allowing for the combined effects of chromosomal and multifactorial sex determination. Such structures may be relevant in some cases of fish, amphibian, reptile, and general invertebrate species. In the following context sex expression depends principally on an XX/XY mechanism with XY invariably male. However, the sex of XX females can be transformed depending on a multifactorial quantitative index variable V, e.g., reflecting on birth size, environment at conception and genetic dispositions (cf. Model VII, Chapter 2). Suppose that an XX female of index value $V = v$ is transformed to a male with probability $p(v)$. Let $\phi_t(v)$ be the distribution of V for XX phenotypes in the population at generation t and $\psi(v)$ that for XY phenotypes, the latter density unchanged from generation to generation.

Let $k_t = \int_{-\infty}^{\infty} p(v)\phi_t(v)\, dv$ be the proportion of the XX karyotypes that become males and q_t the initial fraction of females in the tth generation. The sex ratio among adults is $q_t k_t + 1 - q_t = m_t$ of which the fraction of transformed females is $q_t k_t / m_t$ (of genotype XX) while the fraction of XY males is $(1 - q_t)/m_t$. The frequency of XX (assuming all XX mate) in the next generation is

$$q_{t+1} = \frac{q_t k_t + \left(\dfrac{1 - q_t}{2}\right)}{q_t k_t + 1 - q_t}. \tag{9.31}$$

Two transmission models are proposed for the changes of the index variable over two successive generations.

Model I. We assume that an XX individual in the $(t + 1)$th generation carries index value

$$U_{t+1} = \lambda U_t^{**} + (1 - \lambda)\varepsilon_t \tag{9.32}$$

where ε_t refers to an independent environmental perturbation altering the index value. The density of U_t^{**} is $[1 - p(v)]\phi_t(v)/\int_{-\infty}^{\infty} [1 - p(\xi)]\phi_t(\xi)\, d\xi$ where $p(v)$ is the probability that an XX female of index value v is transformed to a male and $\phi_t(v)$ is the density of U_t.

The transmission (9.32) apart from environmental terms recognizes only the mother contribution (e.g., reflecting on cytoplasmic factors) to an XX progeny. When $p(v)$ is increasing it is expected that the recursion (9.32) has a limiting density $\phi_\infty(v)$ with mean index value less than $E[\varepsilon_1]$ and limiting variance $\sigma_\infty^2 < (1 - \lambda)\,\mathrm{Var}[\varepsilon_1]/(1 + \lambda)$.

280

Model II. The index value distribution of XX individuals in the $t + 1$th generation is that of the mixture of random variables

$$U_{t+1} = \begin{cases} \lambda \dfrac{U_t^* + W_t}{2} + (1 - \lambda)\varepsilon_t \text{ with probability } \dfrac{1 - q_t}{m_t} \\[3mm] \lambda \dfrac{U_t^* + U_t^{**}}{2} + (1 - \lambda)\varepsilon_t \text{ with probability } \dfrac{k_t q_t}{m_t} \end{cases} \quad (9.33)$$

where

U_t^* follows the modified density $\dfrac{p(v)\phi_t(v)}{\int_{-\infty}^{\infty} p(\xi)\phi_t(\xi)\,d\phi}$,

U_t^{**} follows the density $\dfrac{[1 - p(v)]\phi_t(v)}{\int_{-\infty}^{\infty} [1 - p(\xi)]\phi_t(\xi)\,d\xi}$,

W_t is governed by the density $\psi(v)$ independent of t

and

ε_t is an independent environmental random addend.

When k_t is a constant k as occurs if $\lambda = 0$ (for $\lambda = 0$, $k = \int_{-\infty}^{\infty} p(v)e(v)\,dv$ where $e(v)$ is the density function of ε_t), then the recursion (9.31) for q_t converges at a geometric rate (with factor $2k$) to $1/2(1 - k)$ for $k < 1/2$ and at a geometric rate (with factor $1/2k$) to 1 for $k > 1/2$.

When k_t converges to a limit k, then the recursion (9.31) approaches, as $t \to \infty$, the same limit as that associated with (9.34).

If $k_t \to 0$, there are ultimately no transformed XX-females and the sex ratio approaches $1/2$. For the model (9.32) in the extreme case $\lambda = 1$, we have $k_t \to 0$ independent of $\phi_0(v)$.

In general, where the environmental contribution to the XX index value is relatively small, the sex ratio is expected to approximate $1/2$, whereas if there exists a strong environmental influence on whether an XX-individual has sex phenotype transformed or not and the parental determination of the index value is pronounced, a non-even sex ratio is expected to evolve.

9.8. SOME GENERAL QUALITATIVE
IMPLICATIONS OF MSD

The models of Sections 9.2–9.7 suggest that an even population sex ratio is unlikely in the province of multifactorial sex
determination except under very special symmetric (relative to
sex expression) environmental determinants and with respect to
parental contributions. Indeed, when the index variable (or set
of variables) underlying sex determination is (on an average)
asymmetric in affecting male versus female sex function or if
the strength of the parental multifactorial contributions are intrinsically unequal, a non-even sex ratio is predicted.

With deterministic environmental effects the index vari-
able(s) governing MSD tend(s) to coalesce to a single index
value and subsequent convergence to genic or chromosomal sex
determination is expected.

Sex conversion forms (cf. Section 9.7) appear most likely to
be maintained only in the presence of predominantly environmental influences.

MSD models, as we have formulated them in this chapter, are
transmission models, but not properly "polygenic." In particular, the within-family variance intrinsic to polygenic inheritance
is not accommodated in the transformation equations (9.6).

In the tradition of quantitative multifactorial transmission
models (e.g., see Karlin, 1979b and Karlin et al. 1983, app. F)
the index value of a typical male (female) parent is of the form

$$\mathbf{Y} = \hat{\mathbf{Y}} + \xi, \mathbf{Z} = \hat{\mathbf{Z}} + \eta \qquad (9.34)$$

where $\hat{\mathbf{Y}}(\mathbf{Z}4)$ is the genetic part of $\mathbf{Y}(\mathbf{Z})$ whereas $\xi(\eta)$ are individual residual terms. The offspring index value is then deter-
mined in the form

$$\mathbf{X} = \lambda\hat{\mathbf{Y}} + \mu\hat{\mathbf{Z}} + \zeta + \varepsilon \qquad (9.35)$$

where ζ is a special individual residual term and ε an independent environmental addend. The parameters λ and μ scale the
relative contributions coming from the parents.

282

The transformation equations (9.34) and (9.35) are at equilibrium and the ζ random variable embodies a within-family variance term. The qualitative conclusions on sex ratio evolution derived from these more general models are expected to be the same as in the preceding sections.

APPENDIX A. PROOF OF THEOREM 9.1

Suppose $E[X_t] = m_t$. Taking expectations in (9.13) and rearranging them yields

$$m_{t+1} - m_t = \lambda \left\{ \frac{(1 - 2m_t)(s_t - m_t^2)}{2m_t(1 - m_t)} \right\} + (1 - \lambda)(\tfrac{1}{2} - m_t) \quad \text{(A.1)}$$

where $s_t = E[X_t^2]$. Consider the case $m_t \leq 1/2$ (the analysis of the case $m_t \geq 1/2$ is similar reversing all the inequalities). It follows because $s_t \geq m_t^2$ by Schwartz inequality that $m_t < m_{t+1}$, and because $s_t < m_t$ also $m_{t+1} < 1/2$. Thus m_t converges. Because the distribution of $f_t(x)$ cannot degenerate except to $1/2$ and only when the density $e(x)$ is degenerate, we deduce that any limit value $(s_{t_i} \to s^*)$ necessarily has $s^* > (m^*)^2$ where $m^* = \lim_{t \to \infty} m_t$. Proceeding to a limit in (A.1) along an appropriate subsequence, if necessary, we find both terms on the right of (A.1) are positive unless $\lim_{n \to \infty} m_t = 1/2$.

The corresponding equation for the variance after some simplications becomes

$$\sigma_{t+1} = \frac{\lambda^2}{4} \left[\frac{\sigma_t}{1 - m_t} + \frac{m_t^2}{1 - m_t} + \frac{(1 - 2m_t)w_t}{m_t(1 - m_t)} \right.$$
$$\left. - \frac{(\sigma_t + m_t^2)^2}{m_t^2} - \frac{(m_t - \sigma_t - m_t^2)^2}{(1 - m_t)^2} \right]$$
$$+ (1 - \lambda)^2 b \quad \text{(A.2)}$$

where $w_t = E[X_t^3]$ and $b = \text{Var}[\varepsilon_t]$.

Replacing m_t by its limit $1/2$, the recursion (A.2) reduces to $\sigma_{t+1} = (\lambda^2/2)[\sigma_t - 4\sigma_t^2] + (1 - \lambda)^2 b + \eta_t$ where η_t goes to

zero as $t \to \infty$. Since $m_t \to 1/2$ it follows that $\varlimsup_{t \to \infty^-} \sigma_t \leq 1/4$
and because any limit distribution is nondegenerate when
$0 < \lambda < 1$ we deduce that $\varlimsup_{t \to \infty} \sigma_t < 1/4$. (Note that $b < 1/4$ as
$E[\varepsilon_t] = 1/2$.) The mapping $\sigma' = T(\sigma) = (\lambda^2/2)(\sigma - 4\sigma^2) +$
$(1 - \lambda)^2 b$ has a unique positive fixed point σ^* necessarily
$< 1/4$, since $0 < T(0) = T(1/4) = (1 - \lambda)^2 b < 1/4$.

The derivative $T'(\sigma^*) = (\lambda^2/2)(1 - 8\sigma^*)$ since $0 < \sigma^* <$
$1/4$ satisfies $|T'(\sigma^*)| < 1$ and the iterates of T converge geometrically fast to σ^*. Returning to (A.2) we write

$$\sigma_{t+1} = S(\sigma_t) = T(\sigma_t) + \eta_t$$

where η_t converges to zero. It follows that the iterates of S behave as the iterates of T converging to σ^*.

Writing out a recursion formula for the third moment of X_t
with $w_t = E[X_t^3]$ in the form

$$w_{t+1} = \left(\frac{\lambda}{2}\right)^3 \left[\frac{w_t}{1 - m_t} + \frac{3w_t(m_t - 2s_t)}{m_t(1 - m_t)}\right] + c_t + \delta_t \quad \text{(A.3)}$$

where $\delta_t \to 0$ and c_t is a continuous function of m_t, s_t and the
moments of the random variable ε_t, which converges to a limit
c. Inserting the limit values, (A.3) takes the form

$$w_{t+1} = \gamma w_t + c + \eta_t \text{ with } |\gamma| < 1$$

where $\eta_t \to 0$. It follows that $\lim_{t \to \infty} w_t$ exists.

Suppose we have ascertained the convergence of the first
$r - 1$ moments of X_t $(r \geq 3)$. Subtracting the means from (9.13)
we get the relationship

$$X_{t+1} - m_{t+1} = \frac{\lambda}{2}\left[\left(X_t^* - \frac{s_t}{m_t}\right) + \left(X_t^{**} - \frac{m_t - s_t}{1 - m_t}\right)\right]$$
$$+ (1 - \lambda)\left(\varepsilon_t - \frac{1}{2}\right).$$

Taking the rth power and expectation yields, setting $\mu_t^{(r)} = E((X_t - m_t)^r)$,

284

$$\mu_{t+1}^{(r)} = \left(\frac{\lambda}{2}\right)^r \left\{ E\left[\left(X_t^* - \frac{s_t}{m_t}\right)^r\right] + E\left[\left(X_t^{**} - \frac{m_t - s_t}{1 - m_t}\right)^r\right]\right\}$$

$$+ (1 - \lambda)^r E\left[\left(\varepsilon_t - \frac{1}{2}\right)^r\right]$$

$$+ \text{sums of products of the factors } E\left[\left(X_t^* - \frac{s_t}{m_t}\right)^k\right],$$

$$E\left[\left(X_t^{**} - \left(\frac{m_t - s_t}{1 - m_t}\right)\right)^l\right] \text{ and } E\left[\left(\varepsilon_t - \frac{1}{2}\right)^m\right], \quad \text{(A.4)}$$

where k, l, and m are integers not exceeding $r - 2$ since a factor to the $(r - 1)$th power would be multiplied by an expectation of a first central moment which then vanishes.

Let $m_t^{(r)} = E[(X_t)^r]$. Observe that

$$E[(X_t^*)^r] + E[(X_t^{**})^r] = \frac{m_t^{(r)}}{1 - m_t} + \frac{(1 - 2m_t)m_t^{(r+1)}}{m_t(1 - m_t)}$$

which behaves as

$$2m_t^{(r)} + \delta_t$$

with $\delta_t \to 0$ as $t \to \infty$. It follows that

$$E\left[\left(X_t^* - \frac{s_t}{m_t}\right)^r\right] + E\left[\left(X_t^{**} - \frac{m_t - s_t}{1 - m_t}\right)^r\right]$$

$$= 2m_t^{(r)} + \delta_t - r\left(\frac{s_t}{m_t}\right)\left(\frac{m_t^{(r)}}{m_t}\right)$$

$$- r\left(\frac{m_t - s_t}{1 - m_t}\right)\left(\frac{m_t^{(r-1)} - m_t^{(r)}}{1 - m_t}\right)$$

$$+ \text{terms involving lower } (\leq r - 1) \text{ moments of } X_t.$$

$$\text{(A.5)}$$

Combining (A.4)–(A.5) with the fact that $\mu_t^{(r)} = m_t^{(r)} +$ terms involving moments $m_t^{(k)}$, $0 \leq k \leq r - 1$ and inserting the limiting values of $m_t^{(k)}$ for $k = 1, 2, \ldots, r - 1$ we achieve a recursion relation

$$\mu_{t+1}^{(r)} = \gamma_r \mu_t^{(r)} + \beta + \eta_t \quad \text{(A.6)}$$

where $\eta_t \to 0$ with $t \to \infty$, $|\gamma_r| < 1$ and β is a constant. Explicitly

$$\gamma_r = \left(\frac{\lambda}{2}\right)^r [2(1 + r) - 8rs^*] \text{ where } s^* = \lim_{t \to \infty} s_t, r \geq 3.$$

The convergence of $\mu_t^{(r)}$ $(t \to \infty)$ ensues from (A.6) such that $\lim_{t \to \infty} \mu_t^{(r)} = \beta/(1 - \gamma_r)$ Thus we ascertain that all the moments of X_t converge. Since all moments converge and the range is confined to $0 < x < 1$, the distribution of X_t converges. This completes the proof of the first part of Theorem 9.1. The analysis for the remaining statements is easier.

APPENDIX B. PROOF OF THEOREM 9.2

We use the following properties of log concave functions (see Karlin, 1968, chap. 4):

(i) If g and h are log concave then the product $g \cdot f$ is also log concave.

(ii) The convolution of two log concave densities is log concave.

(iii) THEOREM (Karlin, 1982). *Let U be a real random variable that has a log concave density. Then the conditional variances*

$$\text{Var}[U|U \leq b], \text{Var}[U|U \geq a] \tag{B.1}$$

are decreasing as a increases and b decreases.

We turn now to the proof of Theorem 9.2.

Since f_0 is log concave so is f_0^* and f_0^{**} (f^* and f^{**} are defined in (9.2) with $p(x) = 1$ for $x \geq v^*$ and 0 for $x < v^*$) by property (i). On account of property (iii), we have the variance inequality

$$\text{Var}(X_0^*) < \text{Var}(X_0) \text{ and } \text{Var}(X_0^{**}) < \text{Var}(X_0)$$

and therefore

$$\mathrm{Var}(X_1) < \frac{1}{2}\,\mathrm{Var}(X_0) \tag{B.2}$$

Note from property (ii) that X_1 has a log concave density. Now we can iterate the result of (B.2) to obtain

$$\mathrm{Var}(X_t) < \frac{1}{2^t}\,\mathrm{Var}(X_0). \tag{B.3}$$

We next prove that X_t converges to the threshold index value v^* with probability 1. Consider the partition $\mathscr{P}$ of the random variable X_t according as $X_t > v^*$ or $X_t < v^*$. The following variance identity with respect to the partition $\mathscr{P}$ holds

$$\mathrm{Var}[X_t] = E[\mathrm{Var}\,(X_t|\mathscr{P})] + \mathrm{Var}[E(X_t|\mathscr{P})]$$

Since $\mathrm{Var}[X_t] \to 0$, it follows that $\mathrm{Var}(E[X_t|\mathscr{P}])$ goes to zero. But $E(X_t|\mathscr{P})$ is two-valued; namely

$$E[X_t|X_t > v^*] = E[X_t^*] \text{ with probability } P\{X_t > v^*\}$$

and

$$E[X_t|X_t < v^*] = E[X_t^{**}] \text{ with probability } P\{X_t < v^*\}.$$

If $E[X_t^*] - E[X_t^{**}]$ does not converge to zero, then either $X_t^* > v^* + \varepsilon > v^* > X_t^{**}$ or $X_t^* > v^* > v^* - \varepsilon > X_t^{**}$ for some ε with probability bounded away from zero. Since X_t^* and X_t^{**} are independent it follows that $\mathrm{Var}(X_t) = \mathrm{Var}\,((X_t^* + X_t^{**})/2)$ cannot converge to zero. To avert this outcome, we must have $E[X_t^*] - E[X_t^{**}] \to 0$. Since

$$|X_t^* - X_t^{**}| = |[X_t^* - E(X_t^*)] + [E[X_t^*] - E[X_t^{**}]]$$
$$+ (E[X_t^{**}] - X_t^{**})|$$

and each term goes to zero with probability 1. We conclude that $X_t^* - X_t^{**} \to 0$ with probability 1. Accordingly, the distribution of X_t converges to a degenerate distribution concentrating at the threshold value v^*.

CHAPTER TEN

Concluding Observations
and Open Problems

At first reading this book may appear too relentlessly abstract for the richness and complexity of its subject. That may be so, yet there are weighty precedents of formal theoretical studies helping to elucidate evolutionary processes. Theoretical evolutionary biology was dominated in the first half of this century by R. A. Fisher, J.B.S. Haldane, and Sewall Wright, who set forth a variety of basic mathematical analyses concerned with the way the interactive forces of natural selection, mutation, migration, and mating patterns might be supposed to act. In particular, they succeeded in quantifying such concepts as heterozygote advantage, mutation-selection balance, recombination-selection interaction, and the consequences of mixed mating systems. Not only has the theoretical modeling of observed mechanisms provided insights into evolutionary processes, but theoretical analyses have not infrequently suggested mechanisms for biological processes not fully observed or understood. These models also broaden the scope and add perspectives on the nature of the selection forces and possible causal mechanisms contributing to the vast variability that exists on all levels from molecular to population characteristics.

Model formulation is a search for a consistent and instructive analytic framework that approximates natural phenomena. A well-structured model may provide the means to sort disparate evidence into one or several competing coherent frameworks for interpretation. It strives to deepen, qualitatively and quantitatively, our "understanding" of the real phenomena. The process of formulating models often generates new hypotheses and the possibilities of new relationships, motivating new experiments and observations.

A précis of our results was presented in Section 1.7. In this concluding chapter we briefly discuss some important considerations not dealt with in this book.

Progeny Sex Ratio and Brood Size Variation

It is natural to contemplate variations in sex ratio in conjunction with variations in brood size. Differences in the extent to which brood size is dependent on the mating type are equivalent to differential fertility effects. Speith (1974) formulated a general model allowing for both sex ratio and brood size variation. Uyenoyama and Bengtsson (1979, 1981, 1982) focused on cases where linear relationships or additive allelic effects are assumed for the parameters of the model. As we have seen fertility parameters determined by one of the parents can essentially be treated as viability parameters. The analysis of sex-differentiated fertility schemes with general parameters depending on the mating type appears to be difficult, but they are obviously important to a deeper understanding of sex ratio evolution.

Mating Pattern and Sex Ratio

The influence of mating pattern (e.g., sexual selection, assortative mating) on population sex ratio attainments is considered in Fischer (1980), Charnov (1979b), and Queller (1983). The study of theoretical genetic models of mating behavior in relation to mechanisms for sex determination is mostly virgin territory.

Finite Population Size

Population structure and finite population size effects on sex ratio were considered in Bulmer and Taylor (1980a), Taylor and Bulmer (1980), Wilson and Colwell (1981), and Taylor and Sauer (1980), among others. However, these studies are mostly numerical and stipulate an average invariant local population size without adequate consideration of local population sampling fluctuations. A proper stochastic diploid population genetic model of sex ratio variation needs to be investigated. What is necessary is to adapt and refine the theory of diffusion population genetic models—e.g., Kimura (1983) and references therein—in such a way as to focus on the consequences of small population size with respect to sex ratio evolution.

Multilocus Sex Determination

The study of multilocus sex determination is of interest in at least two contexts. First, although some multilocus systems have been documented (see Chapter 1), probably many other such systems for fish, amphibian, reptile, and invertebrate species have yet to be fully worked out. Second, the evolutionary transitions between sex-determining mechanisms can presumably be manipulated through genic substitutions and modifier gene concomitants. For example, the evolution to male or female heterogamety can be conceived in this framework. Bull (1983, chap. 6) considered sex expression to be controlled, a priori, at two independent loci where the genotypic array of one gene corresponds to male heterogamety and the array of the second gene corresponds to female heterogamety. The establishment of male as against female heterogamety then reduces to the evolution of dominant control at the first or second gene. More complex versions based on multiple genes having sex drive effects deserve full analysis (cf. Chapter 5).

The transition from hermaphroditism to dioecy has also been modeled in terms of multigenes of major and modifier genes, e.g., by Ross and Weir (1976).

Sex Ratio and Frequency Dependence

Sex ratio is in many ways a frequency-dependent phenomenon and in this respect plays an important role in the evolution of altruistic traits, especially in haplodiploid species, depending on whether a sister or brother is the recipient of altruism. Two matters to be investigated are the conditions for the evolution of altruism as a function of sex ratio and the consequences for altruism of genetic modification of sex ratio.

The frequency-dependent nature of sex ratio may also offer a clue to circumventing the paradox of the cost of meiosis (see Uyenoyama, 1984), and to a better understanding of nonrandom mating dynamics generally.

General frequency-dependent selection schemes in sexual populations based on two (or more) phenotypes determined by underlying multiallele variants may be accessible to a global investigation of the evolutionary dynamics. Preliminary results indicate two recurrent classes of equilibria: phenotypic equilibrium manifolds characterized by equal mean phenotypic fitnesses (or phenotypic fixations), and pointwise genotypic equilibria arising from the underlying genetic systems, as occurs in sex ratio theory. (See Lessard, 1984.)

Sex Ratio and Sociality

There have been many recent theoretical, experimental, and field studies pertaining to the biological basis of social behavior, among them studies of sterile castes in eusocial Hymenoptera, intra- and intergenerational conflicts, various recognition and communication systems in animal groups, and contrasts between social systems. It seems likely that population sex ratio plays a role in the evolution of dominance hierarchies and social group structures and that the evolution of polygamy and monogamy depends on the representation and distribution of the sexes. Yet almost no analytic models exist for such population genetic phenomena.

General Issues of Sexuality

The theoretical study of sex ratio evolution cannot be insulated from the more general study of sexuality in all its guises. To provide broader perspective we conclude this book by listing some of the most important related issues together with representative references, most of them of a theoretical nature.

The issues of sex evolution broadly considered include: the advantages and disadvantages of sex and recombination (e.g.,

Crow and Kimura, 1965; Maynard Smith, 1968, 1978; Eshel and Feldman, 1970; Bodmer, 1970; Karlin, 1973; Felsenstein, 1974); the benefits of dioecy versus hermaphroditism, and of diploidy versus haplodiploidy (e.g., Hartl and Brown, 1970; Charnov et al., 1976; Charlesworth and Charlesworth, 1978a, b; Lloyd, 1982); the significance of sexual dimorphism as against monomorphism and behavioral covariates (sexual selection) (e.g., O'Donald, 1976, 1977; Maynard Smith, 1978, Ohno, 1979); the variety of mechanisms and controls underlying sex expression, including single-gene determinants, multigene interactions, chromosomal controls, hormonal adjustments, nutritional influences, cytological conditions of the embryo, and endogenous environments (e.g., Crew, 1965; Bacci, 1965; White, 1973; Charnov, 1982; Bull, 1983); isogamy and anisogamy (e.g., Scudo, 1967b; Parker et al., 1972; Bell, 1982); the role of sex ratio and sex allocation in terms of population and individual fitness (e.g., Ghiselin, 1974; Williams, 1979; Charnov, 1982); the evolution of sex chromosomes, Y-degeneration, and dosage compensation (e.g., Ohno, 1967, 1979; Bull, 1983; Jones, 1983); factors of biased sex ratio (e.g., Hamilton, 1967; Maynard Smith, 1978; Uyenoyama and Feldman, 1978; Colwell, 1981); the relation of mating systems to sex determination (e.g., Hamilton, 1967; Leigh, 1970; Fischer, 1980); and the paradox of parthenogenesis (e.g., Williams and Mitton, 1973; Maynard Smith, 1978; Lloyd, 1980; Uyenoyama, 1984).

Fisher (1930) identifies Mendel as "a young mathematician whose statistical interests extended to the physical and biological sciences," who modeled his laws of inheritance to be consistent with his experimental results. The period of 1900-1950 yielded a number of important successes in evolutionary biology when theoretical modeling, mathematical and statistical methodology, field observation, experimentation, and practical breeding were conjoined in this realm of science; see the historical reviews of Dunn (1965) and Provine (1971). Mergers of this sort, we believe, are necessary to advance our understanding of evolutionary processes.

References

Alexander, R. D., and P. W. Sherman. 1977. Local mate competition and parental investment in social insects. *Science* 196:494–500.

Alstad, D. N., and G. F. Edmund, Jr. 1983. Selection, outbreeding depression, and the sex ratio of scale insects. *Science 304*:93–94.

Bacci, G. 1965. *Sex Determination*. Elmsford, N.Y.: Pergamon.

Bar-Anon, R., and A. Robertson, 1975. Variation in sex ratio between progeny groups in dairy cattle. *Theor. Appl. Genet.* 46:63–65.

Bawa, K. S., and J. H. Beach. 1981. Evolution of sexual systems in flowering plants. *Ann. Mo. Bot. Gard. 68*:259–275.

Beiles, A. 1974. A buffered interaction between sex ratio, age difference at marriage, and population growth in humans, and their significance for sex ratio evolution. *Heredity 33*: 265–278.

Bell, G. 1982. *The Masterpiece of Nature: The Evolution and Genetics of Sexuality*. Berkeley and Los Angeles: University of California Press.

Bengtsson, B. O. 1977. Evolution of sex ratio in the wood lemming. In *Measuring Selection in Natural Populations*, ed. F. B. Christiansen and T. M. Fenchel, 333–343. Berlin: Springer-Verlag.

Bodmer, W. F. 1965. Differential fertility in population genetic models. *Genetics 51*:411–424.

————. 1970. The evolutionary significance of recombination in prokaryotes. *Symp. Soc. Gen. Microbiol. 20*:279–294.

————, and A.W.F. Edwards. 1960. Natural selection and the sex ratio. *Ann. Hum. Genet. 24*:239–244.

Borgia, G. 1980. Evolution of haplodiploidy: Models for inbred and outbred systems. *Theor. Pop. Biol. 17*:103–128.

Bridges, C. B. 1916. Non-disjunction as proof of the chromosome theory of heredity. *Genetics 1*:1–52, 107–163.

Bull, J. J. 1979. An advantage for the evolution of male haploidy and systems with similar genetic transmission. *Heredity* 43:361–381.

———. 1980. Sex determination in reptiles. *Quart. Rev. Biol.* 55:3–21.

———. 1981a. Sex ratio evolution when fitness varies. *Heredity* 46:9–26.

———. 1981b. Evolution of environmental sex determination from genotypic sex determination. *Heredity* 47:173–184.

———. 1981c. Coevolution of haplodiploidy and sex determination in the Hymenoptera. *Evolution* 35:568–580.

———. 1983. *Evolution of Sex Determining Mechanisms.* Menlo Park, Calif.: Benjamin-Cummings.

———, and E. L. Charnov. 1977. Changes in the heterogametic mechanism of sex determination *Heredity* 39:1–14.

———, and R. C. Vogt. 1979. Temperature dependent sex determination in turtles. *Science* 206:1186–1188.

———, R. C. Vogt, and M. G. Bulmer. 1982. Heritability of sex ratio in turtles with environmental sex determination. *Evolution* 36:333–341.

Bulmer, M. G. 1983. Models for the evolution of protandry in insects. *Theor. Pop. Biol.* 23:314–322.

———, and J. J. Bull. 1982. Models of polygenic sex determination and sex ratio evolution. *Evolution* 36:13–26.

———, and P. D. Taylor. 1980a. Sex ratio under the haystack model. *J. Theor. Biol.* 86:83–89.

———, and P. D. Taylor. 1980b. Dispersal and the sex ratio. *Nature* 284:448–449.

Callan, H. G., and P. E. Perry. 1977. Recombination in male and female meiocytes contrasted. *Phil. Trans. R. Soc. London* 277:227–233.

Cannings, C., and L. M. Cruz Orive. 1975. On the adjustment of the sex ratio and the gregarious behaviour of animal populations. *J. Theor. Biol.* 55:115–136.

Charlesworth, B. 1977. Population genetics, demography and the sex ratio. In *Measuring Selection in Natural Populations*, ed.

F. B. Christiansen and T. M. Fenchel, 345–363. Berlin: Springer-Verlag.

———. 1978. Model for evolution of Y chromosomes and dosage compensation. *Proc. Nat. Acad. Sci. 75*:5618–5622.

———, and D. Charlesworth. 1978a. A model for the evolution of dioecy and gynodioecy. *Amer. Natur. 112*:975–997.

———, and D. Charlesworth. 1978b. Population genetics of partial male-sterility and the evolution of monoecy and dioecy. *Heredity 41*:137–153.

Charlesworth, D., and B. Charlesworth. 1979. The evolutionary genetics of sexual systems in flowering plants. *Proc. R. Soc. Lond. B., Biol. Sci. 205*:513–530.

———, and B. Charlesworth. 1981. Allocation of resources to male and female functions in hermaphrodites. *Biol. J. Linn. Soc. 14*:57–74.

Charnov, E. L. 1975. Sex ratio selection in an age structured population. *Evolution 29*:366–368.

———. 1978. Sex-ratio selection in eusocial hymenoptera. *Amer. Natur. 112*:317–326.

———. 1979a. The genetical evolution of patterns of sexuality: Darwinian fitness. *Amer. Natur. 113*:465–480.

———. 1979b. Simultaneous hermaphroditism and sexual selection. *Proc. Nat. Acad. Sci. 76*:2480–2484.

———. 1982. *The Theory of Sex Allocation*. Princeton: Princeton University Press.

———, and J. J. Bull. 1977. When is sex environmentally determined? *Nature 266*:828–830.

———, J. Maynard Smith, and J. J. Bull. 1976. Why be an hermaphrodite? *Nature 263*:125–126.

Clark, A. B. 1978. Sex ratio and local resource competition in a prosimian primate. *Science 201*:163–165.

Colwell, R. K. 1981. Group selection is implicated in the evolution of female-biased sex ratios. *Nature 290*:401–404.

Cotterman, C. W. 1953. Regular two-allele and three-allele phenotype systems. *Amer. J. Hum. Genet. 5*:193–235.

Crew, F.A.E. 1937. The sex ratio. *Amer. Natur. 71*:529–559.

————. 1965. *Sex Determination*. 4th ed. New York: Dover.

Crow, J. F., and M. Kimura. 1965. Evolution in sexual and asexual populations. *Amer. Natur.* 99:439–450.

————, and M. Kimura. 1970. *An Introduction to Population Genetics Theory*. New York: Harper and Row.

Darlington, C. D., and K. Mather. 1949. *The Elements of Genetics*. London: Allen and Unwin Ltd.

Dunn, L. C. 1965. *A Short History of Genetics*. New York: McGraw-Hill.

East, E. M., and A. J. Mangelsdorf. 1925. A new interpretation of the heredity behavior of self-sterile plants. *Proc. Nat. Acad. Sci.* 11:166–171.

Edwards, A.W.F. 1961. The population genetics of "sex-ratio" in *Drosophila pseudoobscura*. *Heredity* 16:291–304.

————. 1962. Genetics and the human sex ratio. *Adv. Genet.* 11:239–272.

————.1966. Sex ratio data analyzed independently of family limitation. *Ann. Hum. Genet.* 29:337–347.

————. 1970. The search for genetic variability of the sex ratio. *J. Biosoc. Sci. Suppl.* 2:55–60.

Eshel, I. 1975. Selection on sex ratio and the evolution of sex determination. *Heredity* 34:351–361.

————, and M. W. Feldman. 1970. On the evolutionary effect of recombination. *Theor. Pop. Biol.* 1:88–100.

————, and M. W. Feldman. 1982a. On evolutionary genetic stability of the sex ratio. *Theor. Pop. Biol.* 21:430–439.

————, and M. W. Feldman. 1982b. On the evolution of sex determination and the sex ratio in haplodiploid populations. *Theor. Pop. Biol.* 21:440–450.

Ewens, W. 1979. *Mathematical Population Genetics*. Heidelberg: Springer-Verlag.

Feldman, M. W. 1971. Equilibrium studies of two locus haploid populations with recombination. *Theor. Pop. Biol.* 2:299–318.

Felsenstein, J. 1974. The evolutionary advantage of recombination. *Genetics* 78:737–756.

Finney, D. J. 1952. The equilibrium of a self-incompatible poly-morphic species. *Genetica* 26:33–64.

Fischer, E. A. 1980. The relationship between mating system and simultaneous hermaphroditism in the coral reef fish *Hypoplectrus Nigricans* (Serranidae). *Anim. Behav.* 28:620–633.

Fishelson, L. 1970. Protogynous sex reversal in the fish *Anthias squammipinnis* (Teleostei, Anthiidae) regulated by presence or absence of male fish. *Nature* 227:90–91.

Fisher, R. A. 1941. The theoretical consequences of polyploid inheritance for the mid style form of *Lythrum salicaria*. *Ann. Eugenics* 11:31–38.

———. 1958. *The Genetical Theory of Natural Selection.* 2nd rev. ed. New York: Dover Publications.

Franco, M. G., P. G. Rubini, and M. Vecchi. 1982. Sex determinants and their distribution in various populations of *Musca domestica* of western Europe. *Genet. Res.* 40:279–293.

Frankel, R., and E. Galun. 1977. *Pollination Mechanisms, Reproduction, and Plant Breeding.* Berlin: Springer-Verlag.

Fredga, K., A. Gropp, H. Winking, and F. Frank. 1976. Fertile XX- and XY-type females in the wood lemming *Myopus schisticolor*. *Nature* 261:225–227.

Gantmacher, F. R. 1959. *The Theory of Matrices.* New York: Chelsea.

Gershenson, S. 1928. A new sex ratio abnormality in *Drosophila obscura*. *Genetics* 13:488–507.

Ghiselin, M. T. 1974. *The Economy of Nature and the Evolution of Sex.* Berkeley and Los Angeles: University of California Press.

Gregorius, H. R. 1982. Selection in diplo-haplonts. *Theor. Pop. Biol.* 21:289–300.

Hamilton, W. D. 1967. Extraordinary sex ratios. *Science* 156:477–488.

———. 1972. Altruism and related phenomena, mainly in social insects. *Ann. Rev. Ecol. Syst.* 3:193–232.

———. 1979. Wingless and fighting males in fig wasps and

other insects. In *Sexual Selection and Reproductive Competition in Insects*, ed. M. S. Blum and N. A. Blum, 107–200. New York: Academic Press.

———. 1980. Sex versus non-sex parasite. *Oikos 35*:282–290.

———, and R. M. May. 1977. Dispersal in stable habitats. *Nature 269*:578–581.

Hartl, D. L., and S. W. Brown. 1970. The origin of male haploid genetic systems and their expected sex ratio. *Theor. Pop. Biol. 1*:165–190.

Haskins, C. P., P. Young, R. E. Hewitt, and E. F. Haskins. 1970. Stabilised heterozygosity of supergenes mediating certain X-linked colour patterns in populations of *Lebistes reticulatus*. *Heredity 25*:575–589.

Heslop-Harrison, J. 1972. Sexuality of angiosperms. In *Plant Physiology*. Vol. 6C, ed. F. C. Stewart, 133–290. New York: Academic Press.

Heuch, I. 1979. Equilibrium populations of heterostylous plants. *Theor. Pop. Biol. 15*:43–57.

Hickey, W. A., and G. P. Craig. 1966. Genetic distortion of sex ratio in a mosquito, *Aedes aegypti*. *Genetics 53*:1177–1196.

Hines, W.G.S., and W. S. Moore. 1981. An analysis of sex in random environments, I. *Adv. Appl. Prob. 13*:453–463.

Hodgkin, J. 1983. Two types of sex determination in a nematode. *Nature 304*:267–268.

Iwasa, Y. 1981. Role of sex ratio in the evolution of eusociality in haplo-diploid social insects. *J. Theor. Biol. 93*:125–142.

———, F. J. Odendaal, D. D. Murphy, P. R. Ehrlich, and A. E. Launer. 1983. Emergence patterns in male butterflies: A hypothesis and a test. *Theor. Pop. Biol. 23*:363–379.

Jayakar, S. D. 1982. Sex determination in higher animals: Evolutionary aspects. In *Evolution and the Genetics of Populations*. ed. S. D. Jayakar and L. Zonka. Suppl ATTI ASS. Genet. ITal XXIX pp. 121–140.

———, and Spurway, H. 1966. Reuse of cells and brother-sister mating in the Indian species *Stenodynerus miniatus*. *J. Bomb. Nat. Hist. Soc. 63*:378–398.

————, and Spurway, H. 1968. The nesting activities of the vespoid potter wasp *Eumenes capaniformis esuriens* (Fabr.) compared with the ecologically similar shecoid *Sceliphron madraspatanum* (Fabr.) (Hymenoptera). *J. Bomb. Nat. Hist. Soc. 64*:307–332; *65*:148–181.

Jones, K. W. 1983. Evolution of sex chromosomes. In *Developments in Mammals*, Vol. 5, ed. M. H. Johnson, 297–320. New York: Elsevier Science Publishing Co.

————, and L. Singh. 1982. Conserved sex associated repeated DNA sequences in vertebrates. In *Genome Evolution*, ed. G. A. Dover and R. B. Flavell, 135–154. New York: Academic Press.

Kallman, K. D. 1970. Sex determination and the restriction of sex-linked pigment patterns to the X and Y chromosomes in populations of a poeciliid fish, *Xiphophorus maculatus*, from the Belize and Sibun rivers of British Honduras. *Zoologica 55*:1–16.

————. 1973. The sex-determining mechanism of the platyfish, *Xiphophorus maculatus*. In *Genetics and Mutagenesis of Fish*, ed. J. H. Schroder, 19–28. New York: Springer-Verlag.

————. 1984. A new look at sex determination in poeciliid fishes. In *Evolutionary Genetics of Fishes*, ed. B. J. Turner, 95–171. New York: Plenum Publishing Corp.

Kalmus, H., and C.A.B. Smith. 1960. Evolutionary origin of sexual differentiation and the sex ratio. *Nature 186*:1004–1006.

Karlin, S. 1968a. Equilibrium behavior of population genetics models with non-random mating. *J. Appl. Prob. 5*:231–313, 487–566.

————. 1968b. *Total Positivity*. Stanford: Stanford University Press.

————. 1972. Some mathematical models of population genetics. *Amer. Math. Monthly 79*:699–739.

————. 1973. Sex and infinity: A mathematical analysis of the advantages and disadvantages of genetic recombination. In *The Mathematical Theory of the Dynamics of Biological Popu-*

lations ed. M. S. Bartlett and R. W. Hiorns, 155–194. London: Academic Press.

———. 1977. Gene frequency patterns in the Levene subdivided population model. *Theor. Pop. Biol. 11*:356–385.

———. 1978. Theoretical aspects of multilocus selection balance, I. In *Mathematical Biology, Part II: Populations and Communities*, ed. S. A. Levin, MAA. Studies in Mathematics, Vol. 16, 503–587. Washington, D.C.

———.1979a. Principles of polymorphism and epistasis for multilocus systems. *Proc. Nat. Acad. Sci. 76*:541–545.

———. 1979b. Models of multifactorial inheritance I–IV. *Theor. Pop. Biol. 15*:308–438.

———. 1982a. Classifications of selection-migration structures and conditions for a protected polymorphism. In *Evolutionary Biology*, Vol. 14, ed. M. K. Hecht, B. Wallace, and C.T. Prance, 61–204. New York: Plenum Publishing Corp.

———. 1982b. Some results on optimal partitioning of variance and monotonicity with truncation level. In *Statistics and Probability: Essays in Honor of C. R. Rao*, ed. G. Kallianpur, P. R. Krishnaiah, and J. K. Ghosh, 375–382. Amsterdam: North Holland Publishing Co.

———. 1983. Theoretical aspects of genetic map functions in recombination processes. In *Human Population Genetics*, ed. A. Chakravarti. New York: Van Nostrand Reinhold Co.

———, and H. Avni. 1981. Analysis of central equilibria in multilocus systems: A generalized symmetric viability regime. *Theor. Pop. Biol. 20*:241–280.

———, E. Cameron, and P. T. Williams. 1983. Structured exploratory data analysis applied to mode of inheritance. In *Developments in Statistics*, ed. P. R. Krishnaiah, 185–277. New York: Academic Press.

———, and M. W. Feldman. 1968a. Analysis of models with homozygote x heterozygote matings. *Genetics 59*:105–116.

———, and M. W. Feldman. 1968b. Further analysis of negative assortative mating. *Genetics 59*:117–136.

————, and S. Lessard. 1983. On the optimal sex ratio. *Proc. Nat. Acad. Sci. 80*:5931–5935.

————, and S. Lessard. 1984. On the optimal sex ratio: A stability analysis based on a characterization for one-locus multiallele viability models. *J. Math. Biol. 20*:15–38.

————, and U. Liberman. 1979a. Central equilibria in multilocus systems, I: Generalized nonepistatic selection regimes. *Genetics 91*:777–798.

————, and U. Liberman. 1979b. Central equilibrium in multilocus systems, II: Bisexual generalized nonepistatic selection models. *Genetics 91*:799–816.

————, and U. Liberman. 1979c. Representation of nonepistatic selection models and analysis of multilocus Hardy-Weinberg equilibrium configurations. *J. Math. Biol. 7*: 353–374.

————, and U. Liberman. 1979d. A natural class of multilocus recombination processes and related measures of crossover interference. *Adv. Appl. Prob. 11*:479–501.

Kempthorne, O. 1957. *An Introduction to Genetic Statistics*. New York: John Wiley & Sons.

Kerr, W. E. 1975. Evolution of the population structure in bees. *Genetics 79*:73–84.

Kerr, W. E., Y. Akahira, and C. A. Camargo. 1975. Sex determination in bees, IV: Genetic control of juvenile hormone production in *Melifona quadrifasciata (Apidae)*. *Genetics 81*: 749–756.

Kimura, M. 1983. *The Neutral Theory of Molecular Evolution*. Cambridge: Cambridge University Press.

Kingman, J.F.C. 1961a. A matrix inequality. *Quart. J. Math. 12*:78–80.

————. 1961b. A mathematical problem in population genetics. *Proc. Camb. Phil. Soc. 57*:574–582.

Kolman, W. 1960. The mechanism of natural selection for the sex ratio. *Amer. Natur. 94*:373–377.

Kosswig, C. 1964. Polygenic sex determination. *Experientia 20*: 190–199.

Lancaster, P. 1969. *Theory of Matrices*. New York: Academic Press.

Leigh, E. G., Jr. 1970. Sex ratio and differential mortality between the sexes. *Amer. Natur. 104*:205–210.

———. 1977. How does selection reconcile individual advantage with the good of the group? *Proc. Nat. Acad. Sci. 74*: 4542–4546.

———, E. L. Charnov, and R. R. Warner. 1976. Sex ratio, sex change, and natural selection. *Proc. Nat. Acad. Sci. 73*: 3655–3660.

Lessard, S. 1984. Evolutionary dynamics in frequency-dependent two-phenotype models. *Theor. Pop. Biol. 25*:210–234.

———, and S. Karlin. 1982. A criterion for stability-instability at fixation states involving an eigenvalue one with applications in population genetics. *Theor. Pop. Biol. 22*:108–126.

Leutert, R. 1975. Sex determination in *Bonellia*. In *Intersexuality in the Animal Kingdom*, ed. R. Reinboth, 84–90. Berlin: Springer-Verlag.

Lewis, D. 1979. *Sexual Incompatibility in Plants*. London: Edward Arnold.

Lewontin, R. C. 1968. The effect of differential viability on the population dynamics of *t*-alleles in the house mouse. *Evolution 22*:262–273.

———. 1974. *The Genetic Basis of Evolutionary Change*. New York: Columbia University Press.

Lloyd, D. G. 1974. Theoretical sex ratios of dioecious and gynodioecious angiosperms. *Heredity 32*:11–34.

———. 1980. Benefits and handicaps of sexual reproduction. *Evol. Biol. 13*:69–111.

———. 1982. Selection of combined versus separate sexes in seed plants. *Amer. Natur. 120*:571–585.

———, and K. S. Bawa. 1984. Modification of the gender of seed plants in varying conditions. In *Evolutionary Biology*, Vol. 16, ed. M. K. Hecht, B. Wallace, and C. T. Prance New York: Plenum Publishing Corp. pp. 255–338.

302

Lyon, M. F. 1961. Gene action in the X-chromosome of the mouse (*Mus musculus L*). *Nature 190*:372–373.

———. 1974. Evolution of X chromosome inactivation in mammals. *Nature 250*:651–653.

MacArthur, R. H. 1965. Ecological consequences of natural selection. In *Theoretical and Mathematical Biology*, ed. T. H. Waterman and H. Morowitz, 388–397. New York: Blaisdell.

McKusick, V. A., and F. H. Ruddle. 1977. Status of the gene map of the human chromosomes. *Science 196*:390–405.

MacNair, M. R. 1978. An ESS for the sex ratio in animals with particular reference to the social hymenoptera. *J. Theor. Biol. 70*:449–459.

Maffi, G., and S. D. Jayakar. 1981. A two-locus model for polymorphism for sex-linked meiotic drive modifiers with possible applications to *Aedes aegypti*. *Theor. Pop. Biol. 19*: 19–36.

Maynard Smith, J. 1968. Evolution in sexual and asexual populations. *Amer. Natur. 102*:469–473.

———. 1971. What use is sex? *J. Theor. Biol. 30*:319–335,

———. 1978. *The Evolution of Sex*. Cambridge: Cambridge University Press.

———. 1982. *Evolution and the Theory of Games*. Cambridge: Cambridge University Press.

———, and G. R. Price. 1973. The logic of animal conflict. *Nature 246*:15–18.

———, and N. C. Stenseth. 1978. On the evolutionary stability of the female-biased sex ratio in the wood lemming (*Myopus schisticolor*): The effect of inbreeding. *Heredity 41*: 205–214.

Milani, R., P. G. Rubini, and M. G. Franco. 1967. Sex determination in the housefly. *Genetic Agraria 21*:385–411.

Motro, U. 1982a. Optimal rates of dispersal I: Haploid populations. *Theor. Pop. Biol. 21*:394–411.

———. 1982b. Optimal rates of dispersal II: Diploid populations. *Theor. Pop. Biol. 21*:412–429.

Mrosovsky, N., S. R. Hopkins-Murphy, and J. I. Richardson. 1983. Sex ratio of hatchling loggerhead turtles *Caretta caretta*. *Science 225*:739–740.

Muller, H. J. 1918. Genetic variability twin hybrids and constant hybrids in a case of balanced lethal factors. *Genetics 3*:422–499.

Nagylaki, T. 1977. *Selection in One- and Two-locus Systems*. Lecture Notes in Biomathematics, Vol. 15. Berlin: Springer-Verlag.

———, and Crow, J. F. 1974. Continuous selective models. *Theor. Pop. Biol. 5*:257–284.

Nur, U. 1974. The expected changes in the frequency of alleles affecting the sex ratio. *Theor. Pop. Biol. 5*:143–147.

O'Donald, P. 1976. Mating preferences and their genetic effects in models of sexual selection for colour phases of the Arctic Skua. In *Population Genetics and Ecology*, ed. S. Karlin and E. Nevo, 411–430. New York: Academic Press.

———. 1977. Theoretical aspects of sexual selection. *Theor. Pop. Biol. 12*:298–334.

———. 1980. *Genetic Models of Sexual Selection*. London: Cambridge University Press.

Ohno, S. 1967. *Sex Chromosomes and Sex-Linked Genes*. Berlin: Springer-Verlag.

———. 1979. *Major Sex-Determining Genes*. Berlin: Springer-Verlag.

Orzack, S. H., J. J. Sohn, K. D. Kallman, S. A. Levin, and R. Johnston. 1980. Maintenance of the three sex chromosome polymorphism in the Platyfish. *Evolution 34*:663–672.

Oster, G., I. Eshel, and D. Cohen. 1977. Worker-queen conflict and the evolution of social insects. *Theor. Pop. Biol. 12*:49–85.

Pamilo, P. 1982. Genetic evolution of sex ratios in eusocial Hymenoptera: Allele frequency simulations. *Amer. Natur. 119*:638–656.

Parker, G. A. 1984. Theory and applications of evolutionary stable strategies. In *Behavioral Ecology: An Evolution Approach*, ed. J. R. Krebs and N. B. Bavies, 30–61. Sunderland, Mass: Sinauer Assoc.

————, R. R. Baker, and V.G.F. Smith. 1972. The origin and evolution of gamete dimorphism and the male-female phenomenon. *J. Theor. Biol.* *36*:529–553.

Policansky, D. 1981. Sex choice and the size advantage model in jack-in-the-pulpit (*Arisaema triphyllum*). *Proc. Nat. Acad. Sci.* *78*:1306–1308.

————. 1982. Sex change in plants and animals. *Ann. Rev. Ecol. Syst.* *13*:471–495.

Provine, W. B. 1971. *The Origins of Theoretical Population Genetics.* Chicago: University of Chicago Press.

Queller, D. C. 1983. Sexual selection in a hermaphroditic plant. *Nature 305*:706–707.

Robertson, D. R. 1972. Social control of sex reversal in a coral reef fish. *Science 177*:1007–1009.

Robinson, R. W., H. M. Munger, T. W. Whitaker, and G. W. Bohn. 1976. Genes of the *Cucurbitaceae*. *Hortoscience 11*: 554–568.

Ross, M. D., and Weir, B. 1976. Maintenance of males and females in hermaphrodite populations and the evolution of dioecy. *Evolution 30*:425–441.

Schuster, D. H., and L. Schuster. 1972. Speculative mechanisms affecting sex ratio. *J. Genet. Psychol.* *121*:245–254.

Scudo, F. M. 1964. Sex population genetics. *Ricerca Sci. 34*, *11-B*:93–146.

————. 1967a. Criteria for the analysis of multifactorial sex determination. *Mon. Zool. Ital.* *1*:1–21.

————. 1967b. The adaptive value of sexual dimorphism I. Anisogamy. *Evolution 21*:285–291.

Shaw, R. F. 1958. The theoretical genetics of the sex ratio. *Genetics 43*:149–163.

————, and J. D. Mohler. 1953. The selective advantage of the sex ratio. *Amer. Natur. 87*:337–342.

Speith, P. T. 1971. A necessary condition for equilibrium in systems exhibiting self-incompatible mating. *Theor. Pop. Biol. 2*:404–418.

————. 1974. Theoretical considerations of unequal sex ratios. *Amer. Natur. 87*:337–342.

305

Spurway, H., R. Dronamraji, and S. D. Jayakar. 1964. One nest of *Saliphron madraspatanum* (*Sphecidae*; *hymenoptera*). *J. Bomb. Nat. Hist. Soc. 61*:1–26.

Stenseth, N. C. 1978. Is the female-biased sex ratio in wood leming *Myopus schisticolor* maintained by cyclic inbreeding? *Oikos 30*:83–89.

Stubblefield, J. W. 1980. Theoretical elements of sex ratio evolution. Ph.D. Diss., Harvard University.

Sturtevant, A. H., and T. Dobzhansky. 1936. Geographical distribution and cytology of "sex ratio" in *Drosophila pseudoobscura* and related species. *Genetics 21*:473–490.

Suzuki, V., and Y. Iwasa. 1980. A sex ratio theory of gregarious parasitoids. *Res. Pop. Ecol. 22*:366–382.

Taylor, P. D. 1981. Intra-sex and inter-sex sib interactions as sex ratio determinants. *Nature 241*:64–66.

———, and M. G. Bulmer. 1980. Local mate competition and the sex ratio. *J. Theor. Biol. 86*:409–419.

———, and A. Sauer. 1980. The selective advantage of sex-ratio homeostasis. *Amer. Natur. 116*:305–310.

Teitelbaum, M. S. 1972. Factors associated with the sex ratio in human populations. In *The Structure of Human Populations*, ed. G. A. Harrison and A. J. Boyce, 90–109. London: Oxford University Press.

Thomson, G. J., and M. W. Feldman. 1975. Population genetics of modifiers of meiotic drive, IV: On the evolution of sex-ratio distortion. *Theor. Pop. Biol. 8*:202–211.

Traut, W. 1970. Zur sexualität von *Dinophilus gyrociliatus* III. *Die Geschleschtsbestimmung Biologisisches Zentralblatt 89*:137–161.

Trivers, R. L. 1972. Parental investment and sexual selection. In *Sexual Selection and the Descent of Man, 1871–1971*, ed. B. Campbell, 136–179. Chicago: Aldine.

———. 1974. Parent-offspring conflict. *Amer. Zool. 14*:249–265.

———, and H. Hare. 1976. Haplodiploidy and the evolution of the social insects. *Science 191*:249–263.

————, and D. E. Willard. 1973. Natural selection of parental ability to vary the sex ratio of offspring. *Science 179*:90–92.

Tsukamoto, M., T. Shono, and M. Horio. 1980. Autosomal sex-determining system of the housefly: Discovery of the first-chromosome male factor in Kitakyushu, Japan. *J. Univ. Occup. Env. Health 2*:235–252.

Uyenoyama, M. K. 1984. On the evolution of parthenogenesis: A genetic representation of the "cost of meiosis." *Evolution 38*:87–102.

————, and B. O. Bengtsson. 1979. Towards a genetic theory for the evolution of the sex ratio. *Genetics 93*:721–736.

————, and B. O. Bengtsson. 1981. Towards a genetic theory for the evolution of the sex ratio II: Haplodiploid and diploid models with sibling and parental control of the brood sex ratio and brood size. *Theor. Pop. Biol. 20*:57–79.

————, and B. O. Bengtsson. 1982. Towards a genetic theory for the evolution of the sex ratio III: Parental and sibling control of brood investment ratio under partial sib-mating. *Theor. Pop. Biol. 22*:43–68.

————, and M. W. Feldman. 1978. The genetics of sex ratio distortion by cytoplasmic infection under maternal and contagious transmission: An epidemiological study. *Theor. Pop. Biol. 14*:471–497.

————, and M. W. Feldman. 1981. On relatedness and adaptive topography in kin selection. *Theor. Pop. Biol. 19*:87–123.

Verner, S. 1965. Selection for sex ratio. *Amer. Natur. 99*:419–421.

Werren. J. H. 1980. Sex ratio adaptations to local mate competition in a parasitic wasp. *Science 208*:1157–1159.

————. 1983. Sex ratio evolution under local mate competition in a parasitic wasp. *Evolution 37*:116–124.

Westergaard, M. 1958. The mechanism of sex determination in dioecious flowering plants. *Adv. Genet. 9*:217–281.

White, M.J.D. 1973. *Animal Cytology and Evolution*. Cambridge: Cambridge University Press.

Whiting, P. W. 1943. Multiple alleles in complementary sex determination of *Habrobracon*. *Genetics 28*:365–382.

Williams, G. C. 1975. *Sex and Evolution*. Princeton: Princeton University Press.

———. 1979. The question of adaptive sex ratio in outcrossed vertebrates. *Proc. R. Soc. Lond. B., Biol. Sci. 205*:567–580.

———, and J. B. Mitton. 1973. Why reproduce sexually? *J. Theor. Biol. 39*:545–554.

Wilson, D. S. and R. K. Colwell. 1981. The evolution of sex ratio in structured demes. *Evolution 35*:882–897.

Winge, O. 1932. The nature of sex chromosomes. *Proc. 6th Int. Congr. Genetics 1*:343–355.

———. 1934. The experimental alteration of sex chromosomes into autosomes and vice versa, as illustrated by *Lebistes*. *Compt. Remd. Lab., Carlsburg. Ser. Physiol. 21*:1–49.

Wood, R. J., and M. E. Newton. 1976. Meiotic drive and sex ratio distortion in the mosquito *Aedes aegypti*. *Proc. Int. Congr. Ent*: 97–105.

Workman, P. L. 1964. The maintenance of heterozygosity by partial negative mating. *Genetics 50*:1369–1382.

Wright, S. 1921. Systems of mating III: Assortative mating based on somatic resemblance. *Genetics 6*:144–161.

Yampolsky, E., and H. Yampolsky. 1922. Distribution of sex forms in phanerogamic flora. *Bibl. Genet. 3*:1–62.

Author Index

309

Subject Index

absolute linkage, 173
adaptation and strategy in sex ratio determination, 13
age-dependent sex ratio, 34
age distribution, stable, 234
age-structured populations, 232
arrhenotoky, 8
assortative mating, 22, 289

biased sex ratios, 4, 6, 12, 16, 21, 38, 44
blending transmission, 265
brood sex ratio, 17, 20, 55
brood size variation, 55, 56, 289

central equilibrium, 168, 170, 176
chromosomal sex determination, 24, 25. *See also* dichotomous genotypic sex determination
complementary sex determination, 8
conflict situations for sex ratios, 13

dichotomous (exact) genotypic sex determination, 22, 36, 42, 47, 102, 127, 142, 166. *See also* chromosomal sex determination
dichotomous sex partitions, 48, 127
differential fertilities, viabilities, 65
directed selection, 219, 241
dosage compensation, 25

environmental sex determination (ESD), 11, 23, 29, 260, 267
ESS, 15, 19, 84, 182, 185, 193, 196, 200, 214, 226; ESS (optimal) sex ratio, 17, 19, 38, 103, 198, 199; ESS-mixed, 229; ESS-pure, 288
even sex-ratio equilibria, 70, 78, 79, 82, 102
extrachromosomal factors, 5, 12

family planning and sex ratio, 34
female-biased population sex ratio, 7, 12, 162, 182, 193
female heterogamety, 7, 8, 25, 161, 290

free recombination, 173
frequency dependence, 290

Galtonian blending, 266
gametic selection, 22
gene-dispersal model, 31
genetic sex determination (GSD), 23
genotypic equilibrium, 37, 38
gradient matrix, 217, 236
group selection, 16, 193

haplodiploid populations, 39, 226
haplodiploid sex determination model, 88, 109
haplodiploidy, 5, 8, 27
haploid fertilization, 89
haystack model, 21
hermaphroditism, 4, 9, 31, 184, 195, 212, 227, 233; sequential, 9, 259, 262; simultaneous, 9, 14, 63
heterogamety, 6, 26
homogamety, 6
hymenoptera, 8, 14

inbreeding, 16, 181, 193
inclusive fitness, 19
incompatibility at two loci, 255
incompatibility mechanisms, 3, 10, 252
incompatibility models, 5, 244
initial increase, 19
internal equilibrium, 95
internal stability, 74
isoplethic equilibria, 245

Jacobian matrix, 236

kin selection, 34

linkage disequilibrium, 177
litter size, 13
local mate competition (LMC), 15, 16, 20, 34, 181, 193
local resource competition (LRC), 21, 34, 182, 200
low-density model, 31
Lyapounov function, 52

311

Library of Congress Cataloging-in-Publication Data

Karlin, Samuel, 1923–
    Theoretical studies on sex ratio evolution.

    (Monographs in population biology; 22)
    Bibliography: p.
    Includes indexes.
    1. Sex ratio—Mathematical models.   2. Evolution—
Mathematical models.   I. Lessard, Sabin, 1951–   .
II. Title.   III. Title: Sex ratio evolution.   IV. Series.
QH481.K37   1986   574.5'248   85-43291
ISBN 0-691-08411-4 (alk. paper)
ISBN 0-691-08412-2 (pbk.)